Recovery of Waste Materials: Technological Research and Industrial Scale-Up

Recovery of Waste Materials: Technological Research and Industrial Scale-Up

Editor

Franco Medici

MDPI • Basel • Beijing • Wuhan • Barcelona • Belgrade • Manchester • Tokyo • Cluj • Tianjin

Editor
Franco Medici
"Sapienza" University of Roma
Italy

Editorial Office
MDPI
St. Alban-Anlage 66
4052 Basel, Switzerland

This is a reprint of articles from the Special Issue published online in the open access journal *Materials* (ISSN 1996-1944) (available at: https://www.mdpi.com/journal/materials/special_issues/ Recovery_Waste_Materials).

For citation purposes, cite each article independently as indicated on the article page online and as indicated below:

LastName, A.A.; LastName, B.B.; LastName, C.C. Article Title. *Journal Name* **Year**, *Volume Number*, Page Range.

ISBN 978-3-0365-3062-8 (Hbk)
ISBN 978-3-0365-3063-5 (PDF)

© 2022 by the authors. Articles in this book are Open Access and distributed under the Creative Commons Attribution (CC BY) license, which allows users to download, copy and build upon published articles, as long as the author and publisher are properly credited, which ensures maximum dissemination and a wider impact of our publications.

The book as a whole is distributed by MDPI under the terms and conditions of the Creative Commons license CC BY-NC-ND.

Contents

About the Editor vii

Franco Medici
Recovery of Waste Materials: Technological Research and Industrial Scale-Up
Reprinted from: *Materials* **2021**, *15*, 685, doi:10.3390/ma15020685 1

Vincenzo Santucci and Silvia Fiore
Recovery of Waste Polyurethane from E-Waste. Part II. Investigation of the Adsorption Potential for Wastewater Treatment
Reprinted from: *Materials* **2021**, *14*, 7587, doi:10.3390/ma14247587 5

Vincenzo Santucci and Silvia Fiore
Recovery of Waste Polyurethane from E-Waste—Part I: Investigation of the Oil Sorption Potential
Reprinted from: *Materials* **2021**, *14*, 6230, doi:10.3390/ma14216230 17

Mattia De Colle, Rahul Puthucode, Andrey Karasev and Pär G. Jönsson
A Study of Treatment of Industrial Acidic Wastewaters with Stainless Steel Slags Using Pilot Trials
Reprinted from: *Materials* **2021**, *14*, 4806, doi:10.3390/ma14174806 31

Valentín Gómez Escobar, Celia Moreno González, María José Arévalo Caballero and Ana Ma Gata Jaramillo
Initial Conditioning of Used Cigarette Filters for Their Recycling as Acoustical Absorber Materials
Reprinted from: *Materials* **2021**, *14*, 4161, doi:10.3390/ma14154161 45

Marco Abis, Martina Bruno, Franz-Georg Simon, Raul Grönholm, Michel Hoppe, Kerstin Kuchta and Silvia Fiore
A Novel Dry Treatment for Municipal Solid Waste Incineration Bottom Ash for the Reduction of Salts and Potential Toxic Elements
Reprinted from: *Materials* **2021**, *14*, 3133, doi:10.3390/ma14113133 57

Marcin Małek, Marta Kadela, Michał Terpiłowski, Tomasz Szewczyk, Waldemar Łasica and Paweł Muzolf
Effect of Metal Lathe Waste Addition on the Mechanical and Thermal Properties of Concrete
Reprinted from: *Materials* **2021**, *14*, 2760, doi:10.3390/ma14112760 73

Tova Jarnerud, Andrey V. Karasev and Pär G. Jönsson
Neutralization of Acidic Wastewater from a Steel Plant by Using CaO-Containing Waste Materials from Pulp and Paper Industries
Reprinted from: *Materials* **2021**, *14*, 2653, doi:10.3390/ma14102653 93

Nicolò Maria Ippolito, Franco Medici, Loris Pietrelli and Luigi Piga
Effect of Acid Leaching Pre-Treatment on Gold Extraction from Printed Circuit Boards of Spent Mobile Phones
Reprinted from: *Materials* **2021**, *14*, 362, doi:10.3390/ma14020362 105

Wojciech Kosakowski, Malgorzata Anita Bryszewska and Piotr Dziugan
Biochars from Post-Production Biomass and Waste from Wood Management: Analysis of Carbonization Products
Reprinted from: *Materials* **2020**, *13*, 4971, doi:10.3390/ma13214971 121

Virendra Kumar Yadav, Krishna Kumar Yadav, Vineet Tirth, Govindhan Gnanamoorthy, Nitin Gupta, Ali Algahtani, Saiful Islam, Nisha Choudhary, Shreya Modi and Byong-Hun Jeon
Extraction of Value-Added Minerals from Various Agricultural, Industrial and Domestic Wastes
Reprinted from: *Materials* **2021**, *14*, 6333, doi:10.3390/ma14216333 135

Sadegh Papari, Hanieh Bamdad and Franco Berruti
Pyrolytic Conversion of Plastic Waste to Value-Added Products and Fuels: A Review
Reprinted from: *Materials* **2021**, *14*, 2586, doi:10.3390/ma14102586 165

Rebeca Martínez-García, P. Jagadesh, Fernando J. Fraile-Fernández, Julia M. Morán-del Pozo and Andrés Juan-Valdés
Influence of Design Parameters on Fresh Properties of Self-Compacting Concrete with Recycled Aggregate—A Review
Reprinted from: *Materials* **2020**, *13*, 5749, doi:10.3390/ma13245749 181

About the Editor

Franco Medici is Associate Professor of Materials Science and Technology at the Sapienza University of Roma (2000).

He received his Master's degree in Chemical Engineering at Sapienza University of Roma in 1980. Since 1983, he has been working as researcher in the Applied and Chemical Institute of Engineering Faculty of the l' Aquila University. Then, in 1989, he joined the Department of Chemical Engineering, Materials and Environment of Sapienza University of Roma. Moreover, since 2021, he has been member of the academic team for the PhD in Chemical and Process Engineering.

He was (2006–2009) a scientific coordinator for the research national program (Italian Ministry of Education) titled "Recovery of waste materials: technological research, industrial scale-up and legislation." He is currently head of international cooperation projects in developing countries.

His current research seeks to investigate cement chemistry, solidification/stabilization of industrial wastes and recovery of waste materials. He has published 120 scientific papers exclusively on the above-mentioned topics. His H-index (Scopus) is equal to 20.

Editorial

Recovery of Waste Materials: Technological Research and Industrial Scale-Up

Franco Medici

Department of Chemical Engineering, Materials and Environment, Faculty of Civil and Industrial Engineering, "Sapienza" University of Roma, Via Eudossiana 18, 00184 Roma, Italy; franco.medici@uniroma1.it

Citation: Medici, F. Recovery of Waste Materials: Technological Research and Industrial Scale-Up. *Materials* **2022**, *15*, 685. https://doi.org/10.3390/ma15020685

Received: 2 January 2022
Accepted: 7 January 2022
Published: 17 January 2022

Publisher's Note: MDPI stays neutral with regard to jurisdictional claims in published maps and institutional affiliations.

Copyright: © 2022 by the author. Licensee MDPI, Basel, Switzerland. This article is an open access article distributed under the terms and conditions of the Creative Commons Attribution (CC BY) license (https:// creativecommons.org/licenses/by/ 4.0/).

An increase in population, booming economy, rapid urbanization and the rise in living standards have exponentially accelerated waste production. In this scenario, solid waste management is one of the fundamental problems of our society and constitutes a symbol of its limits and inefficiency.

Currently, 2 billion tons per year of municipal solid waste are produced worldwide and, at least, about 33% of this amount remains uncollected by the different municipalities. Recent research estimated that the production of this waste could reach 3.4 billion tons per year in 2050. However, waste production as a whole concerns different streams and origins other than municipal solid waste, including industrial, agricultural, construction and demolition, hazardous, medical and electronic waste. Production estimates of these latter types of waste are more uncertain because the source of production is differentiated and widespread. However, industrial waste generation is almost 18 times greater than municipal solid waste, and global agricultural waste production is more than 4 times the municipal solid waste. Several solutions have been proposed regarding the problem, for example, the theoretical approach of "zero waste", which is a philosophy that encourages the redesign of resources' life cycles so that all products can be recycled. In a zero-waste system, material flow is circular, and the same materials are used over and over again until the optimum level of consumption is reached. Moreover, the practice of circular economy includes our lifestyles, social organization and the restructuring of industrial production.

In light of this, the main problem that we have posed in the Special Issue "Recovery of waste materials: technological research and industrial scale-up" is to offer a minimum contribution to the solution of a huge problem, by presenting some findings for the recovery and the recycling of industrial waste. Contributions, coming from three continents and several countries (Canada, Germany, Korea, India, Italy, Poland, Saudi Arabia, Spain and Sweden), were collected from the research community illustrating the current direction and innovative advances in the field of waste materials.

Published papers as a whole concern different waste materials such as the recovery of different building materials [1–4], the treatment of waste deriving from electrical and electronic equipment [5–7], the utilization of stainless-steel slags [8,9], agricultural and domestic waste [10,11] and plastics [12].

The largest group of contributions concerns the recovery of waste materials in the construction sector, one paper (Gomez Escobar V. et al.) [1] considers the reuse of cigarette butts as acoustic absorbers in building constructions, through a preliminary process of chemical cleaning to eliminate the major metal ions as well as organic pollutants present in the samples. After the cleaning procedure, samples present higher adsorption coefficients than those of non-cleaned samples, this is due to an increase in the adsorption surface of the filters' fibers after the cleaning procedure.

Abis M. et al. [2] highlight that the main obstacle in reusing the bottom ash from municipal solid waste incinerators as a recycling aggregate to be used as a construction material is the content of salts and potential toxic elements concentrated in a layer that coats the bottom ash particles. In this work, a dry treatment process based on abrasion for the removal of salts and other elements is presented. A third paper (Malek M. et al.) [3] shows that the steel chips generated by lathes and mill machines are difficult to recycle, but this steel waste can be conveniently reutilized as a replacement for fine aggregate to produce concrete. Lastly, Martinez-Garzia R. et al., in a review paper, [4] present an overview of the bibliographic status of the design parameter's influence on the mix proportion of self-compacting concrete with recycled aggregate derived from construction and demolition waste.

Waste deriving from Electrical and Electronic Equipment (WEEE) is growing significantly all over the world; regarding this, Ippolito N.M. et al. [5] considered the recovery of gold from the printed circuit boards of used mobile phones, as well as silver and palladium, making them among the most valuable components of WEEE. Their recycling would also reduce the environmental impact of this waste due to the presence of heavy metals like copper and nickel. Two different hydrometallurgical routes for the recovery of copper and then gold were tested, moreover a flow sheet of the process was proposed. In another study, Santucci V. and Fiore S. proposed two papers [6,7], the first one considered the valorization of polyurethane foam (PUF) deriving from end-of-life refrigerators through a simple sieving process to obtain three fractions to be used as oil adsorbent. Particularly, the obtained fine fraction (d < 0.71 mm) revealed oil sorption performance at least 3–4 times higher than that of commercial products. Furthermore, the second one explored the performance of such waste (PUF) as an adsorbent for wastewater treatment after a pre-treatment sieving and washing, concluding that this waste could be applied "rough cut".

Jarnerud T. et al. [8], more or less in the same topic, have studied how CaO-containing waste from pulp and paper industries such as fly ash and calcined lime mud can be utilized to neutralize and purify wastewaters from the pickling processes in steel mills. De Colle M. et al. [9] proposed the utilization of stainless steel slags for the pH buffering of acidic wastewaters.

Recycling biomass and different types of organic waste constitutes a way of increasing the share of renewable sources in energy production; in fact, the Sustainable Development Goals set out by the United Nations highlight renewable energy as a key to the success of Agenda 2030. In this framework Kosakowsky W. et al. [10] studied the thermo-chemical decomposition of six types of agricultural waste biomass. The biomass conversion process was studied under a condition of limited oxygen in the reactor, and the temperature was raised from 450 to 850 °C for over 30 min, followed by a residence time of 60 min. Moreover, the obtained biochars present a combustion heat and a calorific value much higher compared to the biomass from which they were made and comparable to a good quality coal.

The last two papers are review works. Yadav K. V. et al. [11] consider agricultural, industrial and household waste that have the potential to generate value-added products, more specifically, industrial waste as fly-ash, gypsum waste and red mud can be used for the recovery of alumina, silica and zeolite. Agricultural waste materials, which are mainly organic, are biodegradable and can be used for the development of carbon-based materials and activated carbon. Furthermore, domestic waste, such as incense stick ash and eggshells, which is rich in calcium can be used as a potential source of either calcium oxide or carbonate.

From a sustainability point of view, the conversion of plastic waste to fuel or, better yet, to individual monomers, leads to a much greener waste management compared to landfill. Following this approach Papari S. et al. [12] reviewed the potential of pyrolysis as an effective thermo-chemical conversion method for the valorization of plastic waste. This waste, which can be a source of detrimental problems to terrestrial and marine ecosystems, can be thermochemically converted into valuable products, such as gasoline, diesel and wax.

In conclusion, the published works demonstrate a scientific and technological relevance to the topics dealt with, but the problems addressed in this Special Issue go beyond any solution that the scientific community is able to propose.

In fact, the "Industrial system, at the end of its cycle of production and consumption, has not developed the capacity to absorb and reuse waste and by-products. We have not yet managed to adopt a circular model of production capable of preserving resources for present and future generations, while limiting as much as possible the use of non-renewable resources for present and future generations, moderating their consumption, maximizing their efficient use, reusing and recycling them. A serious consideration of this issue would be one way of counteracting the throwaway culture which affects the entire planet, but it must be said that only limited progress has been made in this regard". (Encyclical Letter, Laudato Sì, Holy Father Francis-Bergoglio J.M., 2015).

Funding: This research received no external funding.

Institutional Review Board Statement: Not applicable.

Informed Consent Statement: Not applicable.

Data Availability Statement: Not applicable.

Acknowledgments: The Guest Editor would like to thank all the contributing authors, the reviewers and the editorial team of the materials. Special thanks go to May Zhang for her kind assistance and for her quick answers to my questions during this hard and long period and to Giacomo Medici (University of Guelph), who strongly encouraged me to be the Guest Editor and kindly revised the language of this Editorial.

Conflicts of Interest: The author declares no conflict of interest.

References

1. Gómez Escobar, V.; Moreno González, C.; Arévalo Caballero, M.J.; Gata Jaramillo, A.M. Initial Conditioning of Used Cigarette Filters for Their Recycling as Acoustical Absorber Materials. *Materials* **2021**, *14*, 4161. [CrossRef] [PubMed]
2. Abis, M.; Bruno, M.; Simon, F.-G.; Grönholm, R.; Hoppe, M.; Kuchta, K.; Fiore, S. A Novel Dry Treatment for Municipal Solid Waste Incineration Bottom Ash for the Reduction of Salts and Potential Toxic Elements. *Materials* **2021**, *14*, 3133. [CrossRef] [PubMed]
3. Małek, M.; Kadela, M.; Terpiłowski, M.; Szewczyk, T.; Łasica, W.; Muzolf, P. Effect of Metal Lathe Waste Addition on the Mechanical and Thermal Properties of Concrete. *Materials* **2021**, *14*, 2760. [CrossRef] [PubMed]
4. Martínez-García, R.; Jagadesh, P.; Fraile-Fernández, F.J.; Morán-del Pozo, J.M.; Juan-Valdés, A. Influence of Design Parameters on Fresh Properties of Self-Compacting Concrete with Recycled Aggregate—A Review. *Materials* **2020**, *13*, 5749. [CrossRef] [PubMed]
5. Ippolito, N.M.; Medici, F.; Pietrelli, L.; Piga, L. Effect of Acid Leaching Pre-Treatment on Gold Extraction from Printed Circuit Boards of Spent Mobile Phones. *Materials* **2021**, *14*, 362. [CrossRef] [PubMed]
6. Santucci, V.; Fiore, S. Recovery of Waste Polyurethane from E-Waste—Part I: Investigation of the Oil Sorption Potential. *Materials* **2021**, *14*, 6230. [CrossRef] [PubMed]
7. Santucci, V.; Fiore, S. Recovery of Waste Polyurethane from E-Waste. Part II. Investigation of the Adsorption Potential for Wastewater Treatment. *Materials* **2021**, *14*, 7587. [CrossRef] [PubMed]
8. Jarnerud, T.; Karasev, A.V.; Jönsson, P.G. Neutralization of Acidic Wastewater from a Steel Plant by Using CaO-Containing Waste Materials from Pulp and Paper Industries. *Materials* **2021**, *14*, 2653. [CrossRef] [PubMed]
9. De Colle, M.; Puthucode, R.; Karasev, A.; Jönsson, P.G. A Study of Treatment of Industrial Acidic Wastewaters with Stainless Steel Slags Using Pilot Trials. *Materials* **2021**, *14*, 4806. [CrossRef] [PubMed]
10. Kosakowski, W.; Bryszewska, M.A.; Dziugan, P. Biochars from Post-Production Biomass and Waste from Wood Management: Analysis of Carbonization Products. *Materials* **2020**, *13*, 4971. [CrossRef] [PubMed]
11. Yadav, V.K.; Yadav, K.K.; Tirth, V.; Gnanamoorthy, G.; Gupta, N.; Algahtani, A.; Islam, S.; Choudhary, N.; Modi, S.; Jeon, B.-H. Extraction of Value-Added Minerals from Various Agricultural, Industrial and Domestic Wastes. *Materials* **2021**, *14*, 6333. [CrossRef]
12. Papari, S.; Bamdad, H.; Berruti, F. Pyrolytic Conversion of Plastic Waste to Value-Added Products and Fuels: A Review. *Materials* **2021**, *14*, 2586. [CrossRef]

Recovery of Waste Polyurethane from E-Waste. Part II. Investigation of the Adsorption Potential for Wastewater Treatment

Vincenzo Santucci and Silvia Fiore *

Department of Engineering for Environment, Land, and Infrastructures (DIATI), Politecnico di Torino, Corso Duca degli Abruzzi 24, 10129 Torino, Italy; vincenzo.santucci@polito.it
* Correspondence: silvia.fiore@polito.it

Abstract: This study explored the performances of waste polyurethane foam (PUF) derived from the shredding of end-of-life refrigerators as an adsorbent for wastewater treatment. The waste PUF underwent a basic pre-treatment (e.g., sieving and washing) prior the adsorption tests. Three target pollutants were considered: methylene blue, phenol, and mercury. Adsorption batch tests were performed putting in contact waste PUF with aqueous solutions of the three pollutants at a solid/liquid ratio equal to 25 g/L. A commercial activated carbon (AC) was considered for comparison. The contact time necessary to reach the adsorption equilibrium was in the range of 60–140 min for waste PUF, while AC needed about 30 min. The results of the adsorption tests showed a better fit of the Freundlich isotherm model (R^2 = 0.93 for all pollutants) compared to the Langmuir model. The adsorption capacity of waste PUF was limited for methylene blue and mercury (K_f = 0.02), and much lower for phenol (K_f = 0.001). The removal efficiency achieved by waste PUF was lower (phenol 12% and methylene blue and mercury 37–38%) compared to AC (64–99%). The preliminary results obtained in this study can support the application of additional pre-treatments aimed to overcome the adsorption limits of the waste PUF, and it could be applied for "rough-cut" wastewater treatment.

Keywords: adsorption; circular economy; wastewater; refrigerator; WEEE

1. Introduction

According to the latest report published by the association of plastic manufacturers Plastic Europe [1], the demand for polyurethane in Europe was equal to 4 Mt in 2019, representing 7.9% of the total plastic demand. The main contributors to polyurethane requirement are the manufacturing of pillows and mattresses (31%), and the construction and building (24.5%), electrical and electronic (21.3%), and automotive (11%) sectors [1]. Of the 4 Mt/y polyurethane requested in Europe in 2019, approximately two thirds are in the form of foams (1.68 Mt flexible foam, 1 Mt rigid foam) [2]. In China, polyurethane output in 2011 reached 7.5 Mt, and polyurethane foam (PUF) accounted for 60% [3]. PUF wastes are product scraps, as the production of rigid polyurethane foam usually creates 15% of waste [3] and post-consumer waste materials. Of the total 29.1 Mt of plastic generated in Europe in 2019, approximately 1.5 Mt are made by PUF, of which one third is recycled, while the rest is incinerated or sent to landfill.

The scientific and technical literature offers several potential perspectives for material recovery from waste PUF, mostly as an oil absorbent [4–7], additive for construction materials [8–11], and adsorbent of pollutants from wastewater [12,13]. Nowadays, the market competition in the field of wastewater treatment technologies is increasing due to the need of achieving effective removal performances with limited costs. The most common adsorbent at the state-of-the-art level is activated carbon (AC), as dust or granular material, suitable for a variety of applications for drinking water, swimming pools, urban and industrial wastewater, etc. Alternatives to AC are oxides and zeolites, polymeric adsorbents (intended for application in industrial wastewater treatments, but their high costs of production and regeneration have prevented a broader application), and, developed more recently, low-cost adsorbents derived from wastes [14]. The literature is rich of studies that investigated the adsorption potential of industrial and agricultural wastes, particularly for the removal of dyes or metals from wastewater (Table 1) [15–24].

Table 1. Overview of studies describing the properties and performances of commercial and novel adsorbents towards different contaminants (SSA: specific surface area; C_{Li}: initial concentration in the liquid phase; q_{eq}: amount adsorbed on the solid phase; t_{eq}: contact time).

Adsorbent Parent Material	SSA (m^2/g)	Adsorbent Dose (g/L)	Contaminant	C_{Li} (mg/L)	q_{eq} (mg/g)	t_{eq}	% Removal	Ref.
commercial activated carbon	698–1281	-	phenol	100–5000	200–270	1 h	99	[15]
biochars from lignocellulose biomass	63–211	-	phenol	100–5000	65–104	5 h	68	[15]
composite lignosulfonate sodium/cotton biochar	-	0.2	Pb	50–100	203.5	3 h	-	[16]
	-	0.2	methylene blue	5–30	109.1	24 h	-	
various bio-waste derived adsorbents	0.67–65.19	1–5	Cd	5–250	7.5–230.5	40–480 min	-	[17]
	-	0.6–15	Cr	5–8000	1.3–249	25–250 min	99.2	
	1.8–105	1–10	Pb	6.35–2000	8.6–909.1	30–300 min	>94	
	0.853–450	0.4–10	Cu	5–100	2.1–19.5	30–360 min	-	
	0.75–17.38	1–5	Ni	23–250	0.3–285.7	20–180 min	-	
	0.75–206.8	1–10	As	2.5–500	0.42–133	60–360 min	-	
	59–450	1–18	Zn	20–5000	2.4–68.5	20–300 min	-	
	0.78–186	0.6–4	Co	10–600	14.8–349.6	3–120 min	-	
maize straw ash	38.3	0.2–1.2	perfluorinated compounds	1–500	811	48 h	-	[18]
chitosan-based polymer	-	-	perfluorinated compounds	20–550	1452	32 h	40–60	[19]
non-ionic resins	-	-	perfluorinated compounds	0.01–5	37–46	10–96 h	-	[20]
industrial by-products (blast furnace residues, fly ash, red mud)	4.5–1740	0.25–8	different commercial dyes	-	1.3–390	2–72 h	-	[21]
	3–1440	0.1–50	Cu, Zn, Cr, As, Ni, Cd, Pb	1–4000	1–140	3–72 h	-	
	69–380	0.2–200	phenols	200–1500	11.4–190.2	2–8 h	-	
physically immobilized PUF	-	4	Cr	10	-	2 h	98.6	[22]
thiazolidinone steroids impregnated PUF	-	1	Cd	5–10	-	1 h	94–96	[23]
candle sooth PUF	-	50	Rhodamine B	50	15.066	150 min	96	[24]

Table 1 provides an overview of the literature data describing the properties and performances of adsorbents deriving from different "parent" materials, categorized per type of contaminant. The most promising experimental applications of low-cost adsorbents were industrial wastewater containing dyes, metals, and halogenated compounds. The dose of adsorbent was in the range of 0.1–20.0 g/L, though it was higher for the removal of phenols. The specific surface area (SSA) directly affects adsorption, and high values are usually desirable to provide many adsorption sites. AC exhibited SSA values between 500 and 1500 m^2/g [15,21]; however, adsorbents with relatively low values (<200 m^2/g) could also achieve good adsorption capacities towards metals such as lead, cadmium, nickel, and cobalt [17]. The application of PUF as an adsorbent material for the removal of several pollutants from wastewaters is a recently investigated perspective [12,24]. PUF-based adsorbents achieved adsorption capacities between 20 and 30 mg/g for copper, cadmium, and chromium [13], making them less performant than commercial products, but still with a good adsorption capacity, higher than other waste-derived materials such as fly ash and hemp. Commercial AC is usually made from non-renewable resources or biomass transported over long distances, resulting in high environmental impacts due to feedstock and transportation, and in relevant energy demand [25]. The estimated impact on climate change of granular AC is 1.44 Kg CO_2/kg adsorbent [20,26]. To sum up, a good adsorbent should: be made from a porous raw material with high SSA; have good affinity for the target contaminants; and have limited costs for raw material procurement-also including transportation, and for its preparation. To limit the environmental impacts, adsorbents with minimal energy consuming pre-treatments are preferable, and the feasibility of their regeneration after adsorption must be considered as well.

The interest of the scientific and industrial worlds is shifting towards waste-derived non-conventional adsorbents, derived from biological, agricultural, or industrial processes, which are available almost free of cost [27,28]. The porous structure of PUF is a desirable feature for an adsorbent because it provides numerous potential sites of adsorption; also, open-cell PUF can be successfully used in columns for the treatment of large volumes of wastewaters [14]. The potential as adsorbent of virgin PUF in combination with different reagents has been previously tested [29,30] but, to our knowledge, there are not many studies specifically exploring the application of "plain" (e.g., without modification of its chemistry) waste PUF as an adsorbent. The main goal of this study is to investigate the adsorption potential of waste PUF in the field of wastewater treatment technologies for the removal of inorganic and organic pollutants. Waste PUF is employed "as such" separated from end-of-life (EoL) refrigerators, after the application of minimal and simple physic treatments to eliminate the impurities (i.e., sieving and washing with water). The perspective explored by this study is coherent with the Circular Economy strategy of the actual European policy and regulations. This solution, if proven effective, can lead to a double potential benefit when costs and environmental burdens are reduced in comparison to the use of conventional adsorbents.

2. Materials and Methods

2.1. Waste PUF Origin and Characteristics

The tested material was waste PUF in a loose granular form derived from the shredding of EoL refrigerators (category 1 WEEE) at a TBD treatment plant managed by AMIAT in the metropolitan area of Turin, Italy. The waste PUF was sampled across 5 weeks (one sample per week) to account for any composition variability. The samples (1 kg each) were collected according to standard methods UNI 10802:2013 and UNI 14899:2006 at the end of the working day. The samples were assumed to be representative, considering that 3300 t/y EoL refrigerators entering the plant roughly correspond to over 300 items shredded per day [4]. The collected samples were quartered to obtain representative secondary samples for the characterization and adsorption tests. A complete characterization of the waste PUF is reported in a previous study [4], describing the investigation of the oil absorption potential of the same material (whole material and selected particle-size fractions). Compared to our previous study [4], this research explored the adsorption potential for wastewater treatment of the fraction of waste PUF with dimensions between 0.71 and 5 mm. The main features of the considered waste PUF are reported in Table 2. Commercial powdered Activated Carbon (AC) FILTERCARB RO, provided by Carbonitalia srl (Livorno, Italy) was chosen as reference material for the adsorption tests (Table 2).

Table 2. Main features of the considered waste PUF and of the reference commercial AC.

Parameter	Measure Unit	Waste PUF	AC
Specific Surface Area	m^2/g	-	>1750
ash at 550 °C	%	10.40 ± 1.60	<3.00
bulk density	kg/m^3	47.57	<350.00
pH in water	pH units	8.02 ± 0.16	5.00 ± 1.00
moisture	%	<0.1	<10.0
particle size distribution	mm	0.710 ÷ 5.000	0.015 ÷ 0.110
electrical conductivity	µS/cm	125.50 ± 12.70	<200.00

2.2. Pre-Treatment

The waste PUF sampled in the WEEE shredding plant contained impurities such as plastic, paper, and metal. Before the adsorption tests, the waste PUF (fraction having dimensions between 0.71 and 5 mm) underwent a washing pre-treatment (0.125 L water/g PUF) aimed at removing the impurities as higher density (sink) fraction after 15 min of shaking at 150 rpm in an ARGOLAB SKI 4 orbital shaker. After a 3 min rest, the floating particles of PUF were collected and wet sieved at 0.71 mm with 0.03 L water/g PUF. The washed samples were drained for 10 days in ambient conditions (21 °C, relative humidity 63%) and stored in a dry container.

2.3. Target Pollutants

Three pollutants were considered in the adsorption tests: methylene blue, an organic compound present in paints used in textile and plastic industries; phenol, an organic pollutant derived from the polymer, chemical, and food industries; and mercury, a carcinogenic metal well known for its bioaccumulation potential in water reservoirs affected by industrial or mining activities [31]. The target pollutants solutions were prepared from the dilution in deionized water of: 1000 mg/L mercury solution Chem-Lab (Zedelgem, Belgium) 99.5+% phenol pellets Chem-Lab (Zedelgem, Belgium); 99.5% methylene blue ($C_{16}H_{18}ClN_3S \cdot 3H_2O$) CarloErba Reagents (Cornaredo, MI, Italy).

The analyses of phenol and methylene blue were performed directly on the aqueous phases through an ONDA UV-30 SCAN UV-VIS spectrophotometer (at 269 and 668 nm, respectively). Mercury was analyzed through an NEX DE VS Rigaku XRF spectrometer.

2.4. Adsorption Tests

All adsorption experiments were performed in an ARGOLAB SKI 4 orbital shaker at 260 rpm and 20 °C. The AC was tested at a solid/liquid ratio equal to 0.75 g/L. All tests were conducted in three replicates.

Firstly, equilibrium tests were necessary to find the equilibrium time (t_{eq}) for each target pollutant and the pre-treated waste PUF. Flasks of 250 mL were filled with 200 mL of 10 mg/L solution of each pollutant and 5 g of PUF (solid/liquid ratio equal to 25 g/L, chosen according to literature studies in Table 1). Three milliliter aliquots of solution were withdrawn after different time intervals, filtered on 0.45 µm cellulose ester syringe filters, and analyzed to measure the residual pollutant concentration. t_{eq} was determined as the time after which no decrease in the residual aqueous concentration was detected. q_{eq}, i.e., the amount of pollutant adsorbed, was calculated as the difference between the initial concentration of pollutant in the liquid phase (C_{Li}) and the residual value (C_{Lf}).

The adsorption tests were performed in 50 mL falcon test tubes filled with 40 mL of pollutants solution and 1 g of pre-treated waste PUF (solid/liquid ratio equal to 25 g/L). The pollutant solutions were as follows: methylene blue: 0.5, 1, 2, 5, 7.5, 10, 15, 18, 20 mg/L; phenol: 6, 8, 10, 12, 14, 16, 19, 24, 30 mg/L; mercury: 2, 3, 4, 5, 6.5, 10, 12, 17, 22 mg/L. The tubes were shaken for an interval equal to the t_{eq} of each pollutant. The supernatant was separated from the solid phase through a Z20A Hermle centrifuge (Labortechnik GmbH, Wehingen, Germany) at 3500 rpm for 5 min, then filtered on 0.45 µm cellulose ester syringe filters and analyzed. The adsorption tests involved three replicates.

2.5. Isotherm Models

At a constant temperature, the process of adsorption can be described by an adsorption isotherm. After the equilibrium state has been reached, the concentrations of the adsorbate on the solid phase are plotted against concentrations of adsorbate in liquid phase. Two models were used for the interpretation of experimental data. The Freundlich model is based on Equation (1) [14]:

$$q_{eq} = K_f \, (C_{eq})^{1/n} \tag{1}$$

where q_{eq} is the amount of adsorbate transferred on the sorbent at equilibrium; K_f is the capacity factor, a parameter that characterizes the strength of adsorption, and it is directly proportional to q_{eq}. The exponent $1/n$ determines the curvature of the isotherm, and it denotes the intensity of adsorption.

The Langmuir model is based on Equation (2) [14]:

$$q_{eq} = \frac{q_{max} \, b}{1 + bC_e} \cdot C_e \tag{2}$$

where q_{eq} is the amount of adsorbate transferred on the sorbent at equilibrium; q_{max} is the maximum capacity of adsorption at saturation (assuming the formation of a single layer of adsorbed molecules); b is the Langmuir constant related to the adsorption energy.

3. Results and Discussion

3.1. Adsorption Equilibrium Tests

Considering the results of the equilibrium tests (Figure 1 and Table 3), for all target contaminants, the equilibrium of adsorption was reached more quickly with AC, so that the test was stopped earlier than for reactors with PUF, since no significant changes in liquid concentration were detectable. Compared to PUF, the much shorter t_{eq} found for AC is reasonably a consequence of its high specific surface area [14] and of the hydrophobic nature of polyurethane, which could make adsorption slower [32]. The pollutants reached the adsorption equilibrium on AC rather quickly (30–35 min), while waste PUF required much longer times: 60 min for methylene blue, and 135–140 min for phenol and mercury. From these preliminary tests and considering the amounts of pollutant transferred on the solid adsorbent (q_{eq}), methylene blue exhibited the highest affinity, compared to phenol and mercury, both for waste PUF and AC (Table 3).

Table 3. Details and results of the adsorption equilibrium tests performed on PUF and AC (C_{Li}: initial concentration in the liquid phase; C_{Lf}: final concentration in the liquid phase; t_{eq}: equilibrium time; q_{eq}: amount of contaminant transferred on the sorbent).

Adsorbent	Adsorbate	Adsorbent Dose (g/L)	C_{Li} (mg/L)	C_{Lf} (mg/L)	t_{eq} (min)	q_{eq} (mg/kg)
Waste PUF	methylene blue	25.00	12.50	7.49	60	0.24
	phenol	25.00	40.00	35.00	140	0.17
	mercury	25.00	6.00	2.97	135	0.13
AC	methylene blue	0.75	55.73	0.14	30	74.11
	phenol	0.75	38.48	9.89	30	38.12
	mercury	0.75	10.90	2.41	35	11.32

3.2. Adsorption Batch Tests

The results of the adsorption batch tests (Figure 2 and Table 4) showed that the Freundlich isotherm model better fitted, compared to Langmuir model, the data related to waste PUF with an adequate correction factor ($R^2 = 0.93$) for all the three pollutants. The adsorption capacity of waste PUF was moderate for methylene blue and mercury (K_f values around 0.02), while it was considerably lower for phenol (K_f around 1×10^{-3}). Indeed, the maximum removal efficiency achieved from the batch tests by waste

PUF (Table 5) was also rather limited: 12.2% for phenol and 37–38% for methylene blue and mercury.

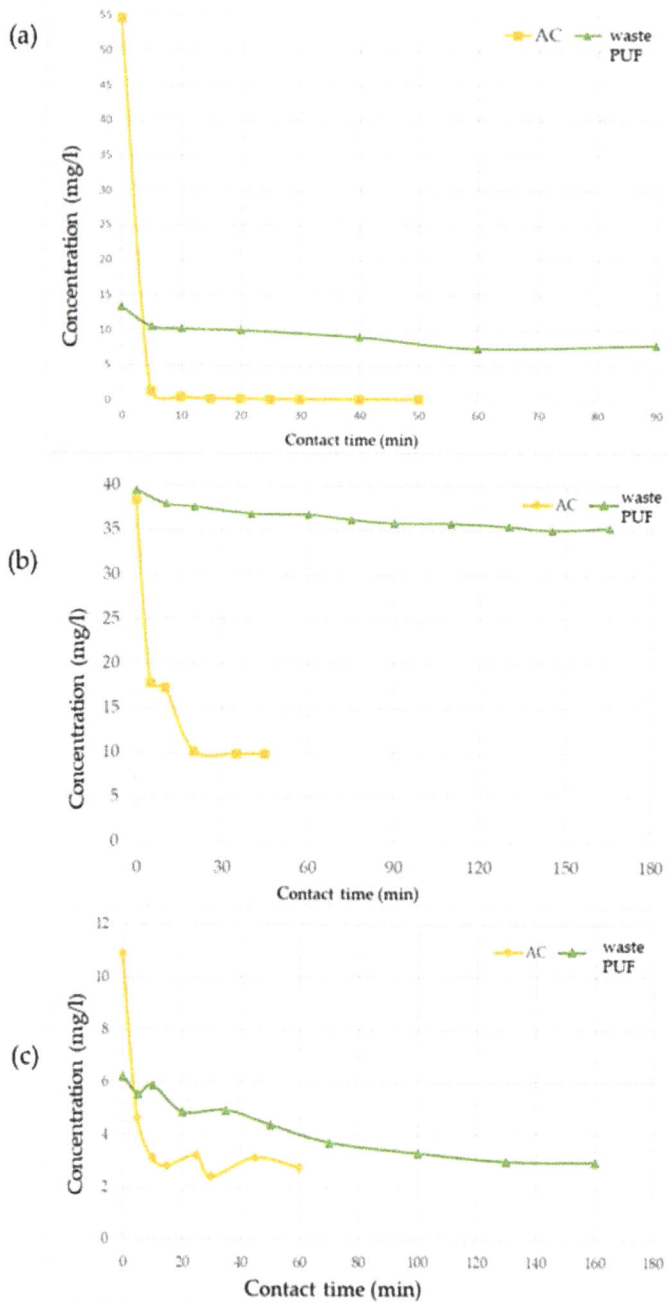

Figure 1. Results of the adsorption equilibrium tests performed on waste PUF and AC with (**a**) methylene blue, (**b**) phenol, and (**c**) mercury.

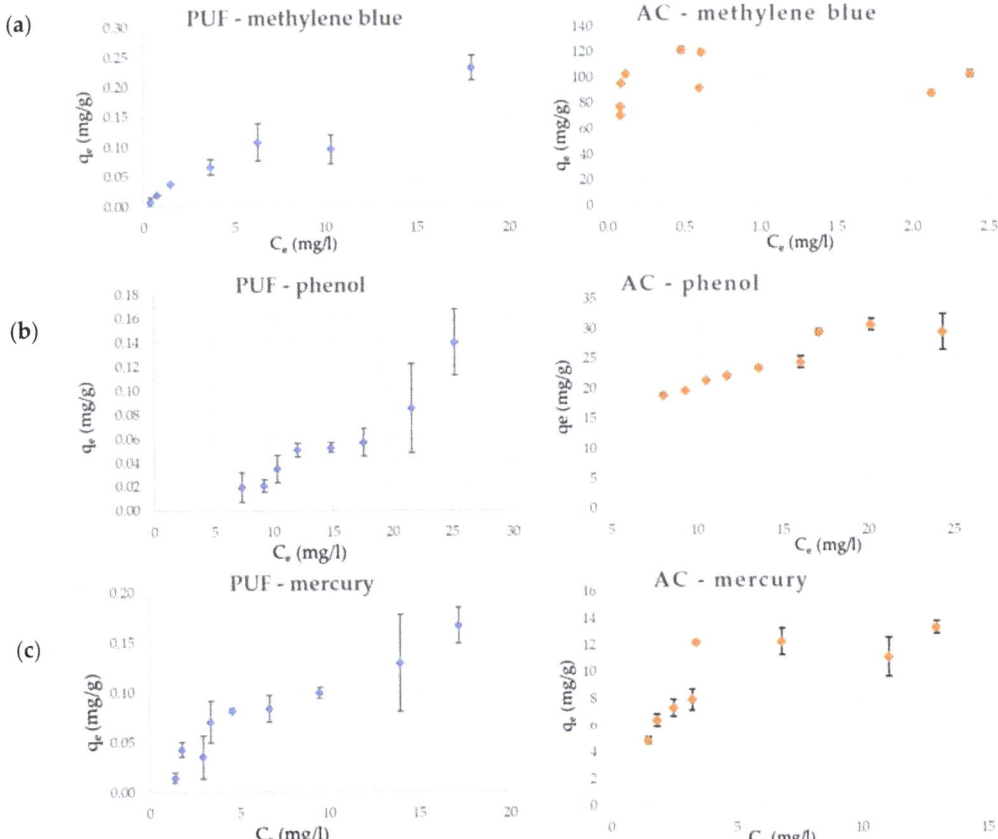

Figure 2. Results of the adsorption batch tests performed with waste PUF and AC in contact with (**a**) methylene blue, (**b**) phenol, and (**c**) mercury (C_e: equilibrium concentration in the liquid phase; q_e: equilibrium concentration in the solid phase).

The results of the adsorption tests performed on AC were described with higher accuracy by the Langmuir model for methylene blue ($R^2 = 0.99$) and mercury ($R^2 = 0.95$). Only in the case of phenol was the Freundlich model more adequate in describing the adsorption by AC ($R^2 = 0.88$) than the Langmuir model ($R^2 = 0.59$). The maximum removal efficiency achieved by AC for methylene blue was 99.9%, leading to very low residual concentrations in the liquid phase ($C_{Lf} = 0.04$ mg/L). The Freundlich model had inadequate experimental results obtained for methylene blue ($R^2 = 0.14$), probably because when the concentrations at equilibrium are much lower than the initial concentrations, the adsorption is generally well described by a linear model. The Freundlich isotherm, which is in exponential form, cannot describe the linear range at very low concentrations. On the contrary, this limit case is well described by Langmuir model and when $b \cdot C_{Lf} \ll 1$, it is equivalent to a linear isotherm. The higher q_{max} found for AC applied to the adsorption of methylene blue (135.13 mg/g), compared to q_{max} of phenol (26.11 mg/g) and mercury (0.05 mg/g), was realistically expected since the considered commercial AC is commonly applied for decolorization purposes.

Table 4. Values of Freundlich (K_f, n) and Langmuir (q_{max}, b) isotherm models' parameters resulting from the interpolation of the experimental data derived from batch adsorption tests with waste PUF and AC.

Pollutant	Adsorption Model	Waste PUF			AC		
	Freundlich	K_f (L/mg)	1/n	R^2	K_f (L/mg)	1/n	R^2
methylene blue		0.022	0.797	0.93	101.110	0.056	0.14
phenol		0.001	1.517	0.93	7.020	0.468	0.88
mercury		0.019	0.784	0.93	5.170	0.39	0.68
	Langmuir	q_{max} (mg/g)	b (L/mg)	R^2	q_{max} (mg/g)	b (L/mg)	R^2
methylene blue		0.363	0.061	0.54	135.130	1.480	0.99
phenol		0.098	0.023	0.57	26.110	0.085	0.59
mercury		0.349	0.048	0.43	0.059	0.410	0.95

Table 5. Maximum removal efficiencies achieved in batch adsorption tests performed with waste PUF and AC (C_{Li}: initial concentration in the liquid phase; C_{Lf}: final concentration in the liquid phase).

Pollutant	C_{Li} (mg/L)		C_{Lf} (mg/L)		% Removal	
	Waste PUF	AC	Waste PUF	AC	Waste PUF	AC
methylene blue	1.27	48.32	0.78	0.04	38.50	99.90
phenol	28.73	22.19	25.24	8.05	12.20	63.70
mercury	2.87	12.53	1.81	3.38	37.00	73.00

Unfortunately, because of the different level of correction factors, a direct comparison of the two adsorbents was not possible. However, since the differences between the values of K_f and q_{max} obtained from waste PUF and AC were of several orders of magnitude almost in every case, it was evident that there was a considerable gap in favor of AC towards the adsorption of the considered target pollutants.

The results of this study were compared to literature data related to other novel and low-cost "non-conventional" (i.e., not commercial) materials tested for the adsorption of mercury (Table 6), phenol (Table 7) and methylene blue (Table 8). These materials, although at an experimental level, all underwent treatments aimed at improving their adsorption performances (e.g., activation for biomass-based sorbents, modification by addition of reagents for other materials). Literature data referred to mercury adsorption (Table 6) exhibited q_{max} in the range of 1.8–13 mg/g from the Langmuir model, and K_f between 0.02 and 19 L/mg from the Freundlich model, with correction factor values exceeding 0.9 for both isotherm models in all studies. Literature data on phenol adsorption (Table 7) found q_{max} values in the range of 38–285 mg/g from the Langmuir model, and K_f between 0.19 and 7.40 L/mg, with correction factor values exceeding 0.9 for both isotherm models in all studies. Methylene blue adsorption literature studies (Table 8) found typical values of q_{max} in the range of 29–2639 mg/g for the Langmuir model, and K_f between 0.82 and 1746 L/mg, with correction factor values around 0.8–0.9 for both isotherm models in all studies.

The fact that waste PUF did not show similar adsorption performances in the present study means that the tested material was not yet ready to provide competitive adsorption performances. Indeed, the gap was not so large when comparing the Freundlich parameters obtained from waste PUF (K_f = 0.019 L/mg) and other non-commercial adsorbents in contact with mercury solutions (K_f mostly in the range 0.02–4.50 L/mg, with one exception).

Table 6. Performances of some non-commercial adsorbents tested for the removal of mercury.

Adsorbent	Langmuir Model			Freundlich Model			Temperature	Ref.
	q_{max} (mg/g)	b (L/mg)	R^2	K_f (L/mg)	n	R^2	°C	
biochar	6.54	0.328	0.995	1.72	2.204	0.987	25	[33]
modified biochar	9.15	0.608	0.992	3.22	1.803	0.949	25	[33]
bentonite	2.01	0.125	0.984	0.29	2.505	0.995	25	[33]
biochar-bentonite composite	11.72	0.749	0.991	4.50	2.482	0.981	25	[33]
hydrated lime	12.93	0.070	0.990	0.02	50	1.00	room	[34]
co-doped molybdenum selenide (nitrogen and sulfur)	-	-	-	18.96	0.40	0.988–0.995	25	[35]
chitosan modified PUF	1.84	0.989	0.888	0.30	0.623	0.942	room	[36]

Table 7. Performances of some non-commercial adsorbents tested for the removal of phenol.

Adsorbent	Langmuir Model			Freundlich Model			Temperature	Ref.
	q_{max} (mg/g)	b (L/mg)	R^2	K_f (L/mg)	n	R^2	°C	
zeolite/AC composite	37.92–40.31	0.022–0.032	0.929–0.944	5.74–7.40	0.20–0.32	0.998	25–40	[37]
modified halloysite nanotubes	-	-	-	0.19	0.99	0.987	25	[38]
biochar from lignocellulose biomass	65.00–104.00	0.00054–0.00094	-	1.10–4.80	0.29–0.52	-	25	[39]
Biochar from sewage sludge	216.76	0.0067	0.998	2.66	0.7635	0.987	35	[40]
carbon pellets from cigarette butts	211.45–285.11	0.0096–0.015	0.976	-	-	-	10–40	[41]

Table 8. Performances of some non-commercial adsorbents tested for the removal of methylene blue.

Adsorbent	Langmuir Model			Freundlich Model			Temperature	Ref.
	q_{max} (mg/g)	b (L/mg)	R^2	K_f (L/mg)	n	R^2	°C	
biochar from soybean	2488.00–2639.00	0.39–1.04	0.999–1.00	1672.00–1746.00	11.65–16.95	0.849–0.912	25	[42]
graphene-oxide-based nanocomposites from rice husks	478.47–632.91	3.66–10.38	0.859–0.985	334.37–422.22	6.18–6.83	0.893–0.929	ambient	[43]
corn husk powder	30.30	0.003	0.949	8.51	2.27	0.827	25–28	[44]
biochar from eucalyptus	114.60	20.68	0.901	86.58	0.085	0.980	30	[45]
zeolite clays combined with $ZnTiO_3/TiO_2$	29.14–49.81	0.43–1.00	0.990	11.98–18.80	0.30–0.38	0.970	ambient	[46]
adsorbent based on magnetic metal–organic compounds	148.80	0.051	0.961	17.40	0.47	0.992	ambient	[47]
biochar from Paulownia wood	255.89	0.003	0.886	0.82	40.27	0.839	20–40	[47]

4. Conclusions

Investigating any possible opportunities for the recovery of plastics is a key step for supporting the European Circular Economy strategies. This research provides preliminary results about the adsorption properties of waste PUF deriving from the shredding of EoL refrigerators. In this study, waste PUF performances for the removal of methylene blue, phenol, and mercury from aqueous phases were compared with the ones of a commercial AC. Adsorption batch tests allowed to determine the adsorption isotherm parameters. The Freundlich isotherm model better fitted ($R^2 = 0.93$), compared to the Langmuir model ($R^2 < 0.60$), the adsorption of methylene blue, phenol, and mercury on waste PUF. In the considered experimental conditions, waste PUF showed a constrained affinity in adsorbing the target pollutants. The obtained Freundlich adsorption parameter K_f was around 0.02 L/mg for mercury and methylene blue, and 0.001 L/mg for phenol. These values were three or four orders of magnitude lower compared to commercial AC, and rather low when compared to the average adsorption capacities of non-commercial adsorbents according to the literature. Moreover, the long time required to reach the adsorption equilibrium (60–140 min depending on the pollutant) in the considered experimental conditions makes waste PUF direct application as an adsorbent rather challenging, especially in fixed-bed columns wherein short equilibrium times are desirable to design columns of reasonable height.

However, summarizing the results obtained in this study, it must be considered that waste PUF is a material deriving from a waste treatment process totally unintended for any adsorption application, and with a minimal preparation consisting only of sieving and washing. The results of this study can support the design of other pre-treatments aimed at overcoming the adsorption limits of the waste PUF "as such". For instance, reducing the particle size of waste PUF, and thus increasing the available specific surface area, would benefit the rate of adsorption. After these additional studies, waste PUF could be applied for "rough-cut" wastewater treatment. When industrial wastewater with high pollution loads is delivered to treatment plants, a rough removal of contamination can be conducted with a relatively low-performant adsorbent such as PUF, prior to a second- more advanced purification process. Additionally, considering the comparison with the performances of other non-conventional (i.e., non-commercial) adsorbents, PUF exhibited the most promising affinity towards mercury. Therefore, further research could be conducted aiming at a feasible application of PUF for mercury removal.

Author Contributions: Conceptualization, methodology, experimental investigation, data curation, and writing—original draft preparation: V.S.; conceptualization, methodology, supervision, writing—review and editing, project administration, and funding acquisition: S.F. All authors have read and agreed to the published version of the manuscript.

Funding: This research was performed in the framework of the project "Material recovery from WEEE" (in Italian: "Recupero di materia da RAEE/R1-R2") funded by the Italian Ministry for Environmental Transition, and ongoing between January 2019 and August 2021. The project was coordinated by Politecnico di Torino (Polytechnic University of Turin) and involved as industrial partners, among others, IREN Group and AMIAT. Specifically, this study was based on the activities of work package 1 of the project, dedicated to the recovery of waste PUF from the shredding of end-of-life refrigerators.

Institutional Review Board Statement: Not applicable.

Informed Consent Statement: Not applicable.

Data Availability Statement: The data presented in this study are available on reasonable request from the corresponding author.

Conflicts of Interest: The authors declare no conflict of interest.

References

1. Plastics Europe-Association of Plastic Manufacturers (Organization). *Plastics—The Facts 2020*; Plastic Europe 16; Plastics Europe: Brussels, Belgium, 2020.
2. The Essential Chemical Industry. Polyurethanes. 2017. Available online: https://www.essentialchemicalindustry.org/polymers/polyurethane.html (accessed on 21 July 2021).
3. Yang, W.; Dong, Q.; Liu, S.; Xie, H.; Liu, L.; Li, J. Recycling and Disposal Methods for Polyurethane Foam Wastes. *Procedia Environ. Sci.* **2012**, *16*, 167–175. [CrossRef]
4. Santucci, V.; Fiore, S. Recovery of Waste Polyurethane from E-Waste—Part I: Investigation of the Oil Sorption Potential. *Materials* **2021**, *14*, 6230. [CrossRef] [PubMed]
5. Keshawy, M.; Farag, R.K.; Gaffer, A. Egyptian crude oil sorbent based on coated polyurethane foam waste. *Egypt. J. Pet.* **2020**, *29*, 67–73. [CrossRef]
6. Ouda, Y.W. The Effect of Fulica Atra Feather on Oil Sorption Capacity of Polyurethane Foam. *Polyurethane* **2015**, *5*, 90–94.
7. Zia, K.M.; Bhatti, H.N.; Ahmad Bhatti, I. Methods for polyurethane and polyurethane composites, recycling and recovery: A review. *React. Funct. Polym.* **2007**, *67*, 675–692. [CrossRef]
8. Gómez-Rojo, R.; Alameda, L.; Rodríguez, Á.; Calderón, V.; Gutiérrez-González, S. Characterization of Polyurethane Foam Waste for Reuse in Eco-Efficient Building Materials. *Polymers* **2019**, *11*, 359. [CrossRef]
9. Gu, L.; Ozbakkaloglu, T. Use of recycled plastics in concrete: A critical review. *Waste Manag.* **2016**, *51*, 19–42. [CrossRef] [PubMed]
10. Jia, Z.; Jia, D.; Sun, Q.; Wang, Y.; Ding, H. Preparation and Mechanical-Fatigue Properties of Elastic Polyurethane Concrete Composites. *Materials* **2021**, *14*, 3839. [CrossRef]
11. Yang, C.; Zhuang, Z.H.; Yang, Z.G. Pulverized polyurethane foam particles reinforced rigid polyurethane foam and phenolic foam. *J. Appl. Polym. Sci.* **2014**, *131*, 1–7. [CrossRef]
12. Dacewicz, E.; Grzybowska-Pietras, J. Polyurethane foams for domestic sewage treatment. *Materials* **2021**, *14*, 933. [CrossRef]
13. Teodosiu, C.; Wenkert, R.; Tofan, L.; Paduraru, C. Advances in preconcentration/removal of environmentally relevant heavy metal ions from water and wastewater by sorbents based on polyurethane foam. *Rev. Chem. Eng.* **2014**, *30*, 403–420. [CrossRef]
14. Catizzone, E.; Sposato, C.; Romanelli, A.; Barisano, D.; Cornacchia, G.; Marsico, L.; Cozza, D.; Migliori, M. Purification of Wastewater from Biomass-Derived Syngas Scrubber Using Biochar and Activated Carbons. *Int. J. Environ. Res. Public Heal.* **2021**, *18*, 4247. [CrossRef] [PubMed]
15. Zhang, N.; Cheng, N.; Liu, Q. Functionalized Biomass Carbon-Based Adsorbent for Simultaneous Removal of Pb^{2+} and MB in Wastewater. *Materials* **2021**, *14*, 3537. [CrossRef] [PubMed]
16. Bilal, M.; Ihsanullah, I.; Younas, M.; Ul Hassan Shah, M. Recent advances in applications of low-cost adsorbents for the removal of heavy metals from water: A critical review. *Sep. Purif. Technol.* **2021**, *278*, 119510. [CrossRef]
17. Chen, X.; Xia, X.; Wang, X.; Qiao, J.; Chen, H. A comparative study on sorption of perfluorooctane sulfonate (PFOS) by chars, ash and carbon nanotubes. *Chemosphere* **2011**, *83*, 1313–1319. [CrossRef] [PubMed]
18. Yu, Q.; Deng, S.; Yu, G. Selective removal of perfluorooctane sulfonate from aqueous solution using chitosan-based molecularly imprinted polymer adsorbents. *Water Res.* **2008**, *42*, 3089–3097. [CrossRef]
19. Mudiyanselage, S.T.; Senevirathna, L.D. Development of Effective Removal Methods of PFCs (Perfluorinated Compounds) in Water by Adsorption and Coagulation. Ph.D. Thesis, Kyoto University, Kyoto, Japan, 2010.
20. De Gisi, S.; Lofrano, G.; Grassi, M.; Notarnicola, M. Characteristics and adsorption capacities of low-cost sorbents for wastewater treatment: A review. *Sustain. Mater. Technol.* **2016**, *9*, 10–40. [CrossRef]
21. El-Shahawi, M.S.; Bashammakh, A.S.; Abdelmageed, M. Chemical speciation of chromium(III) and (VI) using phosphonium cation impregnated polyurethane foams prior to their spectrometric determination. *Anal. Sci.* **2011**, *27*, 757–763. [CrossRef]
22. Makki, M.S.T.; Abdel-Rahman, R.M.; Alfooty, K.O.; El-Shahawi, M.S. Thiazolidinone steroids impregnated polyurethane foams as a solid phase extractant for the extraction and preconcentration of cadmium(II) from industrial wastewater. *E-J. Chem.* **2011**, *8*, 887–895. [CrossRef]
23. Singh, V.P.; Vaish, R. Candle soot coated polyurethane foam as an adsorbent for removal of organic pollutants from water. *Eur. Phys. J. Plus* **2019**, *134*, 419. [CrossRef]
24. Joseph, B.; Kaetzl, K.; Hensgen, F.; Schäfer, B.; Wachendorf, M. Sustainability assessment of activated carbon from residual biomass used for micropollutant removal at a full-scale wastewater treatment plant. *Environ. Res. Lett.* **2020**, *15*, 064023. [CrossRef]
25. Moreira, M.T.; Noya, I.; Feijoo, G. The prospective use of biochar as adsorption matrix—A review from a lifecycle perspective. *Bioresour. Technol.* **2017**, *246*, 135–141. [CrossRef] [PubMed]
26. Jiao, G.J.; Ma, J.; Li, Y.; Jin, D.; Ali, Z.; Zhou, J.; Sun, R. Recent advances and challenges on removal and recycling of phosphate from wastewater using biomass-derived adsorbents. *Chemosphere* **2021**, *278*, 130377. [CrossRef]
27. Jang, S.H.; Min, B.G.; Jeong, Y.G.; Lyoo, W.S.; Lee, S.C. Removal of lead ions in aqueous solution by hydroxyapatite/polyurethane composite foams. *J. Hazard. Mater.* **2008**, *152*, 1285–1292. [CrossRef]
28. Murthy, K.S.R.; Marayya, R. Studies on the Removal of Heavy Metal Ions from Industrial Effluents Using Ammonium Pyrrolidine Dithio Carbamate (APDC) Loaded Polyurethane Foams (PUF). *World Appl. Sci. J.* **2011**, *12*, 358–363.
29. US Geological Survey. Mercury in the Environment. In *Environmental Science and Technology*; US Geological Survey: Reston, VA, USA, 2000. Available online: http://www.usgs.gov/themes/factsheet/146-00/index.html (accessed on 6 September 2021).

30. Fierro, V.; Torné-Fernández, V.; Montané, D.; Celzard, A. Adsorption of phenol onto activated carbons having different textural and surface properties. *Microporous Mesoporous Mater.* **2008**, *111*, 276–284. [CrossRef]
31. Goel, N.K.; Kumar, V.; Dubey, K.A.; Bhardwaj, Y.; Varshney, L. Development of functional adsorbent from PU foam waste via radiation induced grafting I: Process parameter standardization. *Radiat. Phys. Chem.* **2013**, *82*, 85–91. [CrossRef]
32. Bai, Y.; Hong, J. Preparation of a novel millet straw biochar-bentonite composite and its adsorption property of hg^{2+} in aqueous solution. *Materials* **2021**, *14*, 1117. [CrossRef] [PubMed]
33. Ullah, S.; Al-Sehemi, A.G.; Mubashir, M.; Mukhtar, A.; Saqib, S.; Bustam, M.A.; Cheng, C.K.; Ibrahim, M.; Show, P.L. Adsorption behavior of mercury over hydrated lime: Experimental investigation and adsorption process characteristic study. *Chemosphere* **2021**, *271*, 129504. [CrossRef] [PubMed]
34. Long, C.; Li, X.; Jiang, Z.; Zhang, P.; Qing, Z.; Qing, T.; Feng, B. Adsorption-improved MoSe2 nanosheet by heteroatom doping and its application for simultaneous detection and removal of mercury (II). *J. Hazard. Mater.* **2021**, *413*, 125470. [CrossRef]
35. Iqhrammullah, M.; Mustafa, I. The application of chitosan modified polyurethane foam adsorbent. *Rasayan J. Chem.* **2019**, *12*, 494–501. [CrossRef]
36. Cheng, W.P.; Gao, W.; Cui, X.; Ma, J.H.; Li, R.F. Phenol adsorption equilibrium and kinetics on zeolite X/activated carbon composite. *J. Taiwan Inst. Chem. Eng.* **2016**, *62*, 192–198. [CrossRef]
37. Słomkiewicz, P.; Szczepanik, B.; Czaplicka, M. Adsorption of phenol and chlorophenols by HDTMA modified halloysite nanotubes. *Materials* **2020**, *13*, 3309. [CrossRef]
38. Sposato, C.; Catizzone, E.; Romanelli, A.; Marsico, L.; Barisano, D.; Migliori, M.; Cornacchia, G. Phenol Removal from Water with Carbons: An Experimental Investigation. *Tec. Ital. -Ital. J. Eng. Sci.* **2020**, *64*, 143–148. [CrossRef]
39. Dalhat, N.; Zubair, M. Sludge-Based Activated Carbon Intercalated MgAlFe Ternary. *Molecules* **2021**, *26*, 4266. [CrossRef]
40. Medellín-Castillo, N.A.; Ocampo-Pérez, R.; Forgionny, A.; Labrada-Delgado, G.; Zárate-Guzmán, A.; Cruz-Briano, S.; Flores-Ramírez, R. Insights into equilibrium and adsorption rate of phenol on activated carbon pellets derived from cigarette butts. *Processes* **2021**, *9*, 934. [CrossRef]
41. Ying, Z.; Huang, L.; Ji, L.; Li, H.; Liu, X.; Zhang, C.; Zhang, J.; Yi, G. Efficient removal of methylene blue from aqueous solutions using a high specific surface area porous carbon derived from soybean dreg. *Materials* **2021**, *14*, 1754. [CrossRef]
42. Liou, T.H.; Liou, Y.H. Utilization of rice husk ash in the preparation of graphene-oxide-based mesoporous nanocomposites with excellent adsorption performance. *Materials* **2021**, *14*, 1214. [CrossRef]
43. Malik, D.S.; Jain, C.K.; Yadav, A.K.; Kothari, R.; Pathak, V.V. Removal of Methylene Blue Dye in Aqueous Solution by Agricultural Waste. *Int. Res. J. Eng. Technol.* **2016**, *3*, 864–880.
44. Amin, M.T.; Alazba, A.A.; Shafiq, M. Successful application of eucalyptus camdulensis biochar in the batch adsorption of crystal violet and methylene blue dyes from aqueous solution. *Sustainability* **2021**, *13*, 3600. [CrossRef]
45. Jaramillo-Fierro, X.; González, S.; Montesdeoca-Mendoza, F.; Medina, F. Structuring of zntio3 /tio2 adsorbents for the removal of methylene blue, using zeolite precursor clays as natural additives. *Nanomaterials* **2021**, *11*, 898. [CrossRef] [PubMed]
46. Zhang, G.; Wo, R.; Sun, Z.; Hao, G.; Liu, G.; Zhang, Y.; Guo, H.; Jiang, W. Effective magnetic mofs adsorbent for the removal of bisphenol a, tetracycline, congo red and methylene blue pollutions. *Nanomaterials* **2021**, *11*, 1917. [CrossRef] [PubMed]
47. Alam, S.; Khan, M.S.; Bibi, W.; Zekker, I.; Burlakovs, J.; Ghangrekar, M.M.; Bhowmick, G.D.; Kallistova, A.; Pimenov, N.; Zahoor, M. Preparation of activated carbon from the wood of paulownia tomentosa as an efficient adsorbent for the removal of acid red 4 and methylene blue present in wastewater. *Water* **2021**, *13*, 1453. [CrossRef]

Recovery of Waste Polyurethane from E-Waste—Part I: Investigation of the Oil Sorption Potential

Vincenzo Santucci and Silvia Fiore *

Department of Engineering for Environment, Land, and Infrastructures (DIATI), Politecnico di Torino, Corso Duca degli Abruzzi 24, 10129 Torino, Italy; vincenzo.santucci@polito.it
* Correspondence: silvia.fiore@polito.it

Abstract: The shredding of end-of-life refrigerators produces every year in Italy 15,000 tons of waste polyurethane foam (PUF), usually destined for energy recovery. This work presents the results of the investigation of the oil sorption potential of waste PUF according to ASTM F726–17 standard. Three oils (diesel fuel and two commercial motor oils) having different densities (respectively, 0.83, 0.87, and 0.88 kg/dm^3) and viscosities (respectively, 3, 95, and 140 mm^2/s at 40 °C) were considered. The waste PUF was sampled in an Italian e-waste treatment plant, and its characterization showed 16.5 wt% particles below 0.71 mm and 13 wt% impurities (paper, plastic, aluminum foil), mostly having dimensions (d) above 5 mm. Sieving at 0.071 mm was applied to the waste PUF to obtain a "coarse" (d > 0.71 mm) and a "fine" fraction (d < 0.71 mm). Second sieving at 5 mm allowed an "intermediate" fraction to be obtained, with dimensions between 0.71 and 5 mm. The oil sorption tests involved the three fractions of waste PUF, and their performances were compared with two commercial oil sorbents (sepiolite and OKO-PUR). The results of the tests showed that the "fine" PUF was able to retain 7.1–10.3 g oil/g, the "intermediate" PUF, 4.2–7.4 g oil/g, and the "coarse" PUF, 4.5–7.0 g oil/g, while sepiolite and OKO-PUR performed worse (respectively, 1.3–1.6 and 3.3–5.3 g oil/g). In conclusion, compared with the actual management of waste PUF (100 wt% sent to energy recovery), the amount destined directly to energy recovery could be limited to 13 wt% (i.e., the impurities). The remaining 87 wt% could be diverted to reuse for oil sorption, and afterward directed to energy recovery, considered as a secondary option.

Keywords: absorption; circular economy; oil spill; refrigerator; WEEE

1. Introduction

Approximately 97,000 tons of waste temperature exchange equipment were collected in Italy in 2020, mostly consisting of end-of-life (EoL) refrigerators [1]. This type of waste falls into category 1 of waste electrical and electronic equipment (WEEE, or e-waste), as defined by the Directive 2012/19/EU [2]. On average in Italy 90 wt% of collected EoL temperature exchange equipment is made of refrigerators (the rest are air conditioners), whose 15–16 wt% is made of polyurethane foam (PUF) sprayed inside the frame as thermal-insulating material [3,4]. Hence, the expected amount of waste PUF deriving from EoL refrigeration equipment is nearly 15,000 tons every year. In the current Italian WEEE management system, waste PUF is often an unrecovered fraction, since there is no common practice to reuse it as secondary raw material, and it is disposed of in incineration plants. After the Paris agreement, signed in 2015 by members of the United Nations on the reduction in greenhouse gas (GHG) emissions, and the New Circular Economy Action Plan, stated in 2020 by the European Commission, WEEE management optimization become a strategic issue in urgent need of optimization. Therefore, research activity driven by the aims of reducing GHG emissions and keeping the resources in use as long as possible is highly needed. With the aim of exploring the potential perspectives for material recovery from waste PUF, the scientific and technical literature was surveyed according to the following two complementary points of view:

- *A top-down approach* considering the general uses of PUF: the main applications (as raw or secondary material) in the global manufacturing industry were identified. The most common applications of PUF are connected with two purposes: thermal insulation (in buildings and refrigeration equipment) due to its outstanding thermal properties [5–7], and the manufacturing of flexible and porous items such as mattresses, car seats, shoes, sports equipment, etc. due to the facility of molding it into the desired frame and/or applying it as spray foam [8,9].
- *A bottom-up approach* based on the uses of recovered waste PUF: the technical properties required for polyurethane for specific applications were evaluated. Three promising solutions for the valorization of waste PUF were found: application as sorbent material for oil spills [10–12]; additive for construction materials as lightweight mortars or insulant materials [13–16]; application in filters for the adsorption of pollutants from wastewater [17,18].

Matching the results of the literature survey based on the two complementary viewpoints, the research was focused on two main directions and organized into two parts: part I explores the oil sorption potential, which is presented in this work, and part II, investigates the adsorption potential toward inorganic and organic contaminants for wastewater treatment, which will be discussed in another work [19]. Part I of the research, here presented, specifically considers the application of PUF as contingency equipment for absorbing oil spills. Due to its interesting properties (high porosity, hydrophobicity), PUF is applied as the primary raw material in the manufacturing of industrial absorbents, which can absorb crude oil and related by-products. US EPA defines fundamental features for such products to be oleophilic (oil attracting) and hydrophobic (water repellent) [20]. Contingency equipment for oil spills in the form of pads and booms made of PUF can absorb 20–40 times their own weight in oil, whereas synthetic granular sorbents have generally lower capacity of absorption, between 5 and 30 times their weight [21,22]. Since the excessive use of sorbents at a spill scene, especially in granular or particulate forms, can lead to cleanup problems, their applications are intended for industrial workplaces such as warehouses and repair shops [23,24].

Many studies on PUF were devoted to the improvement of the performance of products deriving from primary materials, tested on oils and several waste fluids, through the chemical treatment of the porous surface, or applying special synthesis processes [25–28]. Recently, numerous scientific studies focused on how to recover waste PUF by chemical or thermochemical recycling processes [29–33], but, to our knowledge, there is not yet any consistent research about the absorbing performances of waste PUF that can be achieved without altering its chemical structure. This work aimed at defining an easily applicable valorization opportunity of waste PUF in loose form, based on physical/mechanical treatment processes commonly performed in a WEEE treatment plant. The experimental plan of the here-presented study was designed to assess the effectiveness of waste PUF derived from EoL refrigerators in absorbing oily substances. The purpose of this specific research is that, once the lifespan of the refrigerator is over, a new use phase for the PUF as recovered absorbent would be feasible, with the lowest possible requirements about technical complexity and related economic and environmental costs. Compared with recycling through chemical treatments, this approach can limit energy consumptions and the generation of residual by-products, thus resulting in an easier and cheaper technical solution. The outcomes of this research can provide support in the evaluation of material-recovery opportunities from WEEE consistently with circular economy principles, and in the optimization of the management of waste PUF from EoL refrigerators.

2. Materials and Methods

2.1. Waste Origin and Sampling

The tested material was PUF in loose form, derived from the mechanical treatment of EoL refrigerators (category 1 WEEE) at the TBD treatment plant, a few km from Turin, Italy, and managed by AMIAT SpA. AMIAT oversees municipal solid waste management in Turin (900,000 inhabitants). The catchment area of the TBD plant extends through northwestern Italy, accounting in total for 3300 t of category 1 WEEE in 2018. Several treatments were applied in the TBD plant on the EoL refrigerators: initial disassembling of the refrigerating circuit and depollution from hazardous fluids; shredding of the carcasses to recover ferrous and non-ferrous metals, respectively, through magnetic and eddy current separators, while mixed plastics are processed in another plant. The heterogeneous material resulting from the shredding phase underwent an air separation treatment in a "zig-zag" process, which diverted PUF to a milling phase to achieve particle size below 10 mm. Finally, the milled PUF was briquetted and sent to incineration.

The milled PUF was sampled for the research along 5 weeks, one sample per week, to account for any composition variability. The samples (1 kg each) were collected at the milling unit crate according to standard methods UNI 10802:2013 and UNI 14899:2006. The samples were assumed representative, considering that 3300 t/y EoL refrigerators entering the plant roughly correspond to over 300 items shredded per day (average weight of 1 item: 42 kg) [3]. The collected samples were quartered to obtain representative secondary samples (40–50 g) for the characterization.

2.2. Characterization

Within the project, part of the characterization was performed in Iren Group laboratories, analyzing (the reference methods are detailed in parentheses): the speed of combustion (UNI CEN/TS 16023:2014), the chemical composition (UNI EN 13657:2004 and UNI EN ISO 11885:2009), the density (ASTM D 5057-10), the content of organic compounds, PCBs and pesticides (EPA 3545A 2007 + EPA 8270E 2018, EPA 5021A 2014 + EPA 8260D 2018), and performing UNI EN 12457-2 leaching test to assess the release of contaminants into the environment. The list of measured parameters with the related values is in the Supplementary Materials Section. The rest of the characterization, consisting of the particle-size and component analyses, was performed in the Circular Economy Lab at DIATI, Politecnico di Torino. The particle-size analysis was performed through a Giuliani IG3/EXP siever equipped with 7 sieves of different mesh screens. All analyses were performed on aliquots of five different secondary samples.

2.3. Pre-Treatment

The visual analysis of the waste PUF highlighted the presence of coarse impurities (plastic, paper, aluminum foil) (Figure 1) and of fine particles. Therefore, the samples underwent a sieving pre-treatment with 2 subsequent dimensional cuts (0.71 and 5 mm), aimed at separating the "fine" fraction (dimensions below 0.71 mm) from the "coarse" fraction (dimensions above 0.71 mm), and afterward, to eliminate the impurities (dimensions above 5 mm) and obtain an "intermediate" fraction, having dimensions between 0.71 and 5 mm (Table 1). The pre-treatment was performed on three different secondary samples (see Section 2.2).

Table 1. Features of different types of samples used in oil sorption tests.

Size Range (mm)	Sample ID	Bulk Density (kg/m^3)
d < 0.71	fine	129.65
0.71 < d < 5	intermediate	47.57
d > 0.71	coarse	42.00
2 < d < 4	sepiolite	440
d < 1	OKO-PUR	127.5

Figure 1. Appearance of the waste polyurethane (PUF) foam: (**a**) PUF sample collected at the milling unit; (**b**) PUF sample after the removal of impurities; (**c**) impurities separated by type.

2.4. Oil Sorption Tests

The tests were conducted on the three mentioned particle-size fractions of waste PUF deriving from the pre-treatment (Table 1). They were compared with two commercial products used as loose sorbents for oil spill control: sepiolite and OKO-PUR. Sepiolite is a traditional sorbent material made of clay minerals' granules of magnesium silicate; OKO-PUR is a full saturation oil binder in powder, derived, likewise the tested PUF, from waste plastic foam. OKO-PUR and sepiolite were provided by Amiat and purchased from Airbank company (www.airbank.it, accessed on 8 May 2021). Their particle-size characteristics, as specified by the supplier, are presented in Table 1.

The tests were performed in triplicates according to the international ASTM F726–17 standard, which provides methods for assessing the performance of adsorbents in removing crude oils and related spills [34]. The PUF in loose form falls under the definition of Type II adsorbent in the ASTM F726–17 standard ("an unconsolidated, particulate material without sufficient form and strength to be handled except with scoops and similar equipment"). The oil sorption capacity was assessed as the ratio

$$\frac{S_O}{S_i} \tag{1}$$

where

$$S_O = S_{OT} - S_i \qquad (2)$$

is the net oil absorbed; S_{OT} is the weight of the absorbent at the end of the test after 2 min dripping; S_i is the initial dry absorbent weight (Figure 2). Specifically, the oil sorption capacity represents the amount of sorbed oil normalized by the absorbent's mass. The oil sorption capacity of the three particle-size fractions of waste PUF and of the two commercial competitors was assessed toward three different oils, according to the ASTM F728-17 standard. Specifically, a diesel fuel (Quaser, produced by Q8) and two commercial motor oils (10w40 Prestige and 20w50 Select, produced by Delphi) with different densities (respectively 0.83, 0.87, and 0.88 kg/dm^3) and viscosities (respectively 3, 95, and 140 mm^2/s at 40 °C) were considered. The procedure for the sorption tests is detailed as follows: A 19 cm crystallizing dish was filled up to 2.5 cm height with the oil. The PUF sample (5 g) was placed in a steel mesh basket with 0.1 mm mesh openings and kept in contact with the oil for 15 min. During the contact, the gradual sinking of the whole absorbent mass was visible. Then, the sample was removed from the crystallizing dish, and the soaked absorbent with the mesh basket was drained for 10 min. Within this time, the oil in excess in contact with the absorbent's surface, but not completely absorbed, is released by dripping. According to ASTM F726–17 standard, the weight measurement should be taken 30 s after removing the sample from the crystallizing dish. In case of heavy or weathered oils, a 2 min drain time is recommended. However, if the drainage is not adequately long, the values reported for the oil sorption capacity will be inaccurately high [35]. The weight measurements on the sample were taken after 0.5, 1, 2, 5, and 10 min to evaluate the actual amount of absorbed oil and the tendency to release it. In the case of the heaviest oil (20w50), preliminary tests showed that the sample was not fully saturated, since a part of it floated and did not sink after 15 min of contact with the oil. This behavior is typical of high-viscosity oils that require a longer time to saturate the pores of the tested material. This happened for the "intermediate" and OKO samples. Hence, a long test (24 h contact time) according to the ASTM F726-17 standard was conducted on these two materials.

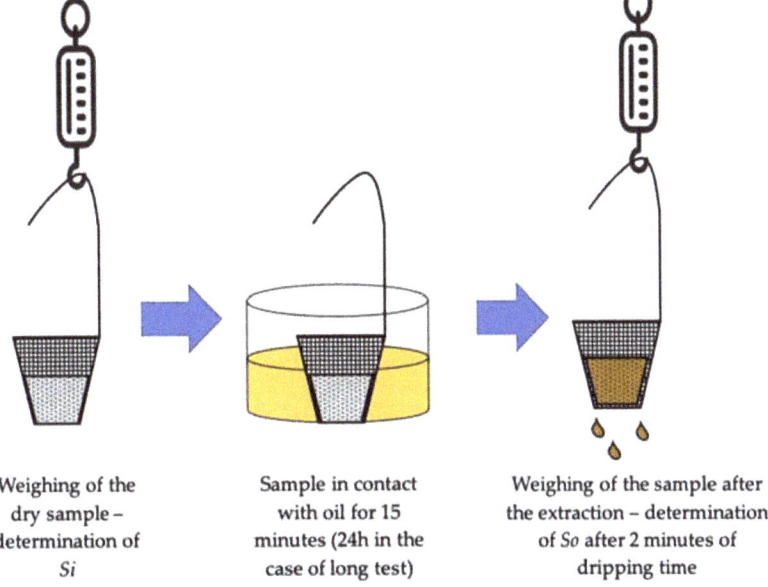

Figure 2. Scheme of the procedure applied for the oil sorption tests.

3. Results and Discussion

3.1. Characterization

The results of the waste PUF characterization are reported in full detail in the Supplementary Materials Section. To summarize, the preliminary characterization revealed that the waste PUF in granular form had residual water content equal to zero, low bulk density (about 50 kg/m³), pH equal to 8, and 10 wt% ash. As expected, it was easily flammable, with LHV equal to 26,900 kJ/kg; this feature requires additional caution due to fire risk when storing and handling the material. Regarding this aspect, recent literature presents some studies on the synthesis of polyurethane with reduced flammability [36]. The chemical analyses highlighted the content of total carbon equal to 65.12%, and the presence of several metals (0.65% Al, 0.32% Fe, 0.23% Zn, 0.66% Ca), whereas for most organic compounds, PCBs, and pesticides, the concentrations were minimal (sum of aromatic compounds 1.95 mg/kg and hydrocarbons C1-C40 0.16%) or under the detection limit of the analytical methods (<0.1–0.2 mg/kg). The results of the UNI EN 12457-2 leaching test were compared with the maximum thresholds limits allowed by the Italian regulation for the recovery of secondary raw materials from non-hazardous waste [37]. Values above the thresholds were found for copper (0.14 mg/L compared with a 0.05 mg/L limit) and fluorides (55 mg/L compared with a 1.5 mg/L limit) only.

The particle-size distribution of the waste PUF (Figure 3) showed that only 0.33 wt% of the sample had dimensions above 10 mm, and it was made of rigid plastic particles. The finest fraction (dimensions below 0.2 mm) was less than 3.5 wt%. The impurities (about the 25 wt% of waste PUF) were made of four main materials: rigid plastic (20 wt%), paper (3.6 wt%), aluminum (1.5 wt%), and polystyrene (0.4 wt%).

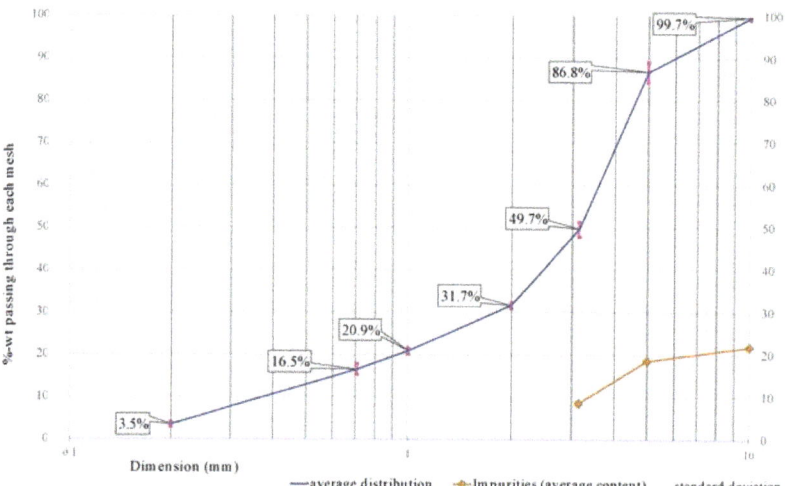

Figure 3. Particle-size distribution of the waste polyurethane foam.

3.2. Pre-Treatment

The two-step sieving pre-treatment resulted in three particle-size fractions from the waste PUF: after the first sieving at 0.71 mm, a "fine" fraction, having dimensions below 0.71 mm (16.5 wt%) and a "coarse" fraction, having dimensions above 0.71 mm (83.5 wt%). After the second sieving at 5 mm to eliminate the impurities, the "intermediate" fraction having dimensions between 0.71 and 5 mm (70 wt%) was obtained.

3.3. Oil Sorption Tests

The sorption tests were performed on the fine and intermediate fractions obtained from the pre-treatments (Sections 2.3 and 3.2), and on the two commercial products—Sepiolite and OKO-PUR. The results of the sorption tests (Figure 4) showed that for both commercial products the lowest sorption capacity was found with diesel fuel (Figure 4a), which has lower density compared with motor oils, which, in turn, directly resulted in lower oil sorption capacity. Sepiolite was characterized by oil sorption capacities between 1.3 and 2 g oil/g sorbent, depending on the type of oil. Similarly, OKO PUR showed oil sorption capacity ranges between 3.2 and 5.2 g oil/g sorbent. In these tests, since still an abundant oil release was observed at 30 s dripping time, the oil sorption capacity was calculated considering the weight of the sample after 2 min of dripping time. After this time interval, variations in oil sorption capacity were negligible, leading to more precise results. Hence, Table 2 reports the oil sorption capacities (relative to 2 min of dripping time) measured for the considered materials.

Table 2. Oil sorption capacities of the tested materials (values expressed in g of oil/g of sorbent). The results are expressed as mean of three replicates with related standard deviation.

Sample Description	Diesel Fuel	Motor Oil 10w40	Motor Oil 20w50
PUF coarse (d > 0.71 mm)	4.51 ± 0.64	5.11 ± 0.50	7.02 ± 1.49
PUF intermediate (0.71 < d < 5 mm)	4.17 ± 0.31	7.36 ± 0.50	6.41 ± 2.41
PUF fine (<0.71 mm)	7.07 ± 0.23	7.72 ± 0.80	10.30 * ± 0.89 *
OKO-PUR	3.26 ± 0.24	4.60 ± 0.30	5.27 * ± 0.11 *
Sepiolite	1.30 ± 0.01	1.34 ± 0.14	1.64 ± 0.06

* long test (24 h contact time).

The best oil sorption performance (Table 2) was achieved by the fine fraction of the waste PUF, which was capable of adsorbing oil for over 7 times the sample weight in all tests. Considering the results of the short test, the highest oil sorption capacity (7.72 g oil/g sorbent) was achieved for the fine PUF with 10w40 motor oil. However, part of the fine PUF and OKO samples did not sink totally in contact with 20w50 oil, resulting in a significant gap in the saturation level of the absorbent. Considering the long test conducted for the fine PUF and OKO samples, the highest oil sorption capacity (10.30 g oil/g absorbent) was achieved for the fine PUF with 20w50 motor oil. Furthermore, the oil sorption capacity curves obtained from the fine PUF (Figure 4) showed that the release of oil after removing the absorbent from the crystallizing dish is significantly constrained. The curves related to the fine PUF are flatter in comparison with the ones obtained from the coarser fractions. One reason for this behavior can be the higher surface area of the fine PUF, which provides a higher number of sites in the micropores of the absorbent where the oil can create bonds. Indeed, it is reasonable to expect that the fine PUF has a higher surface area in comparison with the other samples. Additionally, a practical convenience of the fine powders, compared with coarser absorbents, is the physical state and compactness of the material soaked with oil. The fine PUF loose particles are bonded together by capillarity, resulting in a waste handy to collect after its application (Figure 5). After oil absorption, the PUF can be destined for thermal recovery. Compared with actual management (direct thermal recovery of 100% waste PUF), the scenario would change into 13 wt% waste PUF (i.e., coarse fraction) directly sent to thermal recovery, and 87 wt% reused as oil absorbent and afterward directed to thermal recovery as a secondary option.

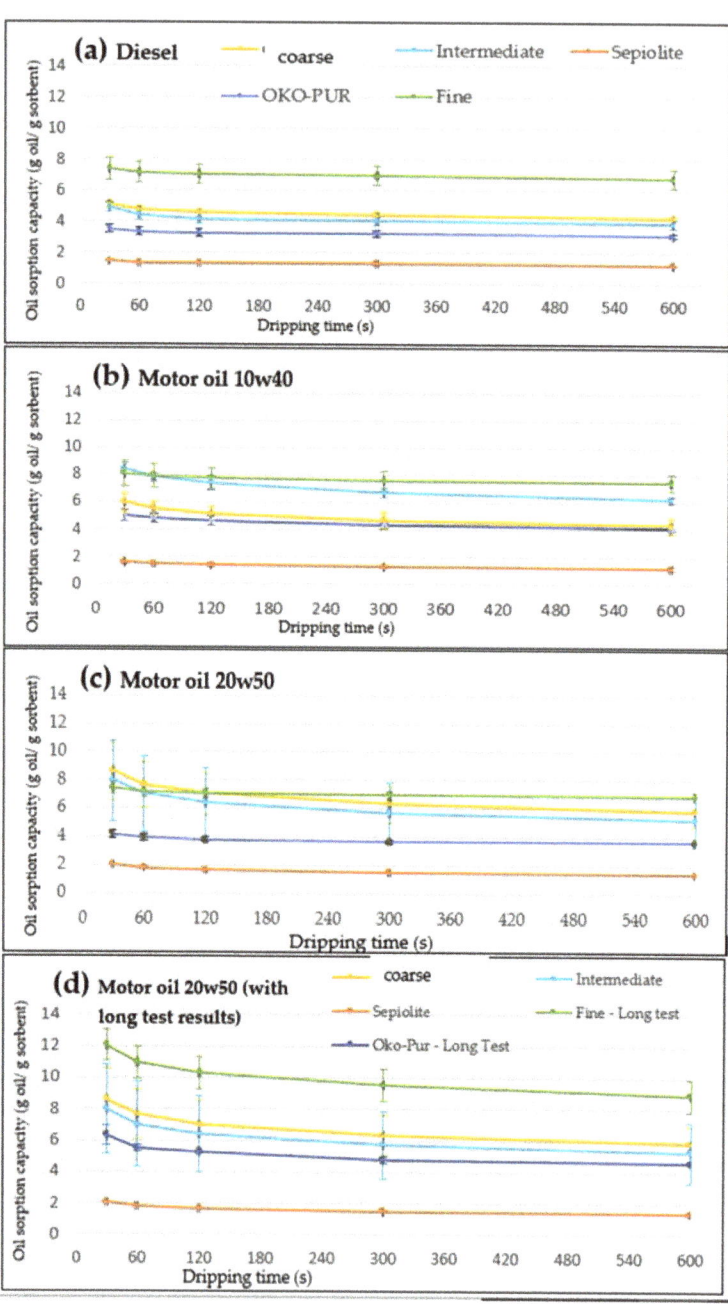

Figure 4. Results of the oil sorption capacity tests performed with (**a**) diesel, (**b**) 10w40 oil, (**c**) 20w50 oil, and (**d**) 20w50 oil in long test (contact time 24 h). The results obtained for the coarse PUF are indicated in the legend as "over 0.71".

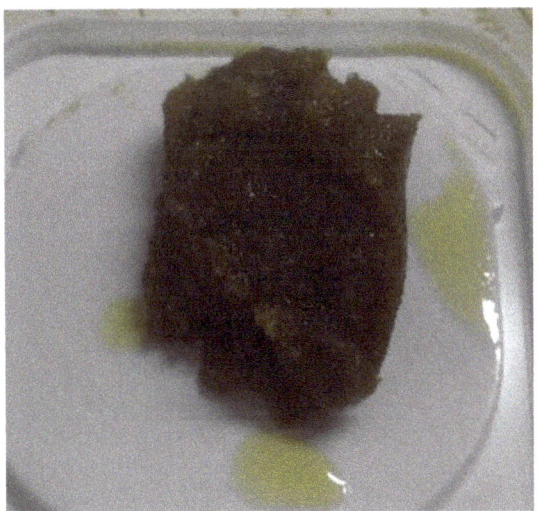

Figure 5. Fine sample of polyurethane foam soaked with 20w50 oil after oil sorption long test.

The other particle-size fractions of the waste PUF were characterized by good oil sorption capacities, with oil sorption values higher for the motor oils than the diesel fuel (Table 2). Specifically, the values achieved were, respectively, 4.51, 5.11, and 7.02 g oil/g for the coarse PUF, and 4.17, 7.36, and 6.41 for the intermediate PUF. These results, considered in comparison with the ones achieved for fine PUF, highlighted a clear improvement in the oil sorption performance when coarser fractions were removed from the waste PUF. Finally, considering the commercial oil sorbents, OKO-PUR demonstrated performances (3.26–5.27 g oil/g) that can be considered comparable with the coarse and intermediate PUF, while sepiolite sorption capacity was worse (1.3–1.6 g oil/g).

To better understand whether the oil sorption capacities of tested PUF samples are satisfactory, the results from similar experimental studies conducted on both commercial and alternative (i.e., waste-derived) materials were compared in Table 3. The absorbent materials derived from recovered wastes, classified in Table 3 according to their origin (industrial, mineral, organic/vegetable), were characterized by oil sorption capacities between 3 and 28.5 g of oil/g of absorbent. Commercial sorbents, instead, were characterized by oil sorption capacities between 3 and 31 g of oil/g of absorbent. Therefore, the performances achieved in this work by waste PUF can be assumed consistent with the literature.

Table 3. Oil sorption capacities reported in the literature for commercial and alternative absorbent materials.

Absorbent Type	Description	Oil Sorption Capacity (g/g) by Short Test	Oil Sorption Capacity (g/g) by Long Test	Dripping Time (s)	Pre-Treatment	Source
Commercial organic sorbents	Reo Amos. Commercial oil sorbent, mixture of loose sorbents in a polypropylene matrix (0.5–5 mm)	Oil 10w40 9.24	-	30	Hydrophobic treatment	[21]
	ACME cellular synthetic sorbent	Diesel 5.8 Medium-density oil 12.3 High-density oil 9.7	Diesel 6.2 Medium-density oil 12 High-density oil 14.3	30 120 for heavy oil	Not available	[38]
	Osprey cellulose-based sorbent	Diesel 18.06 Medium-density oil 21.85 High-density oil 29.96	Diesel 18.11 medium-density 24.60 high-density oil 30.66	30	Not available	[39]
Commercial mineral sorbents	Expanded perlite, from amorphous aluminous silicate (0.5–2.5 mm)	Oil 10w40 3.33	-	30	Industrial treatments	[21]
	Absodan Plus, made from clay minerals (1.5–3 mm)	Oil 10w40 1.06	-	30	Industrial treatments	[21]
	Eco-dry plus, from mixture of rock-based minerals (0.5–2.5 mm)	Oil 10w40 1.37	-	30	Industrial treatments	[21]
Industrial waste-derived sorbents	Aerogels from waste-tire derived textile fibers (recycled)	Oil 5W-30 10.3 (2 h of contact)	-	300 s	Chemical and physical treatment	[40]
	Carton and paper scraps	Diesel 9.6 Oil 0w30 12 Oil 10w30 12	-	30	Surface modification, chemical	[41]
Organic sorbents at experimental level	Microplastics (size 8.6 µm)	Crude oil 1.73 (10 min of contact)	-	60	Sample inserted in an envelope made of polypropylene	[42]
	Polyurethane foam	Diesel 8 Gasoline 18	-	60	Surface modification, chemical	[20]
Sorbents derived from mineral waste	Generic soil (0.1–4 mm)	Oil 10w40 0.45–3.82	-	30	Drying	[21]

Table 3. *Cont.*

Absorbent Type	Description	Oil Sorption Capacity (g/g), by Short Test	Oil Sorption Capacity (g/g) by Long Test	Dripping Time (s)	Pre-Treatment	Source
Sorbents derived from organic and vegetable waste	Sugarcane bagasse (average size 0.2 mm)	Crude oil 3.3–8	-	30	Comminution, Drying	[43]
	Fibers from phragmites australis	Crude oil 5.5	-	30	Comminution, Drying	[43]
	Sugarcane leaves straw	Crude oil 4.5	-	30	Comminution, Drying	[43]
	Switchgrass (average particles size mm)	Oil 10W-30 (2 h of contact) 3.0	-	1800	Grounding	[44]
	Needles from larch, fir, and pine trees	Oil 10w40 4.5	-	30	Drying	[21]
	Beech sawdust	Oil 10w40 7.01	-	30	Drying	[21]
	Spruce sawdust	Oil 10w40 6.54	-	30	Drying	[21]
	Leaf residues	Oil 10w40 15.47	-	30	Drying	[21]
	Moss	Oil 10w40 28.47	-	30	Drying	[21]
	Coconut coir	Vegetable oil 7.2 Diesel fuel 6.5	-	30	Drying	[45]
	Banana peels (average particle size 0.36 mm)	Diesel 6.8 Vegetable oil 7.2	-	30	Comminution, Drying	[46]

4. Conclusions

The idea at the base of this work was to obtain, in the same WEEE treatment plant producing the waste PUF, secondary PUF ready for a new life, which could be then considered a *by-product* according to EU regulations (e.g., "destined for a use for which there is a market and without any harm for the environment and human health, after the application of processes common in the industrial practice") instead of a waste. A simple sieving process performed on waste PUF with two-dimensional cuts (0.71 and 5 mm) allowed the impurities to be separated (dimensions above 5 mm, directly sent to thermal recovery), while the rest could be diverted to reuse as an oil absorbent. The oil sorption performances of waste PUF were promising, compared with commercial mineral (sepiolite) and organic (OKO-PUR) products commonly used to control oil spills. Particularly, the fine fraction of waste PUF (dimensions below 0.71 mm) revealed oil sorption performances at least 2–3 times higher than the commercial products. The performances achieved for waste PUF in this work can be assumed interesting also considering the literature data on commercial and alternative absorbents. This work may pave the way for further research in the field, which is needed to develop "end-of-waste" guidelines for waste PUF, e.g., the operations and requirements defined to convert waste PUF into a secondary raw material according to EU regulations. The approach of this work was consistent with circular economy principles in a difficult sector as WEEE recycling and recovery and allowed to demonstrate that material recovery can be possible for waste PUF and that thermal recovery could be a secondary option.

Supplementary Materials: The following are available online at https://www.mdpi.com/article/10.3390/ma14216230/s1, Table S1: Characterization of the waste PUF.

Author Contributions: Conceptualization, methodology, experimental investigation, data curation, and writing—original draft preparation: V.S.; conceptualization, methodology, supervision, writing—review and editing, project administration, and funding acquisition: S.F. All authors have read and agreed to the published version of the manuscript.

Funding: This research was performed in the framework of the project "Material recovery from WEEE" (in Italian: "*Recupero di materia da RAEE/R1-R2*") funded by the Italian Ministry for Environmental Transition, and ongoing between January 2019 and August 2021. The project was coordinated by Politecnico di Torino and involved as industrial partners, among others, IREN Group and AMIAT. Specifically, this study was based on the activities of work package 1 of the project, dedicated to the recovery of waste PUF from EoL refrigerators.

Conflicts of Interest: The authors declare no conflict of interest.

References

1. CDC RAEE. *Rapporto Annuale 2019-Ritiro E Trattamento Dei Rifiuti Da Apparecchiature Elettriche Ed Elettroniche in Italia*; Centro di Coordinamento RAEE: Milano, Italy, 2020.
2. *European Union Directive on Waste Electrical and Electronic Equipment (WEEE)*; Publications Office of the European Union: Luxembourg, Luxembourg, 2012; pp. 38–71.
3. CDC RAEE. *Manuale Per Le Aziende Di Trattamento Del CDC RAEE-Frigoriferi E Congelatori*; Centro di Coordinamento RAEE: Milano, Italy, 2020; pp. 1–12.
4. Fiore, S.; Ibanescu, D.; Teodosiu, C.; Ronco, A. Improving waste electric and electronic equipment management at full-scale by using material flow analysis and life cycle assessment. *Sci. Total Environ.* **2019**, *659*, 928–939. [CrossRef]
5. Briones-Llorente, R.; Calderón, V.; Gutiérrez-González, S.; Montero, E.; Rodríguez, Á. Testing of the integrated energy behavior of sustainable improved mortar panels with recycled additives by means of energy simulation. *Sustainability* **2019**, *11*, 3117. [CrossRef]
6. Hýsek, Š.; Neuberger, P.; Sikora, A.; Schönfelder, O.; Ditommaso, G. Waste utilization: Insulation panel from recycled polyurethane particles and wheat husks. *Materials* **2019**, *12*, 3075. [CrossRef] [PubMed]
7. Łach, M. Geopolymer Foams—Will They Ever Become a Viable Alternative to Popular Insulation Materials?—A Critical Opinion. *Materials* **2021**, *14*, 3568. [CrossRef]
8. Sonnenschein, M.F. *Polyurethanes: Science, Technology, Markets, and Trends*, 2nd ed.; Grossman, R.F., Domasius, N., Eds.; John Wiley &amp: Hoboken, NJ, USA, 2021; ISBN 9781119669418.

9. The Essential Chemical Industry Polyurethanes. Available online: https://www.essentialchemicalindustry.org/polymers/polyurethane.html (accessed on 21 July 2021).
10. Keshawy, M.; Farag, R.K.; Gaffer, A. Egyptian crude oil sorbent based on coated polyurethane foam waste. *Egypt. J. Pet.* **2020**, *29*, 67–73. [CrossRef]
11. Ouda, Y.W. The Effect of Fulica Atra Feather on Oil Sorption Capacity of Polyurethane Foam. *Polyurethane* **2015**, *5*, 90–94.
12. Zia, K.M.; Bhatti, H.N.; Ahmad Bhatti, I. Methods for polyurethane and polyurethane composites, recycling and recovery: A review. *React. Funct. Polym.* **2007**, *67*, 675–692. [CrossRef]
13. Gómez-Rojo, R.; Alameda, L.; Rodríguez, Á.; Calderón, V.; Gutiérrez-González, S. Characterization of polyurethane foam waste for reuse in eco-efficient building materials. *Polymers* **2019**, *11*, 359. [CrossRef]
14. Gu, L.; Ozbakkaloglu, T. Use of recycled plastics in concrete: A critical review. *Waste Manag.* **2016**, *51*, 19–42. [CrossRef] [PubMed]
15. Jia, Z.; Jia, D.; Sun, Q.; Wang, Y.; Ding, H. Preparation and mechanical-fatigue properties of elastic polyurethane concrete composites. *Materials* **2021**, *14*, 839. [CrossRef]
16. Yang, C.; Zhuang, Z.H.; Yang, Z.G. Pulverized polyurethane foam particles reinforced rigid polyurethane foam and phenolic foam. *J. Appl. Polym. Sci.* **2014**, *131*, 1–7. [CrossRef]
17. Dacewicz, E.; Grzybowska-Pietras, J. Polyurethane foams for domestic sewage treatment. *Materials* **2021**, *14*, 933. [CrossRef] [PubMed]
18. Teodosiu, C.; Wenkert, R.; Tofan, L.; Paduraru, C. Advances in preconcentration/removal of environmentally relevant heavy metal ions from water and wastewater by sorbents based on polyurethane foam. *Rev. Chem. Eng.* **2014**, *30*, 403–420. [CrossRef]
19. Part II—Investigation of the adsorption potential for wastewater treatment. In *Recovery of Waste Polyurethane from E-Waste*; In preparation; 2021.
20. US Environmental Protection Agency Booms|Emergency Response|US EPA. Available online: https://www.epa.gov/emergency-response/booms (accessed on 21 July 2021).
21. Liu, H.D.; Wang, Y.; Yang, M.B.; He, Q. Evaluation of Hydrophobic Polyurethane Foam as Sorbent Material for Oil Spill Recovery. *J. Macromol. Sci. Part A Pure Appl. Chem.* **2014**, *51*, 88–100. [CrossRef]
22. Mojžiš, M.; Bubeníková, T.; Zachar, M.; Kačíková, D.; Štefková, J. Comparison of natural and synthetic sorbents' efficiency at oil spill removal. *BioResources* **2019**, *14*, 8738–8752. [CrossRef]
23. Adebajo, M.O.; Frost, R.L.; Kloprogge, J.T.; Carmody, O.; Kokot, S. Porous Materials for Oil Spill Cleanup: A Review of Synthesis and Absorbing Properties. *J. Porous Mater.* **2003**, *10*, 159–170. [CrossRef]
24. Fingas, M. *The Basics of Oil Spill Cleanup*, 3rd ed.; CRC Press: Boca Raton, FL, USA, 2012; ISBN 9781439862476.
25. Tamsilian, Y.; Ansari-Asl, Z.; Maghsoudian, A.; Abadshapoori, A.K.; Agirre, A.; Tomovska, R. Superhydrophobic ZIF8/PDMS-coated polyurethane nanocomposite sponge: Synthesis, characterization and evaluation of organic pollutants continuous separation. *J. Taiwan Inst. Chem. Eng.* **2021**, *125*, 204–214. [CrossRef]
26. Wang, Z.; Ma, H.; Chu, B.; Hsiao, B.S. Super-hydrophobic polyurethane sponges for oil absorption. *Sep. Sci. Technol.* **2017**, *52*, 221–227. [CrossRef]
27. Borreguero, A.M.; Zamora, J.; Garrido, I.; Carmona, M.; Rodríguez, J.F. Improving the hydrophilicity of flexible polyurethane foams with sodium acrylate polymer. *Materials (Basel)* **2021**, *14*, 2197. [CrossRef]
28. Cortez, J.S.A.; Kharisov, B.I.; Quezada, T.E.S.; García, T.C.H. Micro- and nanoporous materials capable of absorbing solvents and oils reversibly: The state of the art. *Pet. Sci.* **2017**, *14*, 84–104. [CrossRef]
29. Deng, Y.; Dewil, R.; Appels, L.; Ansart, R.; Baeyens, J.; Kang, Q. Reviewing the thermo-chemical recycling of waste polyurethane foam. *J. Environ. Manage.* **2021**, *278*, 111527. [CrossRef] [PubMed]
30. Shin, S.R.; Kim, H.N.; Liang, J.Y.; Lee, S.H.; Lee, D.S. Sustainable rigid polyurethane foams based on recycled polyols from chemical recycling of waste polyurethane foams. *J. Appl. Polym. Sci.* **2019**, *136*, 1–9. [CrossRef]
31. Zahedifar, P.; Pazdur, L.; Vande Veld, C.M.L.; Billen, P. Multistage chemical recycling of polyurethanes and dicarbamates: A glycolysis–hydrolysis demonstration. *Sustainability* **2021**, *13*, 3583. [CrossRef]
32. Gu, X.; Luo, H.; Lv, S.; Chen, P. Glycolysis recycling of waste polyurethane rigid foam using different catalysts. *J. Renew. Mater.* **2021**, *9*, 1253–1266. [CrossRef]
33. Godinho, B.; Gama, N.; Barros-Timmons, A.; Ferreira, A. Recycling of different types of polyurethane foam wastes via acidolysis to produce polyurethane coatings. *Sustain. Mater. Technol.* **2021**, *29*, e00330. [CrossRef]
34. ASTM International Designation: F726-17. In *Standard Test Method for Sorbent Performance of Adsorbents for Use on Crude Oil and Related Spills*; ASTM International: West Conshohocken, PA, USA, 2017; pp. 1–6. [CrossRef]
35. Bazargan, A.; Tan, J.; McKay, G. Standardization of Oil Sorbent Performance Testing. *J. Test. Eval.* **2015**, *43*, 20140227. [CrossRef]
36. Kuranska, M.; Prociak, A.; Beneš, H.; Sałasinska, K.; Malewska, E.; Polaczek, K. Development and Characterization of " Green Open-Cell Polyurethane Foams" with Reduced Flammability. *Materials* **2020**, *13*, 5459. [CrossRef] [PubMed]
37. Ministro dell'Ambiente Documentazione DECRETO MINISTERIALE 5 febbraio 1998, Gazzetta Ufficiale n.88 del 16-4-1998 - Suppl. Ordinario n. 72. Available online: https://www.gazzettaufficiale.it/atto/vediMenuHTML?atto.dataPubblicazioneGazzetta=1998-04-16&atto.codiceRedazionale=098A3052&tipoSerie=serie_generale&tipoVigenza=originario (accessed on 21 July 2021).
38. Emergencies Engineering Technologies Office. *ASTM Sorbent Test Program 1999–2000 Interim Report*; SAIC (Science Application International Corporation) Canada: Gloucester, ON, Canada, 2000; pp. 1–24.

39. Galicki, S. *Sorbency Evaluation*; Millsaps College Sorbent & Environmental Laboratory: Jackson Mississippi, United States; Osprey Spill Control: Virginia Beach, VA, USA, 2014.
40. Thai, Q.B.; Le-Cao, K.; Nguyen, P.T.T.; Le, P.K.; Phan-Thien, N.; Duong, H.M. Fabrication and optimization of multifunctional nanoporous aerogels using recycled textile fibers from car tire wastes for oil-spill cleaning, heat-insulating and sound absorbing applications. *Colloids Surfaces A Physicochem. Eng. Asp.* **2021**, *628*, 127363. [CrossRef]
41. Demirel Bayık, G.; Altın, A. Conversion of an industrial waste to an oil sorbent by coupling with functional silanes. *J. Clean. Prod.* **2018**, *196*, 1052–1064. [CrossRef]
42. Martins, L.S.; Zanini, N.C.; Botelho, A.L.S.; Mulinari, D.R. Envelopes with microplastics generated from recycled plastic bags for crude oil sorption. *Polym. Eng. Sci.* **2021**, *61*, 2055–2065. [CrossRef]
43. Sutar, M.S. Oil Spillage Treatment Using Sorbent. In Proceedings of the the International Conference on Global Trends in Engineering, Technology and Management (ICGTETM), Bambhori, Jalgaon, India, 22–24 December 2017; pp. 302–311.
44. Tripathi, J.; Arya, A.; Ciolkosz, D. Switchgrass as oil and water-spill sorbent: Effect of particle size, torrefaction, and regeneration methods. *J. Environ. Manage.* **2021**, *281*, 111908. [CrossRef] [PubMed]
45. Osamor, A.; Momoh, Z. An Evaluation of the Adsorptive Properties of Coconut Husk for Oil Spill Cleanup. In Proceedings of the Conference on Advances in Applied science and Environmental Technology—ASET 2015, G Tower Hotel, Kuala Lumpur, Malaysisa, 11–12 April 2015; Volume 1, pp. 33–37.
46. Alaa El-Din, G.; Amer, A.A.; Malsh, G.; Hussein, M. Study on the use of banana peels for oil spill removal. *Alexandria Eng. J.* **2018**, *57*, 2061–2068. [CrossRef]

Article

A Study of Treatment of Industrial Acidic Wastewaters with Stainless Steel Slags Using Pilot Trials

Mattia De Colle *, Rahul Puthucode, Andrey Karasev and Pär G. Jönsson

Department of Materials Science and Engineering, School of Industrial Engineering and Management, KTH Royal Institute of Technology, SE-100 44 Stockholm, Sweden; rahulpu@kth.se (R.P.); karasev@kth.se (A.K.); parj@kth.se (P.G.J.)
* Correspondence: mattiadc@kth.se; Tel.: +46-765768409

Citation: De Colle, M.; Puthucode, R.; Karasev, A.; Jönsson, P.G. A Study of Treatment of Industrial Acidic Wastewaters with Stainless Steel Slags Using Pilot Trials. *Materials* **2021**, *14*, 4806. https://doi.org/10.3390/ma14174806

Academic Editor: Franco Medici

Received: 5 July 2021
Accepted: 23 August 2021
Published: 25 August 2021

Publisher's Note: MDPI stays neutral with regard to jurisdictional claims in published maps and institutional affiliations.

Copyright: © 2021 by the authors. Licensee MDPI, Basel, Switzerland. This article is an open access article distributed under the terms and conditions of the Creative Commons Attribution (CC BY) license (https://creativecommons.org/licenses/by/4.0/).

Abstract: Different stainless steel slags have been successfully employed in previous experiments, for the treatment of industrial acidic wastewaters. Although, before this technology can be implemented on an industrial scale, upscaled pilot experiments need to be performed. In this study, the parameters of the upscale trials, such as the volume and mixing speeds, are firstly tested by dispersing a NaCl tracer in a water bath. Mixing time trials are used to maintain constant mixing conditions when the volumes are increased to 70, 80 and 90 L, compared to the 1 L laboratory trials. Subsequently, the parameters obtained are used in pH buffering trials, where stainless steel slags are used as reactants, replicating the methodology of previous studies. Compared to laboratory trials, the study found only a minor loss of efficiency. Specifically, in previous studies, 39 g/L of slag was needed to buffer the pH of the acidic wastewaters. To reach similar pH values within the same time span, upscaled trials found a ratio of 43 g/L and 44 g/L when 70 and 90 L are used, respectively. Therefore, when the kinetic conditions are controlled, the technology appears to be scalable to higher volumes. This is an important finding that hopefully promotes further investments in this technology.

Keywords: stainless steel slag; recycling; acidic wastewater treatment; upscale trials; mixing time

1. Introduction

The mining and metal industry is considered to be a mature sector, in terms of technology development and R&D spending. In 2011, Filippou and King [1] estimated that less than 1% of the revenues of the whole sector were used to finance R&D projects. The authors claimed that this declining trend started around the 1980s. There are several compounding factors that hinder innovation in this sector. The most important one is the prohibitive startup costs that often require big capital investments. Opening new mines or metallurgical plants, as well as changing industrial processes or feedstock materials, are inevitably connected to big financial uncertainties. Therefore, a conversative thinking has usually been preferred. The result is a phenomenon called "technological lock-in", which is defined as the tendency to maintain the status quo, due to several barriers impeding disruptive changes from happening [2].

Although, the mining and metals industry is already under pressure by several environmental constraints that require immediate actions. For example, the reduced quality of ores increases the percentage of waste generated [3]. Moreover, the rising carbon emissions pricing [4] is a serious threat to the profitability of the whole sector. Governments are tightening the restrictions around the generation of waste and the opening of new landfills, nudging companies for a change [5]. The public is also pushing for a greener and more sustainable production of goods, adding pressure to the companies to strive for an improved sustainability.

Metallurgical slags are among the most abundant by-products that are generated by the metals industry. Luckily, in many European countries, the technology to recycle those materials is mature enough to avoid excessive landfilling, mostly by employing slags as construction materials [6]. Although, in the case of stainless steel slags, such applications are not viable, due to the composition of the material, forcing their producers to dispose most of their by-products in landfills. In fact, the presence of high concentrations of heavy metals, such as chromium, impedes the use of these slags in such applications [7]. The production of slag in 2015, in Sweden alone, accounted for 1.35 Mton, of which 0.27 Mton came from stainless steel production, which is roughly 20%, as highlighted by the Swedish steel agency Jernkontoret in its latest report [8]. Although, the report shows that blast-furnace (BF) and basic oxygen furnace (BOF) slags are almost entirely valorized. The same applies for low-alloyed electrical arc furnace (L-EAF) slags. The most critical slags to valorize are high-alloyed electrical arc furnace (H-EAF) slags and argon oxygen decarburization (AOD) slags, which are both derived from processes that are associated with the production of stainless steel. By looking at the total of landfilled output, rather than the total amount produced, these slags constitute more than 70% of the total. It is evident that a solution for these kinds of materials is needed. Despite stainless steel being a niche production of a larger subset, it is responsible for most of the waste generation of the whole category.

The authors of this study have previously proposed a new application of stainless steel slags, which can contribute to the overall reduction in the landfilled output. Several stainless steel slags have been successfully tested as lime replacements for the treatment of industrial acidic wastewaters, generated in situ during the pickling process [9]. Lime is frequently used for the treatment of acidic wastewaters, but the reaction products are also often landfilled. Substituting slag with lime is projected to decrease the material input of stainless steel producers, thus decreasing the tons of by-products that are generated as a result. Moreover, if slag can be used as an effective reagent for the treatment of all kinds of industrial wastewaters, the sustainability of the whole stainless steel sector largely increases.

Validating new technologies, especially for mature sectors, where the R&D budget is limited and a large capital is required, can be particularly difficult. Laboratory tests can only provide limited knowledge for the adoption of new technologies. Therefore, the aim of this study is to strengthen the technology validation, by replicating the same experiments that are conducted in laboratory settings in upscaled pilot trials. The primary focus of this study is to test whether the pH buffering of industrial acidic wastewaters, using stainless steel slag, can be replicated for bigger volumes than the one used in precedent laboratory trials [9,10]. Moreover, it is of interest to this study to determine the relationship between the volume of wastewaters to treat and the amount of slag needed to do so, when the kinetic conditions are controlled.

2. Materials and Methods

The first part of the study was to design a physical model that could control and replicate the same kinematic conditions when different volumes of wastewaters are tested. In fact, if the goal is to compare chemical reactions happening when using different volumes, the mixing conditions need to be the same to ensure a good comparison of the results. A physical model that relied on the dispersion of NaCl in a water bath was used to determine the relevant parameters to maintain a similar mixing performance across different volumes. Successively, the pH buffering trials were performed with the use of wastewaters and slags, utilizing the parameters determined by the physical model. The same methodology applied in precedent laboratory trials [9,10] was replicated to evaluate the amount of slag needed to buffer the pH to a value of approximately 9. The pH value of 9 was decided as a target to replicate the industrial processes, which use lime as a reactant to rise the pH level of the treated wastewaters.

2.1. Physical Model Design

How fast added slag can spread in the volume of wastewaters is one of the most important parameters to control when volumes are upscaled from laboratory conditions.

When using small beakers (~1 L), the spreading of the slag can be considered to be instantaneous, given the small size of the container and the speed of the vortex generated by the magnetic stirrer. A homogenous spreading ensures that pH measurements performed at any given point of the volume are representative of the volume's pH level. On the contrary, if the mixing phase is not instantaneous, until the slag is homogeneously distributed, it will react unevenly with the wastewaters across the whole volume. This means that the pH measured in any point during that phase of the upscale trial will likely not represent the pH value of the whole volume. This concern is the highest at the beginning of the trials, and it becomes less so the more the trials continue, since the slag has more time to spread in the volume. Thus, it is important to estimate how fast slag can homogenously distribute in the volume and try to minimize this time in comparison to when the first pH measurement occurs. In conclusion, to achieve kinematic conditions comparable across different volumes of wastewaters, an assessment of the homogenization speed of the solution needs to be performed.

To evaluate the time needed to reach a homogenous distribution of slag, the authors relied on a common method used in the field of metallurgy, which uses a physical water model to assess the kinematic conditions of AOD converters [11,12]. In particular, to determine the mixing performance of the stirring methods applied in agitated tanks, a parameter called "mixing time" has been extensively used. According to a literature survey conducted on the topic [13], there are two types of methods to determine the mixing time: the first is using one or multiple probes to measure local quantities (such as pH, conductivity or temperature). The second relies on global methods that are based on chemical reactions or optical analyses. Each method presents different challenges and limitations, so several sub-categories among these two types of models have been designed to circumvent some of them. However, no method imposed itself as the standard for this kind of investigation. In fact, for AOD converter simulations, several options have been used in the years. Wupperman et al. [14] used a photometer to detect the variation in color when a blue tracer is injected in a transparent tank filled with water. Moreover, Samuelsson et al. [15] used a local method of investigation instead, measuring the variation in pH when CO_2 is injected in a tank filled with a $NaOH-H_2O$ solution. Other studies relied on models that measured the dispersion of a tracer (usually NaCl or KCl solutions) instead, in a tank filled with water [16–20]. The tracer changes the electric conductivity of pure water, thus conductivity probes can be used to evaluate its variation over time. This last method, specifically with using a NaCl tracer, was also used in this study.

To minimize the effect of local measurements, after the tracer is inserted, two probes placed at two different positions were used to measure the local conductivities of the solution over time. In accordance with the majority of the studies analyzed in the literature survey reported above, the water bath is considered to be homogenized when the conductivity of the probe C_i is equal to its final value $C_{final} \pm 5\%$. The last time when C_i falls outside the range $C_{final} \pm 5\%$ is called the "mixing time" (hereafter denoted as T_m). In other words, the "mixing time" is defined as the time after which the conductivity (C_i) differs by no more than 5% of its final value. When two or multiple probes are used, T_m is defined as the greatest value between all the individual mixing times. A graphical example of how T_m can be determined from the conductivity measurement of a single probe is shown in Figure 1, where the ratio C_i/C_{final} is plotted as a function of time. In the example, T_m (36 s) is calculated as being the last time when the curve exceeds the 1.00 ± 0.05 range.

Trial 1: Mixing Time Trials

As previously mentioned, the T_m value will be used to determine a set of parameters that will ensure that the same kinematic conditions are reached, when different volumes are tested. Moreover, if the T_m value is sufficiently smaller compared to the time when the first pH measurement occurs, the conditions in an upscale environment can be compared to the ones obtained in laboratory conditions. In previous experiments [9,10], a 1 L Erlenmeyer flask was used to treat the wastewaters with slag. It is assumed that for the low volumes and high mixing speeds selected in those trials, the kinematic conditions were shape-invariant. Therefore, a standard 200 L cylindrical plastic drum was chosen to contain the wastewaters. An overhead electric engine, paired with a stainless steel 3-blade impeller, was chosen as the stirring mechanism. The engine was secured on top of a steel frame, which also surrounded the drum.

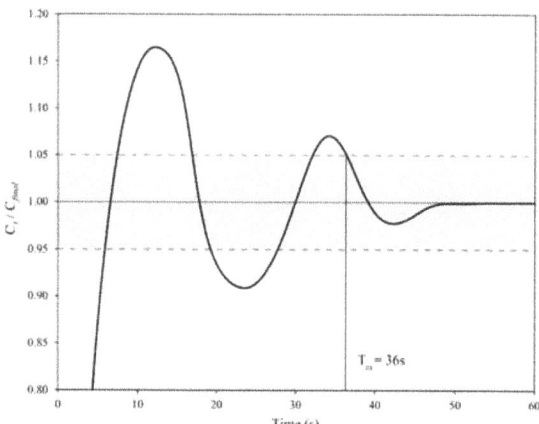

Figure 1. Graphical example of how mixing time can be calculated given the ratio C_i/C_{final} over time.

The engine provided three rotational speeds of 225, 200 and 175 rpm, while the volumes chosen for the upscaled trials were 70, 80 and 90 L. The following several factors restricted the choice of the volumes: first, as the container remains the same, the ratio between height and radius of each volume changes. In addition, the position of the impeller relative to the ground was fixed too, meaning that its distance from the top of the volume increased with the increase in the volume itself. These two factors combined caused the vortex created by the engine to be different for each volume. Therefore, such differences needed to be minimized to maintain comparable T_m values for different volumes. Another restriction was set by the engine torque. In fact, the bigger the volume of water, the harder it was for the engine to spin the mass of liquid. This could result in excessive overheating and potential long-term failures.

Another important factor to consider was that the method chosen to evaluate the T_m value relies vastly on the position of the conductivity probes. Therefore, several positions were tested to have a proper assessment of the parameter. Since the drum can be approximated to a cylinder, thanks to its symmetrical properties there are only a limited number of combinations to test in order to cover the entirety of the volume. By assuming that the points that are harder to reach for the tracer are the ones at the borders of the volume, rather than at the center, 5 points were selected to test different probe positions. The points chosen are shown in Figure 2. Different combinations of rotational speeds of the engine, volumes of water and probe positions were tested. For almost all combinations, triplicate measurements were made to ensure good replicability. In total 30 trials were performed, and they were grouped according to Table 1.

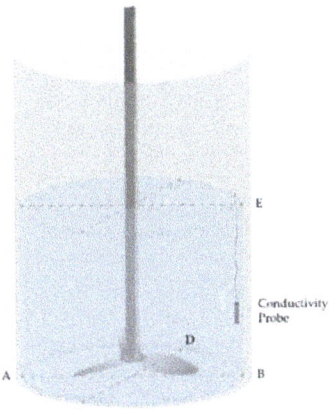

Figure 2. Schematic representing the mixing time trial setup. The probe positions are classified by the letters A, B, C, D and E.

Table 1. Summary of the mixing time trials classified by the rotational speed, volume of water and probe positions.

Trial No.	Volume (L)	Probe Positions	Speed (rpm)
1–3	90	A & B	225
4–6	80	A & B	225
7–9	70	A & B	225
10–12	80	A & B	200
13–15	70	A & B	175
16–18	90	A & B	175
19–21	90	C & D	225
22–24	80	C & D	200
25–27	70	C & D	175
28	90	A & E	225
29	80	A & E	225
30	70	A & E	225

The primary focus of the investigation was understanding whether the rotational speed needed to be adjusted per volume, or if the T_m value was speed-invariant. In this study the rotational speed was considered to be the most important factor in changing the kinematic conditions of the trial, so it was the parameter for which most of the measurements were carried out. For those trials, the two probes were positioned in an A & B configuration, as observed in Figure 2. The bottom of the drum, especially in its peripheral positions, was assumed to be the hardest part to reach by the added tracer. Therefore, the A & B and C & D configurations were preferred. In fact, it was assumed that the T_m values measured when the probes were placed at the bottom of the barrel would be higher than the ones obtained if the probes were placed on top of the volume of water. To test this hypothesis, the probes were also positioned in an A & E configuration.

For each trial, a 20 mL of 20 wt% NaCl solution was used as a tracer. The trial started by activating the engine. Thereafter, the conductivity of the water bath was measured throughout the experiment. Specifically, at t = 0 s the tracer was inserted, and the change in conductivity was measured by the two probes every second. The trials stopped once both conductivity measures plateaued on a stable value. The T_m value was then calculated for each probe individually, and the value for the trial was selected as the highest between the two measured values.

2.2. pH Buffering Trials

The final goal of this study was to compare the results of the upscaled trials with previous laboratory experiments, to check whether the use of the same metallurgical slags as reactants for the treatment of acidic wastewaters can be replicated in an upscaled environment. Moreover, the trials aim at finding the relationship between the volume of treated wastewaters and the amount of slag being used, when the mixing conditions are kept constant. Therefore, the pH buffering methods chosen for the wastewater treatment and the material properties were kept aligned with those used in previous trials, to ensure a proper comparison between them.

2.2.1. Sample Preparation

Given the large amount of wastewater needed per trial, the industrial upscale experiment was conducted in situ at Outokumpu Stainless, in Avesta (SWE). Therefore, compared to previous studies performed on four different slags, only two where available on site, but only one (namely, slag type "O1") did not require being crushed [9]. Although, before using the material as retrieved from the slag yard, some operations were still needed. First, the slag provided was quite wet since it is usually water cooled when it is being disposed. Moreover, it is also stored in an outdoors environment. Slag wetness is not a problem intrinsically. However, to have a reliable estimation of the weights used, the experiments require a dry sample. Additionally, there were several impurities present in the slag, such as gravel or residual rock pieces, so sieving the material was required to remove the impurities. Therefore, the material was dried at 105 °C for 24 h and later sieved through a mesh of 350 μm for 15 min. This did not precisely replicate the conditions used in previous studies (which used a mesh of 1 mm). However, according to previous particle size analysis conducted on the same slag [9], 87% of the volume of the slag was made of particles smaller than 350 μm, ensuring, in principle, a good comparison between the two powders being used.

2.2.2. Trial 2: Replication of Stepwise and Single-Step Dosing Methodology

Despite the constraints posed by the new environment, to effectively compare the results from the current upscale trial to the ones obtained by previous studies [9,10], the same methodology needs to be replicated too. Therefore, the "step-wise dosing" and "single-step dosing" methodologies developed previously, are used once again in this study. Since the single-dosing method operates by trial and error, by adjusting the quantity of reactant necessary to perform the trials, it usually requires a lot of attempts before the optimal quantity to reach pH 9.0 ± 0.2 is found. Therefore, a preliminary stepwise dosing methodology was used to narrow the range of investigation and trial numbers. In previous experiments [9,10], this method has been useful in correctly identifying the order of magnitude of the slag mass to use, avoiding unnecessary waste of materials during the single-step dosing method. Nine trials have been performed during this part of the study and their characteristics are listed in Table 2.

Table 2. Summary of stepwise and single-step dosing trials classified by volume, rotational speed and number of replications performed.

Type	Volume (L)	Speed (rpm)	Replications
Stepwise	90	175	2
Stepwise	70	175	1
Single-step	90	175	3
Single-step	70	175	3

The experimental procedure for the stepwise dosing trials was as follows:

1. At t = 0 a very small amount of reactant of weight w_1 was added to the wastewaters. The reactant was inserted at the center of the vortex to promote good mixing.

2. At t = 10 min and t = 20 min the pH value of the volume was measured.
3. At t = 30 min the pH value was measured again, if $|pH_{30} - pH_{20}| \geq 0.3$ the pH was measured again at intervals of 10 min until $|pH_{i+10min} - pH_i| \leq 0.3$. Once the previous condition was met, if the pH value was not equal to 9.0 ± 0.2 the procedure was repeated from step 1 adding a new weight w_2.
4. When the pH value was 9.0 ± 0.2, the trial was stopped. The total amount of reactant w_{tot} was calculated, by summing all the weights used in the various cycles performed to reach the final pH value.

A schematic representation of the pH evolution over time of a stepwise dosing trial is shown in Figure 3.

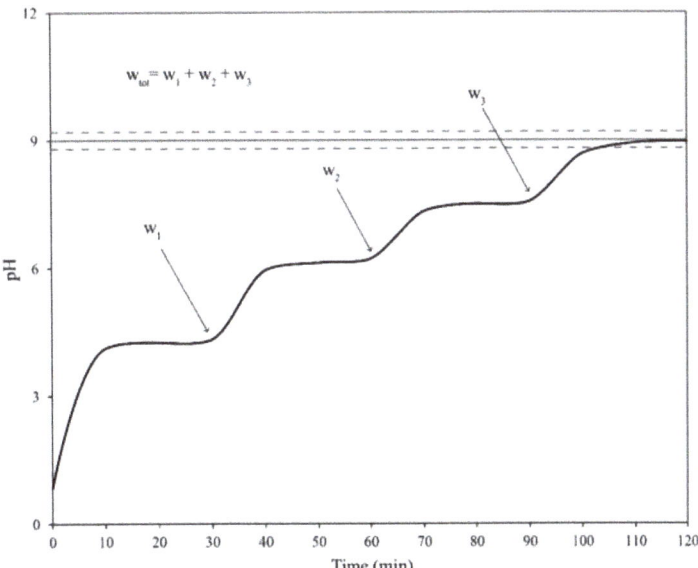

Figure 3. A schematic representation of a stepwise dosing trial with three additions of reactant.

In the case of single-step dosing trials, the procedure was also replicated from previous studies [9,10], changing only the times when the pH values were measured. A schematic representation of the pH evolution over time of a single-step dosing trial is shown in Figure 4. The experimental procedure for the single-step dosing trials was as follows:

1. By using the quantities measured during the stepwise dosing trials as a benchmark, an appropriate amount of reactant of known weight was mixed with the wastewaters at t = 0 min. The reactant was inserted at the center of the vortex to promote good mixing.
2. The pH value was measured at t = 10, 20, 30, 40, 50, and 60 min.
3. If the pH reached the value of 9.0 ± 0.2 at the 30 min mark, the trial was considered to be successful.

For both stepwise dosing and single-step dosing methods, all pH measurements were performed by extracting a wastewater sample of approximately 20 mL from the surface of the agitated tank, close to the vortex generated by the impeller. The pH value of the sample was registered when a stable measurement could be obtained. Afterwards, the liquid inside the sample was poured back into the agitated tank.

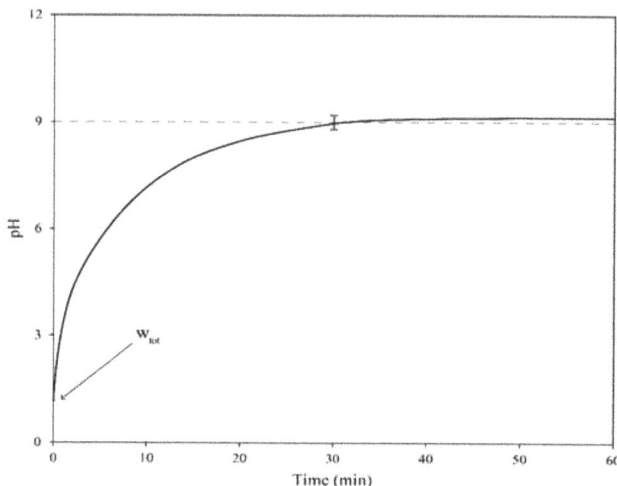

Figure 4. A schematic representation of a single-step dosing trial.

3. Results and Discussion

3.1. Trial 1: Mixing Time Trials

The results of the first set of 18 trials, with the probes in the A & B configuration, are compounded in Table 3. Per combination of rotational speed and volume of water used, triplicate measurement trials were produced. The T_m values of each trial were averaged together, to provide a more statistically significant measure. In the first nine trials, volumes of 90, 80, and 70 L were stirred at 225 rpm. These trials were conducted to test how much changing the volume could influence the T_m value, when the stirring speed remains unchanged. The average T_m value slightly decreased with the decrease in the volume, from 10.3 s, when 90 L were used, to 9.7 s when 80 L were used, and finally 9.3 s when 70 L were used. The variation is in the range of 1 s.

Table 3. Mixing time trials with probe positions in A & B.

Trials #	Volume (L)	Probe Positions	Speed (rpm)	T_m (s)	Average (s)
1	90	A & B	225	11	
2	90	A & B	225	10	
3	90	A & B	225	10	10.3
4	80	A & B	225	10	
5	80	A & B	225	9	
6	80	A & B	225	10	9.7
7	70	A & B	225	9	
8	70	A & B	225	10	
9	70	A & B	225	9	9.3
10	80	A & B	200	10	
11	80	A & B	200	12	
12	80	A & B	200	9	10.3
13	70	A & B	175	11	
14	70	A & B	175	10	
15	70	A & B	175	10	10.3
16	90	A & B	175	10	
17	90	A & B	175	11	
18	90	A & B	175	10	10.3

To test the relationship between different volumes of water and rotational speeds even further, nine other trials were conducted. Here, the tank was filled with 80 L of water and the volume was stirred three more times at 200 rpm, while the volumes of 90 L and 70 L were also stirred three times at 175 rpm. Compared to the 225 rpm trials, when a volume of 80 L was stirred at 200 rpm, there was a slight increase in the T_m value, with an average value of 10.3 s. The same average T_m value of 10.3 s can be found in the 70 L triplicates that were stirred at 175 rpm. At first glance, it might seem appropriate to decrease the rotational speed to accommodate for the decreased volume. However, the same average T_m value of 10.3 s was also found when a volume of 90 L was stirred at 175 rpm. The most plausible interpretation is that the precision of the measurements is not sufficient to detect a meaningful variation in the T_m values, when parameters such as volume and rotational speed are changed in the ranges chosen for this study.

Twelve additional trials were performed, to test whether different probe positions could alter the T_m values. The results are compounded in Table 4. A water volume of 90 L was stirred at 225 rpm and the probes were positioned at the bottom, at a 90° angle, compared to the configuration A & B, namely, the positions that are indicated as C & D in Figure 2. Once again, the average T_m value was comparable to previous results. The variations detected were attributed to the imprecision of the test, rather than being correlated with the probe positions. The same can be said when the trials are performed using the probe configuration C & D, applied to volumes of 80 L and 70 L, stirred a 200 and 175 rpm, respectively.

Table 4. Mixing time trials with probe positions C & D and A & E.

Trials #	Volume (L)	Probe Positions	Speed (rpm)	T_m (s)	Average (s)
19	90	C & D	225	10	
20	90	C & D	225	9	
21	90	C & D	225	9	9.3
22	80	C & D	200	9	
23	80	C & D	200	9	
24	80	C & D	200	10	9.3
25	70	C & D	175	9	
26	70	C & D	175	10	
27	70	C & D	175	10	9.6
28	90	A & E	225	10	
29	80	A & E	225	8	
30	70	A & E	225	7	

Three more trials were performed with one probe located at the bottom of the tank, in position A, and one located in radial symmetry to it, but on the top of the tank, in position E. These trials were performed to test whether the probe at the bottom of the tank could measure a faster T_m value than the one at the top. In those trials, the contrary happened: in the three trials that were tested, the T_m values when the probes were positioned in A & E of 10 s, 8 s, and 7 s were measured. Therefore, when two probes are positioned at the bottom of the tank, either one of the two takes more time to reach its final conductivity value, compared to when one is positioned at the bottom and one at the top.

To summarize, a certain imprecision in the obtained results was expected to appear. This was also indicated in a study by Wupperman et al. [14], which detected a lower accuracy of local measurements compared to global ones. Although, for the scope of this study, the precision obtained during these trials is more than enough to ensure similar kinematic conditions when using different volumes. In fact, it is important to consider the context in which this physical model has been used. The experiments were designed to evaluate the time needed to achieve a homogenous spreading of the NaCl tracer in water, as a proxy for the spreading of slag. Although, the physical model is not very representative of what happens during the pH buffering trials. The ratio between the slag and wastewaters is greater than the one between the tracer and water. Moreover, slag does not fully dissolve as salt does, and it remains in a suspension in the liquid phase. Although, the time needed to homogenize the water bath after the NaCl injection, with the parameters chosen, is an order of magnitude lower compared to the time when the first pH measurement occurs. Thus, it is safe to assume that by the time the first pH measurement occurs, the slag is also homogenously distributed in the wastewaters. Moreover, a variation in the mixing time, between 9 and 10 s, can be considered to be negligible.

3.2. Trial 2: Replication of Stepwise and Single-Step Dosing Methodology

The pH buffering trials were separated in two different investigation methods. As for previous studies [9,10], the stepwise dosing method was used to narrow the range of reactant needed to buffer the pH to a value of 9.0 ± 0.2. The results are collected in Table 5. All the trials successfully reached the target of pH 9.0 ± 0.2. The quantity of slag needed for the stepwise dosing was 3 kg when a volume of 90 L was tested, and 2 kg when the volume was 70 L. Although, it is experimentally proven that the quantity of slag needed in the stepwise dosing method is always lower than the single-step dosing method [9,10]. This is because the latter method sets a time limit for the chemical reactions to happen (≤30 min, in accordance with the industrial requirements), which influences the amount of reactant to buffer the pH value to the desired target. Thus, a higher amount of slag is needed to reach the target pH value in 30 min, compared to the same quantity that can dissolve in a longer amount of time. This phenomenon did not happen when lime was used, which shows that lime has a higher reactivity than slag. In the case of lime, the quantity that was measured during the stepwise dosing was roughly the same as single-step dosing, due to how fast the lime powders reacted with the wastewaters [9].

Table 5. Stepwise dosing trials liters, rpm, total mass of added slag, and final pH obtained.

Trial No.	Volume (L)	Speed (rpm)	m (kg)	Final pH
I	90	175	3	8.8
II	90	175	3	8.9
III	70	175	2	8.8

Given the results from the stepwise dosing tests, 4000 g of slag was chosen as the starting quantity for the first single-step dosing trial, performed with 90 L of wastewaters (44.4 g/L). The target for the trial (IV-1) was to obtain a pH value of 9.0 ± 0.2 at the 30 min mark. The trial IV-1 succeeded in reaching pH_{30min} = 8.8. Thus, the same quantity was used again, to replicate the results and guarantee better reliability of the measurements. The second trial (IV-2) was also successful in reaching pH_{30min} = 9.1, but the third trial (IV-3) missed the mark, reaching only pH_{30min} = 7.7. The pH measurements of all the trials are shown in Figure 5a. It is interesting to point out that, despite the pH_{30min} value for the third trial widely differs from the other two, the pH_{60min} value is much more aligned. In fact, while trials IV-1 and IV-2 quickly plateau to a pH value of approximately nine, trial IV-3 reaches the same value with a slower rate. The same exact situation was obtained during the trials V-1, V-2, and V-3, which were performed with 70 L of wastewater and 3000 g of slag (43 g/L). In fact, trials V-1 and V-2 successfully hit the target value at the 30th minute, while trial V-3 lagged to reach the final pH value. Trial V-3 reached pH_{30min} = 7.0, compared to 9.2 and 9.0 of the two precedent trials, as shown in Figure 5b. A possible explanation to this phenomenon can be found in the variation in the wastewater composition. In fact, the wastewaters were extracted each time with a pump from the continuous flow of the industrial processes. Hence, the composition was not controlled during the trials. In both cases, using volumes of 90 L and 70 L, the first and second trials were conducted consecutively in the morning, whereas the third trials were conducted in the afternoon of the same day. If the wastewaters composition fluctuates between the morning and the afternoon trials, this can be an explanation for the differences in the pH buffering capacity of the slag being used. In fact, a different acid composition in the wastewaters might influence the dissolution rate of the slag. Therefore, an acidic environment that favors the slag dissolution results in a faster reaction. On the contrary, if the slag dissolution is slower, due to the different acidic content, the slag will take longer to dissolve and react with the acids. A confirmation of this hypothesis can be found by looking at the fact that the initial pH values are roughly the same in all the trials (between 1.49 and 1.57), and so are the final pH values, once the situation is closer to the reaction equilibrium. In fact, when a volume of 90 L is treated, the pH values are 9.1 (IV-1), 9.4 (IV-2), and 8.8 (IV-3), at minute 60. At minute 20, instead the values are 8.3, 8.9, and 6.8 for the respective trials. When a volume of 70 L is treated, at minute 60, the pH values are 9.5 (V-1), 9.3 (V-2), and 8.5 (V-3). At minute 20, the pH values are 8.9, 8.7, and 6.2 for the respective trials.

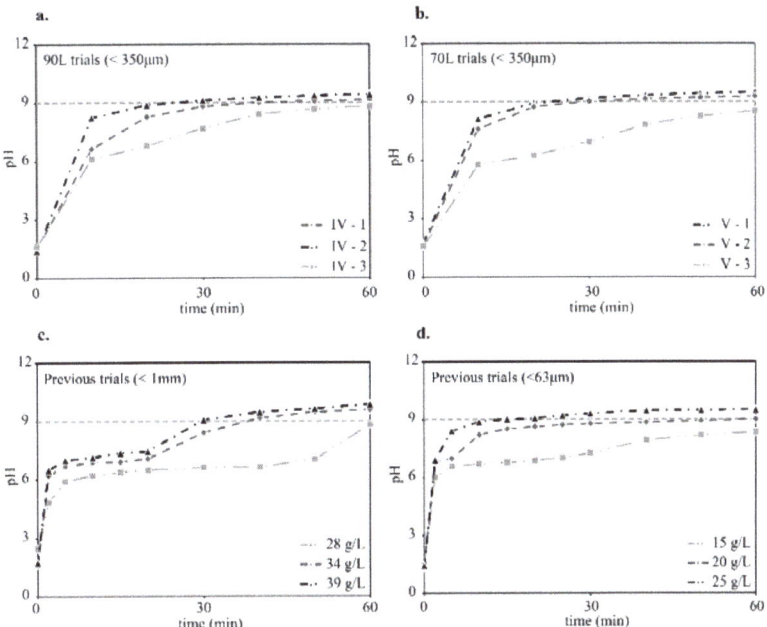

Figure 5. pH buffering obtained with the single-step methodology for volumes of wastewaters of 90 L (**a**) and 70 L (**b**). The results of previous trials conducted on 1 L, with the same slag type and the same company's industrial wastewaters, are also presented. Trials with <1 mm powders are shown in (**c**) [9], while the ones with <63 µm powders are in (**d**) [10]. Adapted with permission from ref. [10]. Copyright 2019. Proceedings of the 6th International Slag Valorisation Symposium.

The results of these trials were also compared to similar experiments that were conducted in the past [9,10] to analyze both the amount of slag being used and their corresponding pH curves. In Figure 5c,d, two set of trials are presented that were conducted with the same kind of slag, extracted from the slag yard of the same steel-making factory. Compared to the pH curves that were obtained when using coarser slags, the current trials seem to reach the final pH value faster. The lack of pH measurements at 2 and 5 min makes it hard to draw a stronger conclusion, but the trends obtained by both trials, when wastewater volumes of 90 and 70 L were tested, seem to resemble the one obtained when using finer slags (d) more. In fact, those trials are characterized by a fast rise in the pH values, followed by a long plateau, even though the quantity of slag used per liter is almost halved compared to the upscale ones. In addition, according to the results shown in Figure 5c, the quantity of slag is slightly higher, compared to the previous results shown in Figure 5d [10], which suggested that a reduced particle size also reduced the amount of slag needed to reduce the wastewaters. Although, the composition of the slag is variable and the effect of different minerals, regarding its capacity to buffer the pH values of the wastewaters, is still unknown. Similarly, a different composition of the wastewaters can alter the quantity needed to buffer its pH value. The different particle size is also a key factor that is hard to account for without being able to carry out proper measurements. Nonetheless, when only comparing the 90 L and 70 L trials to each other, the amount of slag needed to buffer the pH to a value of 9.0 ± 0.2 is roughly the same, while the pH levels reached are quite comparable, both at the 30 min and 60 min marks. Evidence of this can be observed in Figure 6, where a linear regression has been calculated with the amount of slag employed during the trials, with 90 and 70 L. The amount of slag from precedent trials has been added to the graph and compared to the calculated linear regression. As it is possible to notice in the enlarged area, the amount of slag that was measured with particle size <1 mm, falls very close to the regression line. In fact, the calculated amount of slag to buffer the pH of 1 L of wastewater, using the linear regression equation, should be 43.8 g. From empirical evidence, we know it to be 39 g [9], while the amount that is correspondent to the trial with a particle size <63 µm is 25 g [10]. In conclusion, the upscale trials proved to be successful in buffering the pH of bigger volumes of wastewaters, despite a different setup compared to the one used in the laboratory trials. When the particle size, composition of both the wastewaters and slag, as well as the kinematic conditions are controlled, the experiments seem to suggest that there is a linear relationship between the volume of wastewaters and the amount of slag needed to buffer their pH value.

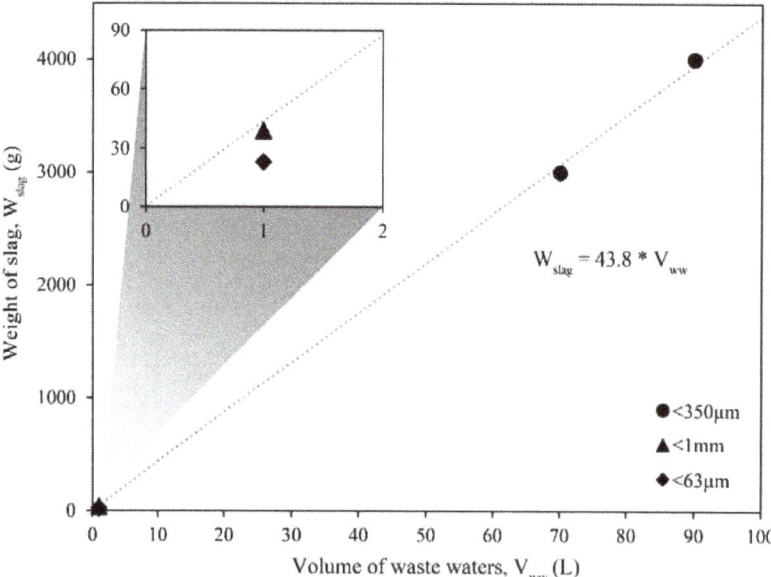

Figure 6. Linear regression of the 90 L and 70 L trials. Precedent trials have been added for comparison [9,10].

The weight of slag needed to buffer the pH of the wastewaters with this setup is likely still too high for its successful use in an industrial implementation. Nonetheless, it is observed, from the laboratory trials, that the quantity per liter can be reduced by almost half, by decreasing the particle size distribution of the powders used [10]. During the current experiments, it was impossible to decrease the particle size of such large quantities of slags to replicate these findings. Although, given the comparable results between the pilot-scale trials and the first laboratory trials [9], it is expected that for upscaled volumes of wastewaters also, reducing the particle size of the powders would optimize the amount of slag needed. Another important factor to consider is that the pilot-scale trials have been conducted as batch tests, whereas the industrial case is a continuous flow process.

Therefore, the optimal weight of the slag per liter of wastewaters, obtained during these trials, may not be indicative of the amount of slag needed in an industrial setting.

Finally, a consideration about the environmental aspect is due, given the nature of this study. Utilizing slag as a substitute for lime in the pH buffering of the wastewaters within the steel industry itself, could spare raw materials from entering the manufacturing process. Depending on the production process and internal reuse of generated by-products, this should constitute in a reduction in the landfilled output. In some cases, depending on the wastewater treatment process, spent lime is landfilled along with slag. Therefore, in that case, the use of slag constitutes a reduction in waste, since the lime is no longer needed. Sometimes, instead, the spent lime is recycled internally and used as a flux agent, thus generating more uncertainties in the possibility of substituting the material. In any case, the reduction in landfilled waste seems modest, as the volumes of spent lime, compared to the slag ones, are quite small. Also, the increase in reactant weight might be incompatible with the current wastewater treatment process, modifying the costs and composition of the landfilled output.

The use of slag as an acidic water treatment agent can be expanded to other industrial processes, translating in a larger volume of substituted lime. Along with the reduction in waste, the reduced need of lime translates in a reduction in CO_2, since the material is produced through the calcination of limestone. Thus, the more lime substituted, the less carbon dioxide produced. Moreover, the composition of the solid residues after the pH buffering highly depends on the treated acid composition and the composition of the slag used for the pH buffering. Therefore, it is quite hard to predict what use can be conducted with the spent slag and how those residues can be further valorized for other uses. Additionally, the toxicity levels of the treated wastewaters were not analyzed during this study, since the validation of the slag used in this regard was already conducted in a more thorough experiment in a precedent study [9].

4. Conclusions

The aim of this study was to test an upscaled environment for the treatment of industrial acidic wastewaters with slag, and to compare the results with previous results from laboratory experiments. Moreover, since the aim was to compare the results between the 70 L and 90 L trials, the kinematic conditions were maintained constant. Specifically, a physical water model was used to determine the kinematic conditions, to find a set of parameters that could ensure similar mixing performances across the two different volumes. The study found that for almost all the combinations of rotational speeds of the impeller, the volumes tested, and the probe positions, the mixing times were 10 ± 2 s.

The results from the stepwise and single-step dosing methods of addition of the slag, developed in previous studies, have been replicated for the current experiments. The results showed that 44 g/L of slag and 43 g/L were needed to reach pH values of 9.0 ± 0.2 in 30 min, when 90 L and 70 L of wastewaters were tested. With the same method of slag addition, previous results estimated that 39 g/L (grain size <1 mm) and 25 g/L (grain size <63 µm) were the adequate quantities. A linear regression, calculated with the data collected during the upscaled trials, predicted that the quantity that is necessary to buffer the pH value to 9.0 ± 0.2 should be approximately 43.8 g/L.

In conclusion, the results of this study deepen the knowledge regarding the use slag for the pH buffering of the acidic wastewaters derived by the pickling process. More specifically, it provides reliable results to show that the material can provide an adequate treatment of the wastewaters, even when their testing volume is increased by 90 times, albeit when the mixing conditions are kept constant. Also, a relationship between the amount of slag and the liters of acidic wastewater is found. Although, the relationship is highly dependent on the properties and compositions of the slags used, as well as the kinetic conditions of the experiments.

Author Contributions: Conceptualization, M.D.C. and R.P.; methodology, M.D.C. and R.P.; formal analysis, M.D.C. and R.P.; investigation, M.D.C. and R.P.; resources, M.D.C. and R.P.; data curation, M.D.C. and R.P.; writing—original draft preparation, M.D.C.; writing—review and editing, M.D.C., A.K., P.G.J.; visualization, M.D.C.; supervision, A.K., P.G.J.; project administration, P.G.J.; funding acquisition, P.G.J. All authors have read and agreed to the published version of the manuscript.

Funding: Funding for this project were provided by VINNOVA and Outokumpu Stainless AB.

Institutional Review Board Statement: Not applicable.

Informed Consent Statement: Not applicable.

Data Availability Statement: No new data were created or analyzed in this study. Data sharing is not applicable to this article.

Acknowledgments: The authors would like to thank Gunnar Ruist and Jyri Kaplin for helping to arrange all the trials at Outokumpu Stainless AB and for the precious feedback.

Conflicts of Interest: The authors declare no conflict of interest. The funders had no role in the design of the study; in the collection, analyses, or interpretation of data; in the writing of the manuscript, or in the decision to publish the results.

References

1. Filippou, D.; King, M.G. R&D prospects in the mining and metals industry. *Resour. Policy* **2011**, *36*, 276–284. [CrossRef]
2. Foxon, T.J. Technological Lock-In. In *Encyclopedia of Energy, Natural Resource, and Environmental Economics*; Elsevier: Amsterdam, The Netherlands, 2013; pp. 123–127. [CrossRef]
3. Mudd, G.M. Sustainable Mining: An Evaluation of Changing Ore Grades and Waste Volumes. In Proceedings of the International Conference on Sustainability Engineering & Science, Auckland, New Zealand, 6–9 July 2004.
4. Ramstein, C.; Dominioni, G.; Ettehad, S.; Lam, L.; Quant, M.; Zhang, J.; Mark, L.; Nierop, S.; Berg, T.; Leuschner, P.; et al. *State and Trends of Carbon Pricing 2019*; The World Bank: Washington, DC, USA, 2019; p. 97. [CrossRef]
5. Hultman, J.; Corvellec, H. The European Waste Hierarchy: From the Sociomateriality of Waste to a Politics of Consumption. *Environ. Plan. A* **2012**, *44*, 2413–2427. [CrossRef]
6. Piatak, N.M.; Parsons, M.B.; Seal, R.R. Characteristics and environmental aspects of slag: A review. *Appl. Geochem.* **2015**, *57*, 236–266. [CrossRef]
7. Autelitano, F.; Giuliani, F. Electric arc furnace slags in cement-treated materials for road construction: Mechanical and durability properties. *Constr. Build. Mater.* **2016**, *113*, 280–289. [CrossRef]
8. Pålsson, K.; Sweden, O.; Stemne, J.; Ruist, G.; Blixt, E. *Stålindustrin Gör Mer än Stål. Handbok för Restprodukter 2018*; Jernkontoret: Stockholm, Sweden, 2018.
9. De Colle, M.; Jönsson, P.; Karasev, A.; Gauffin, A.; Renman, A.; Renman, G. The Use of High-Alloyed EAF Slag for the Neutralization of On-Site Produced Acidic Wastewater: The First Step Towards a Zero-Waste Stainless-Steel Production Process. *Appl. Sci.* **2019**, *9*, 3974. [CrossRef]
10. De Colle, M.; Jönsson, P.; Gauffin, A.; Karasev, A. Optimizing the use of EAF stainless steel Slag to neutralize acid baths. In Proceedings of the 6th International Slag Valorisation Symposium, Mechelen, Belgium, 2–4 April 2019.
11. Figueira, R.M.; Szekely, J. Turbulent Fluid Flow Phenomena in a Water Model of an AOD System. *Metall. Trans. B* **1985**, *16*, 67–75. [CrossRef]
12. Bjurström, M.; Tilliander, A.; Iguchi, M.; Jönsson, P. Physical-modeling study of fluid flow and gas penetration in a side-blown AOD converter. *ISIJ Int.* **2006**, *46*, 523–529. [CrossRef]
13. Cabaret, F.; Bonnot, S.; Fradette, L.; Tanguy, P.A. Mixing Time Analysis Using Colorimetric Methods and Image Processing. *Ind. Eng. Chem. Res.* **2007**, *46*, 5032–5042. [CrossRef]
14. Wuppermann, C.; Giesselmann, N.; Rückert, A.; Pfeifer, H.; Odenthal, H.-J.; Hovestädt, E. A Novel Approach to Determine the Mixing Time in a Water Model of an AOD Converter. *ISIJ Int.* **2012**, *52*, 1817–1823. [CrossRef]
15. Samuelsson, P.; Ternstedt, P.; Tilliander, A.; Appell, A.; Jönsson, P.G. Use of physical modelling to study how to increase the production capacity by implementing a novel oblong AOD converter. *Ironmak. Steelmak.* **2018**, *45*, 335–341. [CrossRef]
16. Wei, J.H.; Ma, J.C.; Fan, Y.Y.; Yu, N.; Yang, S.L.; Xiang, S.H.; Zhu, D.P. Water modelling study of fluid flow and mixing characteristics in bath during AOD process. *Ironmak. Steelmak.* **1999**, *26*, 363–371. [CrossRef]
17. Wei, J.-H.; Zhu, H.-L.; Chi, H.-B.; Wang, H.-J. Physical Modeling Study on Combined Side and Top Blowing AOD Refining Process of Stainless Steel: Fluid Mixing Characteristics in Bath. *ISIJ Int.* **2010**, *50*, 26–34. [CrossRef]
18. Wei, J.-H.; Zhu, H.-L.; Chi, H.-B.; Wang, H.-J. Physical Modeling Study on Combined Side and Top Blowing AOD Refining Process of Stainless Steel: Gas Stirring and Fluid Flow Characteristics in Bath. *ISIJ Int.* **2010**, *50*, 17–25. [CrossRef]
19. Zhou, X.; Ersson, M.; Zhong, L.; Yu, J.; Jönsson, P. Mathematical and Physical Simulation of a Top Blown Converter. *Steel Res. Int.* **2014**, *85*, 273–281. [CrossRef]
20. Ternstedt, P.; Tilliander, A.; Jönsson, P.G.; Iguchi, M. Mixing Time in a Side-Blown Converter. *ISIJ Int.* **2010**, *50*, 663–667. [CrossRef]

Article

Initial Conditioning of Used Cigarette Filters for Their Recycling as Acoustical Absorber Materials

Valentín Gómez Escobar [1,*], Celia Moreno González [1], María José Arévalo Caballero [2] and Ana Mª Gata Jaramillo [2]

[1] Departamento de Física Aplicada, Escuela Politécnica, Universidad de Extremadura, Avda. de la Universidad s/n, 10003 Cáceres, Spain; celiamg@unex.es

[2] Departamento de Química Orgánica e Inorgánica, Escuela Politécnica, Universidad de Extremadura, Avda. de la Universidad s/n, 10003 Cáceres, Spain; arevalo@unex.es (M.J.A.C.); anamariagj@unex.es (A.M.G.J.)

* Correspondence: valentin@unex.es; Tel.: +34-927-257596; Fax: +34-927-257203

Abstract: Used cigarette butts represent a major and problematic form of waste, due to their abundance, toxicity, and durability. Moreover, the few proposals for their recycling are clearly insufficient, and new ones are welcome. For a new proposal regarding the reuse of used cigarette butts as acoustical absorbers in building construction, previous conditioning of the used butts is performed. This conditioning includes the elimination of moisture and toxic products accumulated in the filter of the cigarettes. Thus, in this work, the moisture content effect in acoustical absorption was analyzed, and a proposal for elimination is made. Moreover, a chemical cleaning procedure is proposed, and its influence on the acoustical behavior of the samples was also analyzed.

Keywords: sound absorber; cigarette butts; sustainable material; recycling; chemical cleaning

Citation: Gómez Escobar, V.; Moreno González, C.; Arévalo Caballero, M.J.; Gata Jaramillo, A.M. Initial Conditioning of Used Cigarette Filters for Their Recycling as Acoustical Absorber Materials. *Materials* **2021**, *14*, 4161. https://doi.org/10.3390/ma14154161

Academic Editor: Neven Ukrainczyk

Received: 1 July 2021
Accepted: 25 July 2021
Published: 27 July 2021

Publisher's Note: MDPI stays neutral with regard to jurisdictional claims in published maps and institutional affiliations.

Copyright: © 2021 by the authors. Licensee MDPI, Basel, Switzerland. This article is an open access article distributed under the terms and conditions of the Creative Commons Attribution (CC BY) license (https://creativecommons.org/licenses/by/4.0/).

1. Introduction

Cigarette butts (mainly the filters of the millions of cigarettes consumed) are among the most problematic types of global waste. The importance of the problem derived from this waste can be determined by the four components described below.

Firstly, as smoking is commonly practiced all around the world, trillions of cigarettes are consumed each year on a global scale [1,2]. This amount is even expected to increase by more than 50% by 2025 due both to the increase in the world population and in tobacco production [3]. Almost all these smoked cigarettes incorporate a filter as a consequence of studies carried out in the middle of the twentieth century demonstrating that its use reduces the risk of lung cancer in smokers. Thus, trillions of used filter cigarettes are generated annually.

Secondly, unfortunately, cigarette butts are commonly thrown on the ground [4]. Thus, cigarette butts can be found in almost all environments throughout the world. In fact, cigarette butts are usually the major element in terms of quantity in garbage and even in terms of weight in some cases [5–7]. This second component is even more relevant when linking it with the two components presented below.

Thirdly, butts thrown on the ground are carried by rain or river water to coastal areas, increasing their environmental impact. This environmental impact of cigarette butts is related to both their chemical composition [8] and the difficulty of degradation of cellulose acetate. Thus, some 130 chemical substances present in cigarette butts have been described, chemicals that can be easily leached into water, especially toxic heavy metals, nicotine, and additives [9,10], thereby making these waters toxic to different organisms [8,11,12].

Fourthly, almost all cigarette filters are made of cellulose acetate, a polymer that undergoes a very slow biological degradation and takes several months to photodegrade (only partial photodegradation occurs since the polymer chains are cleaved into smaller pieces) [13]. Therefore, this waste can remain in different environments for a long time.

From the above components, it can be concluded that used cigarette butts constitute a serious environmental and public health problem, as has been stated in a recent study by the World Health Organization [14].

These four components are associated with a fifth, which is the fact that despite the described problem of this waste, there are few initiatives for its recycling. Some authors have developed several proposals to recycle this residue (for example, in the manufacture of supercapacitors [15], as chemical corrosion inhibitors [16], and as a component of bricks [17]). Most of the initiatives proposed in recent years are compiled in two recently published papers [18,19]. The existing proposals, however, are scarce and very limited because of the number of cigarette butts that they require, and they are clearly insufficient, considering the continuous generation of this waste.

As a new proposal for recycling this waste, over the past few years our research group has been working on the elaboration of an acoustical absorber material obtained from used cigarette butts. In these studies [20–23], absorption of the prepared samples (some of them desegregating the cigarette butt filters, but mainly manually inserting them) was fairly satisfactory, being comparable to or even better than those of some other materials conventionally used for sound absorption.

In these previous studies, however, some questions were not solved and are thus addresses in the present work. Therefore, this study examines the case of previous conditioning of samples and identifies the adequate chemical treatment in order to perform previous cleaning of filters.

Both issues (previous conditioning of samples and cleaning chemical treatment) are relevant when first considering the fact that the start samples (used cigarette butts) came from different places and that their moisture contents can differ. As such, we are interested in determining the influence of this moisture content on the samples' acoustical properties, as well as establishing a standard procedure for their elimination. Moreover, the toxic character of this waste (due to the previously mentioned accumulation of toxic residues during the combustion) has made it necessary to develop a proposal of chemical cleaning of used cigarette butts.

Regarding this chemical cleaning, only a few of proposals consider the separation of the butts and the entrapped chemicals in them. For instance, cellulose acetate has been extracted and purified by soaking cigarette butts in distilled water and in ethanol to construct a cellulose-based membrane separator for a high-performance lithium-ion battery [24]; to prepare highly porous carbons as CO_2 super absorbents [25]; or to obtain cellulose [26]. Further research is necessary to analyze the degree of chemical cleaning and to study possible alternatives to this cleaning.

Thus, the aims of this work are (1) to analyze the influence of moisture content on acoustical absorption and the procedure for moisture elimination, and (2) to propose an adequate chemical treatment process in order to carry out previous cleaning of filters and to characterize the influence of this chemical cleaning on the acoustical properties of the samples.

2. Materials and Methods

2.1. Preparation of Samples

Smoked cigarette butts from different brands were picked from ashtrays or from the ground in buildings from the Campus of the University of Extremadura and surroundings. They formed a very heterogeneous mixture; each butt had its original tobacco and blend of additives and different amounts of remaining unburnt tobacco, as well as other possible contaminants, such as lipstick and saliva residue. The remaining non-smoked tobacco and the external wrapping paper were manually separated, and only the cigarette butt filter were taken.

The remaining cigarette butt filters were nonhomogeneous. The length, diameter, presence of burnt regions, squashing degree, humidity, etc. differed for each butt. The most common lengths of filters were 12, 15, 20, and 26 mm, while the most common diameters were 6 and 8 mm.

Two different samples were used in the present study. For the chemical cleaning study, used cigarette butt filters (only without external wrapping paper, as previously mentioned) were used. However, in the analysis of the influence of moisture content and the drying conditions, in order to acquire a structure similar to that of the commercial acoustical products that are also included in the study, cigarette butts were prepared, disaggregating the filters and making a fiber mass. Figure 1 shows some pictures of the prepared samples.

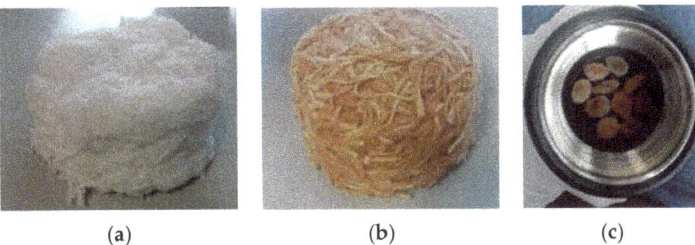

Figure 1. Example pictures of some of the prepared samples: (**a**) sample from disaggregated non-used cigarette butts; (**b**) sample from disaggregated used cigarette butts; (**c**) sample from used cigarette butts before the cleaning treatment.

2.2. Reagents and Solutions for Cleaning Used Cigarette Filters

Absolute ethanol, sulfuric acid (95–97% pure), and sodium chloride were purchased from Scharlab S. L. (Sentmenat, Spain). Ultrapure water was produced by an automatic water distillation apparatus (Barnstend Nanopure, Thermo Scientific, Waltham, MA, USA).

Aqueous solutions, 5% NaCl (w/v) and 0.02% H_2SO_4 (w/v), were prepared by adding the solid or the dilute solute to ultrapure water.

2.3. Cleaning Used Cigarette Filters

Samples of 15 butt filters (without the external the wrapping paper, as described previously) were extracted sequentially under stirring at room temperature with 125 mL of a 0.02% (w/v) H_2SO_4 aqueous solution (five times), with 125 mL of a 5% (w/v) NaCl aqueous solution (four times), and with 100 mL of ethanol (six times). Each extraction was carried out for 1 h. Cleaned butts were dried in an oven at 80 °C for 48 h. Then, they were characterized by nuclear magnetic resonance. ^{1}H-NMR spectra were recorded on a Bruker Avance 500 MHz spectrometer (Billerica, MA, USA), using $CDCl_3$ as solvent. The aqueous leachates were analyzed for heavy metals using an Agilent 7900 ICP mass spectrometer (Santa Clara, CA, USA).

2.4. Instrumentation for Acoustic Absorption Determination

Measurements of the absorption coefficient of the different samples were conducted using the Brüel & Kjær Impedance Tube Kit (Type 4206, Copenhagen, Denmark), equipped with two one-quarter-inch condenser microphones (Type 4187) (Figure 2). The signals were analyzed using a portable Brüel & Kjær PULSE System, with four input data channels (Type 3560-C). Two sample holders, with diameters of 29 mm (with a validity in the frequency range of 500 Hz to 6.4 kHz) and 100 mm (with a validity in the frequency range of 50 Hz to 1600 kHz) were used.

The determination of the sound absorption coefficient of different samples was carried out using an impedance tube following the two-microphone transfer function method described in the ISO 10534-2 standard [27].

The impedance tube was also used for the determination of the airflow resistivity of some of the samples. This property is related to the resistance of the air to penetrate a porous material and, thus, is related to the absorption capacity of the material, as in porous material, a major part of the absorption is produced inside it. The units of this property are Pa·s/m² or Rayls/m. For the determination of airflow resistivity, the Ingard and Dear method was used [28], with a configuration of sample inside the tube as is shown in Figure 3.

Figure 2. Impedance tube disposition used for the measurements.

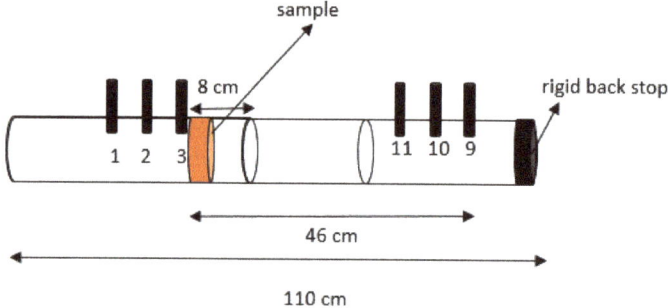

Figure 3. Scheme of the sample configuration for airflow resistivity determinations.

3. Results and Discussion

3.1. Previous Conditioning of Used Cigarette Butts Filters

As previously mentioned, the cigarette butt filters came from ashtrays or directly from the ground. Thus, their initial statuses differed; for instance, their humidity degrees were considerably different. As such, to homogenize the initial humidity degree of the used filters, they were dried.

In order to establish adequate drying conditions, six different samples (two made with used butt filters—samples UBF1 and UBF2; two made with non-used cigarette butts—samples NUBF1 and NUBF2; and two made with commercial absorption products—Mineral Wool 1 and Mineral Wool 2) were prepared. Firstly, they were dried for 48 h at 80 °C. Then, they were placed in a saturated environment for 50 days. The samples were weighted and acoustically characterized during the process.

Variation in weight of the prepared samples is presented in Table 1. As it can be seen, during the 50 days, all of their weights increased by around 13–21% due to the adsorption of water.

Is important to note that the change in weight of the samples produce a variation in their absorption, as can be seen in Figure 4 for three of the samples analyzed.

As can be seen in Figure 4, there is a clear reduction in the absorption capacity of the samples when they are saturated. Another way to quantify this reduction is by calculating the octave band coefficient absorption, which is represented in Figure 5. The average values of the absorption coefficient were 0.42 and 0.35 for the sample UBF1 (dried and wet, respectively); 0.43 and 0.35 for the sample NUBF1 (dried and wet, respectively); and 0.14 and 0.09 for the sample Mineral Wool 1 (dried and wet, respectively).

Table 1. Variation in weight of samples after initial drying and after remaining in a saturated environment for 50 days.

Sample	Initial Weight (g)	Dried Weight (g)	Weight after 50 Days in a Saturated Environment (g)	Increase in Weight (%)
UBF1	21.4	19.4	22.9	17.9
UBF2	20.9	18.9	22.4	18.8
NUBF1	22.2	20.2	24.0	18.9
NUBF2	21.8	19.6	22.7	15.5
Mineral Wool 1	4.1	4.0	4.8	21.0
Mineral Wool 2	5.0	4.9	5.6	13.4

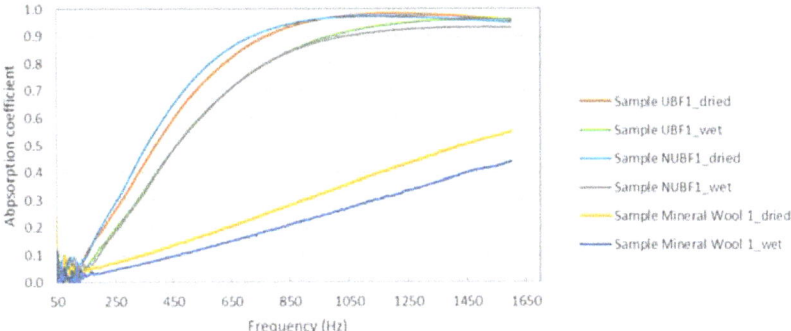

Figure 4. Variation in the absorption coefficient with the degree of humidity in the analyzed samples.

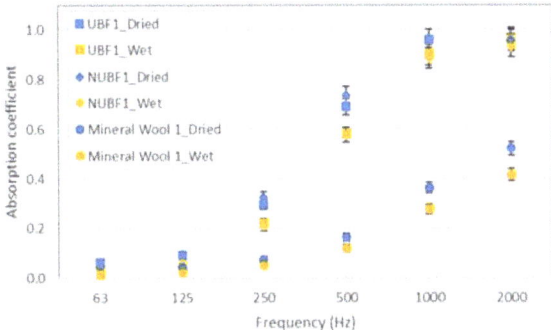

Figure 5. Variation in absorption coefficient with the degree of humidity in the analyzed samples (octave bands).

This observed reduction in the absorption coefficient of the samples with moisture is probably due to the increase in the difficulty of the penetration of waves inside the material due the increase of the size of the fibers with moisture. This hypothesis is confirmed by the measured values of the airflow resistivity of the samples. As can be observed in Figure 6, there is an increase in the airflow resistivity values in the wet samples with respect to those of the dried ones, indicating increased difficulty of air to penetrate the materials and, thus, coherence with the reduction in the absorption coefficient values observed in Figures 4 and 5.

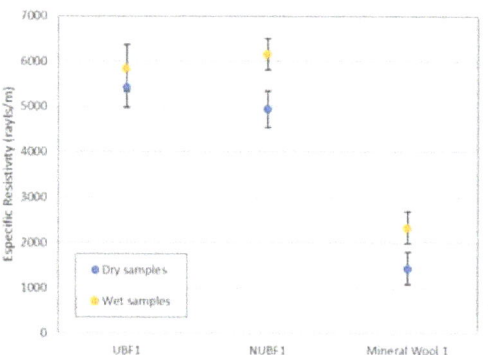

Figure 6. Variation of airflow resistivity of the samples shown in Figure 4.

In the second step, the wet samples were dried, three of them at 105 °C as indicated by the ISO 585 standard [29] and the other three at 80 °C. Variations in the weight of these samples, in percentage, are presented in Table 2.

Table 2. Variation of weight of samples (%) after first drying at 105 °C and 80 °C.

Hours of Drying	Dried at 105 °C			Dried at 80 °C		
	UBF1	NUBF1	Mineral Wool 1	UBF2	NUBF2	Mineral Wool 2
3	−15.5	−15.9	−16.6	−17.69	−13.68	−10.47
24	−0.0	−0.2	0.0	0.0	−0.1	−0.2
48	−0.3	−0.2	0.0	0.0	0.0	0.0
72	−0.2	0.0	0.0	−0.2	−0.1	−0.2
144	−0.2	−0.1	0.0	0.0	0.0	0.0

As can be seen in Table 2, almost all the moisture content is eliminated, both for drying at 105 °C and at 80 °C, in the first three hours. Only a residual fraction (lower than 0.8% in the highest sample) is eliminated after the first three hours.

Thus, according to the results shown in Table 2, a drying period of three hours is sufficient (for drying at both 105 °C and 80 °C) in order to eliminate the moisture content of used cigarette butts from the presented samples.

3.2. Chemical Cleaning of the Used Cigarette Butts Filters

As previously mentioned, used cigarette butt filters have a number of toxic compounds inside that are generated during smoking. Moreover, their color and their smell are generally deemed to be unpleasant. Thus, a second use of this waste must be associated with previous cleaning and even an improvement to its image.

In order to identify an adequate procedure for the cleaning of the samples, several proofs in an attempt to use diluted reactants in order to reduce the posterior problem associated with the treatment of used reactive solutions were carried out. Furthermore, it was important that the used cigarette butts maintained the good acoustical behavior observed for the non-cleaned samples.

After several attempts, a good balance was found for a cleaning procedure based on sequential solid–liquid extractions with 0.02% H_2SO_4 (w/v), 5% NaCl (w/v), and absolute ethanol [30]. Sulfuric acid and sodium chloride aqueous solutions were used to extract heavy and trace metals that can be found in cigarette butts. Table 3 shows the concentration of different metals in the aqueous extracts. Either sulfuric acid or sodium chloride solutions allow the extraction of metals; the former appears to perform better for the extraction of Al, while NaCl is better at eliminating Pb, Mn, Sr, and Ba. Copper and cadmium were not found in the extracts.

On the other hand, extractions with ethanol allow the removal of the organic pollutants that are present in butts. Thus, the ^{1}H-NMR spectrum of butts without extraction (Figure 7a) shows peaks at 2 and 4 ppm, corresponding to protons of triacetin used as adhesive in butts. Moreover, there are two peaks at 8.5 ppm that are due to the aromatic protons of nicotine. These peaks are not found in the cleaned butts spectrum (Figure 7b).

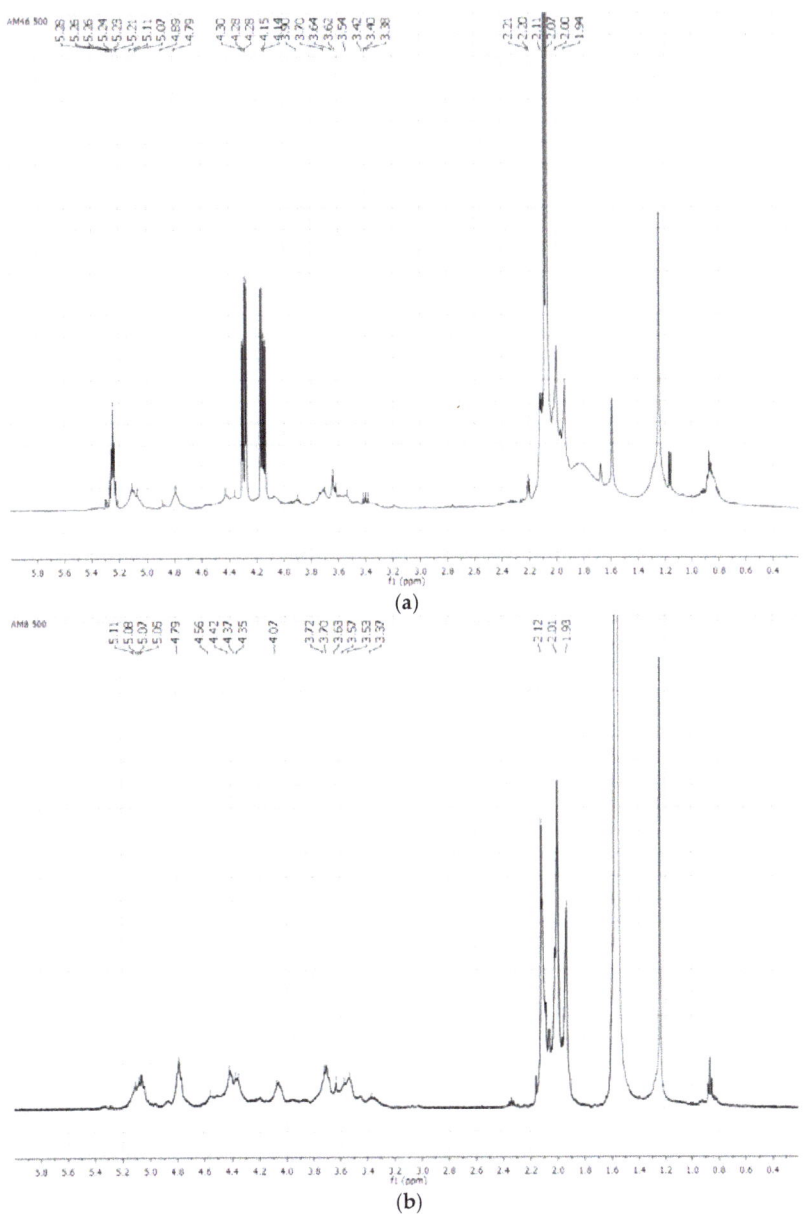

Figure 7. ^{1}H-NMR spectrum of butts before (**a**) and after (**b**) being submitted to the cleaning extractions.

Table 3. Concentrations of metal ions obtained by dissolution tests on cigarette butts.

Metal Ion	Dissolution Test	
	H_2SO_4 aq	NaCl aq
Al (µg/L)	573 ± 1.0	35 ± 5
Mn (µg/L)	26.8 ± 0.4	173 ± 0.8
Fe (µg/L)	22.9 ± 1.1	26 ± 3
Cu (µg/L)	<10	<10
Zn (µg/L)	54.2 ± 1.1	42.7 ± 1.1
Sr (µg/L)	29.5 ± 0.4	2008 ± 1
Cd (µg/L)	<10	<10
Ba (µg/L)	117.3 ± 0.8	657 ± 1.0
Pb (µg/L)	<10	11.3 ± 1.4

An important challenge regarding the cleaning procedure is that, importantly, it might not alter the good acoustical properties of the samples observed before cleaning. To analyze this aspect, 4 samples of 10 used cigarette filters were used. First, they were dried and measured. Then, the cleaning procedure was applied to the different used cigarette filter groups, and, finally, the cleaned filters were dried again. In Figure 8, photographs of the four samples, before and after the application of the cleaning procedure, are shown. As can be seen, there is a clear improvement in the appearance of the cigarette filters.

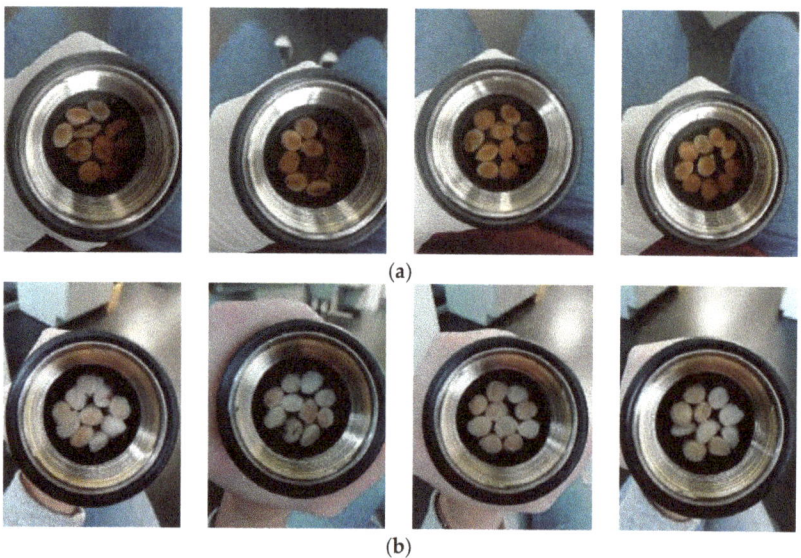

Figure 8. Effect of chemical cleaning on the appearance of the cigarette (**a**) before and (**b**) after cleaning.

In Figure 9, values of the absorption coefficients before and after chemical cleaning are shown.

As can be seen in Figure 9, samples after the cleaning procedure present higher absorption coefficients than those of the non-cleaned samples. This is a surprising result when considering that, after the cleaning procedure, a decrease in the weight of the samples (15 ± 3% in average) can be observed. A possible explanation for this absorption increase could be that, after the cleaning procedure, the filters are less rigid and occupy a larger volume of the impedance tube holder and that there is less airflow resistance (unfortunately, in a tube holder of this size, it is not possible to measure airflow resistivity), meaning that waves can more effectively penetrate the material. Moreover, there can also be an increase in the absorption surface of the filters' fibers after the cleaning procedure.

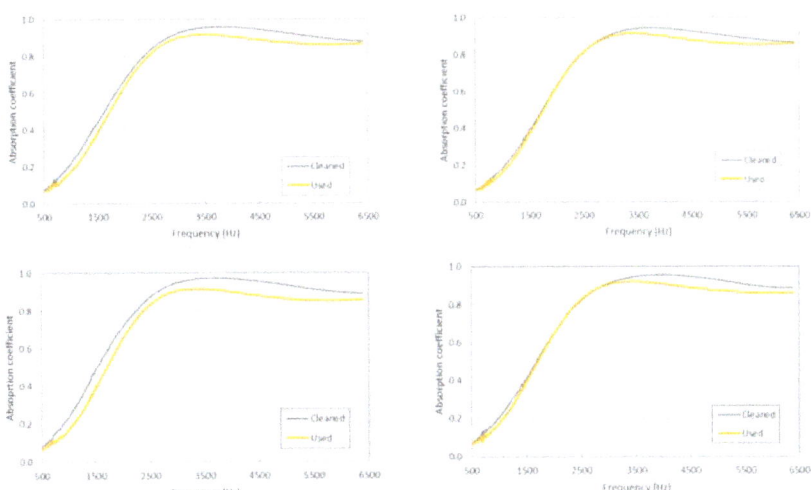

Figure 9. Effect of chemical cleaning in the absorption coefficient values.

Increases in absorption are also observed in octave bands (see Figure 10) and in the average values (in the 500–6400 Hz range) of 0.45, 0.44, 0.45, and 0.45 for the non-cleaned samples and 0.48, 0.46, 0.50, and 0.47 for the cleaned ones.

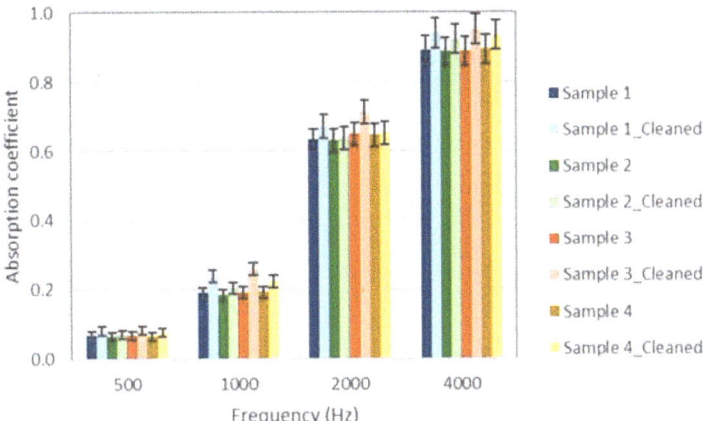

Figure 10. Effect of chemical cleaning on the absorption coefficient values (octave bands).

4. Conclusions

From the results and analysis, the following conclusions can be drawn:

- The moisture content of the samples has an influence on both their weight and acoustical properties (in the values of the absorption coefficient) due to the increase in the difficulty of the sound wave to penetrate the material. A drying period of three hours at 80 °C or 105 °C seems sufficient to complete the drying of the samples.
- A chemical cleaning procedure based on diluted aqueous solutions and ethanol was proposed to eliminate the major metal ions present in the samples, as well as

organic pollutants. With the proposed procedure, an improvement in the samples' appearance was observed. The acoustical behavior of the samples improved following the chemical procedure.

- An improvement in the absorption capacity of the cigarette butts was observed when the total fiber content of the samples is more homogeneously distributed in the sample (not associated to each filter).
- With the proposed previous conditioning of used cigarette butts (cleaning and drying them), the new prepared samples keep high absorption coefficient values for medium and high frequencies, showing their potential application as acoustic absorbers. In any case, further studies related to the different influences on absorption—for instance, of density or thickness—and the possibility of the establishment of a standard method for preparing large samples would be desirable.

Finally, it is worth mentioning that the main goal of this chemical cleaning study was to eliminate pollutants from cigarette butts in order to research their acoustical behavior once they are clean. However, bearing in mind the 3R's—recover, recycle, reuse—further research is needed (and is under development in our laboratories) addressing the elimination of the extracted metals and organic compounds, from either the aqueous or organic solvents, in order to recover them for additional use.

Author Contributions: V.G.E. and C.M.G. designed and performed the acoustical measurements. M.J.A.C. and A.M.G.J., designed and performed the chemical procedure and related determinations. All the authors contributed in a similar way in the original drafting preparation and in the review and editing of the final document. V.G.E. was the responsible of the funding acquisition. All authors have read and agreed to the published version of the manuscript.

Funding: This work was partially supported by Altadis (Imperial Tobacco), by the European Regional Development Fund (ERDF) and also by Junta de Extremadura, Consejería de Economía, Ciencia y Agenda Digital (Project IB18033).

Institutional Review Board Statement: Not applicable.

Informed Consent Statement: Not applicable.

Data Availability Statement: Data sharing not applicable. No new data were created or analyzed in this study. Data sharing is not applicable to this article.

Conflicts of Interest: The authors declare no conflict of interest.

References

1. Euromonitor International. Passport: Global Market Information Database. 2016. Available online: https://tobaccoatlas.org/topic/consumption/ (accessed on 1 June 2021).
2. Available online: https://es.statista.com/estadisticas/636021/consumo-mundial-de-cigarrillos/#statisticContainer (accessed on 1 June 2021).
3. Mackay, J.; Eriksen, M.; Shafey, O. *The Tobacco Atlas*, 2nd ed.; American Cancer Society: Atlanta, GA, USA, 2000; pp. 32–33.
4. Novotny, T.E.; Slaugther, E. Tobacco Product Waste: An Environmental Approach to Reduce Tobacco Consumption. *Curr. Environ. Health Rep.* **2014**, *1*, 208–216. [CrossRef] [PubMed]
5. Ariza, E.; Jiménez, J.A.; Sardá, R. Seasonal evolution of beach waste and litter during the bathing season on the Catalan coast. *Waste Manag.* **2008**, *22*, 2604–2613. [CrossRef]
6. Martínez-Ribes, L.; Basterretxea, G.; Palmer, M.; Tintoré, J. Origin and abundance of beach debris in the Balearic Islands. *Sci. Mar.* **2007**, *71*, 305–314. [CrossRef]
7. Moriwaki, H.; Kitajima, S.; Katahira, K. Waste on the roadside, 'poi-sute' waste: Its distribution and elution potential of pollutants into environment. *Waste Manag.* **2009**, *29*, 1192–1197. [CrossRef]
8. Slaughter, E.; Gersberg, R.M.; Watanabe, K.; Rudolph, J.; Stransky, C.; Novotny, T.E. Toxicity of cigarette butts, and their chemical components, to marine and freshwater fish. *Tob. Control* **2011**, *20*, i25–i29. [CrossRef] [PubMed]
9. Lima, H.H.C.; Maniezzo, R.S.; Kupfer, V.L.; Guilherme, M.R.; Moises, M.P.; Arroyo, P.A.; Rinaldi, A.W. Hydrochars based on cigarette butts as a recycled material for the adsorption of pollutants. *J. Environ. Chem. Eng.* **2018**, *6*, 7054–7061. [CrossRef]
10. Scott Blankenship, L.; Mokaya, R. Cigarette butt-derived carbons have ultra-high surface area and unprecedented hydrogen storage capacity. *Energy Environ. Sci.* **2017**, *10*, 2552. [CrossRef]
11. Register, K.M. Cigarette Butts as Litter- Toxic as Well as Ugly. *Bull. Am. Littoral Soc.* **2000**, *25*, 23–29.

12. Micevska, T.; Warne, M.S.J.; Pablo, F.; Patra, R. Variation in, and Causes of, Toxicity of Cigarette Butts to a Cladoceran and Microtox. *Arch. Environ. Contam. Toxicol.* **2006**, *50*, 205–212. [CrossRef] [PubMed]
13. Puls, J.; Wilson, S.A.; Hölter, D. Degradation of cellulose acetate-based materials: A review. *J. Polym. Environ.* **2011**, *19*, 152–165. [CrossRef]
14. World Health Organization. *Tobacco and Its Environmental Impact: An Overview*; World Health Organization: Geneva, Switzerland, 2017.
15. Lee, M.; Kim, G.-P.; Don Song, H.; Park, S.; Yi, J. Preparation of energy storage material derived from a used cigarette filter for a supercapacitor electrode. *Nanotechnology* **2014**, *25*, 345601. [CrossRef]
16. Zhao, J.; Zhang, N.; Qu, C.; Wu, X.; Zhang, J.; Zhang, X. Cigarette Butts and Their Application in Corrosion Inhibition for N80 Steel at 90 °C in a Hidrochloric Acid Solution. *Ind. Eng. Chem. Res.* **2010**, *49*, 3986–3991. [CrossRef]
17. Mohajerani, A.; KADIR, A.A.; Larobina, L. A practical proposal for solving world's cigarette butt problem: Recycling in fired clay bricks. *Waste Manag.* **2016**, *52*, 228–244. [CrossRef] [PubMed]
18. Torkashvand, J.; Farzadkia, M. A systematic review on cigarette butt management as a hazardous waste and prevalent litter: Control and recycling. *Environ. Sci. Pollut. Res.* **2019**, *26*, 11618–11630. [CrossRef]
19. Marinello, S.; Lolli, F.; Gamberini, R.; Rimini, B. A second life for cigarette butts? A review of recycling solutions. *J. Hazard. Mater.* **2020**, *384*, 121245. [CrossRef]
20. Gómez Escobar, V.; Maderuelo-Sanz, R. Acoustical performance of samples prepared with cigarette butts. *Appl. Acoust.* **2017**, *125*, 166–172. [CrossRef]
21. Maderuelo-Sanz, R.; Gómez Escobar, V.; Meneses-Rodríguez, J.M. Potential use of cigarette filters as sound porous absorber. *Appl. Acoust.* **2018**, *129*, 86–91. [CrossRef]
22. Gómez Escobar, V.; Rey Gozalo, G.; Pérez, C.J. Variability and performance study of the sound absorption of used cigarette butts. *Materials* **2019**, *12*, 2584. [CrossRef]
23. Maderuelo-Sanz, R. Characterizing and modelling the sound absorption of the cellulose acetate fibers coming from cigarette butts. *J. Environ. Health Sci. Eng.* **2021**. [CrossRef] [PubMed]
24. Huang, F.; Xu, F.; Peng, B.; Su, Y.; Jiang, F.M.; Hsieh, Y.-L.; Wei, Q. Coaxial Electrospun Cellulose-Core Fluoropolymer-Shell Fibrous Membrane from Recycled Cigarette Filter as Separator for High Performance Lithium-Ion Battery. *ACS Sustain. Chem. Eng.* **2015**, *3*, 932–940. [CrossRef]
25. Sun, H.; La, P.; Yang, R.; Zhu, Z.; Liang, W.; Yang, B.; Li, A.; Deng, W. Innovative nanoporous carbons with ultrahigh uptakes for capture and reversible storage of CO_2 and volatile iodine. *J. Hazard. Mater.* **2017**, *321*, 210–217. [CrossRef] [PubMed]
26. Abu-Danso, E.; Bagheri, A.; Bhatnagar, A. Facile functionalization of cellulose from discarded cigarette butts for the removal of diclofenac from water. *Carbohydr. Polym.* **2019**, *219*, 46–55. [CrossRef] [PubMed]
27. ISO 10534-2. *Acoustics: Determination of Sound Absorption Coefficient and Impedance in Impedances Tubes. Part 2: Transfer-Function Method*; International Organization for Standardization: Geneva, Switzerland, 1998.
28. Ingard, K.U.; Dear, T.A. Measurement of acoustic flow resistance. *J. Sound Vib.* **1985**, *103*, 567–572. [CrossRef]
29. ISO 585. *Plastics—Unplasticized Cellulose Acetate—Determination of Moisture Content*; International Organization for Standardization: Geneva, Switzerland, 1990.
30. Benavente, M.J.; Arévalo Caballero, M.J.; Silvero, G.; López-Coca, I.; Gómez Escobar, V. Cellulose acetate recovery from cigarette butts. In Proceedings of the Environment, Green Technology and Engineering International Conference (EGTEIC), Cáceres, Spain, 18–20 June 2018; Volume 2, p. 1447.

Article

A Novel Dry Treatment for Municipal Solid Waste Incineration Bottom Ash for the Reduction of Salts and Potential Toxic Elements

Marco Abis [1,*], Martina Bruno [2], Franz-Georg Simon [3], Raul Grönholm [4], Michel Hoppe [5], Kerstin Kuchta [1] and Silvia Fiore [2,*]

1. SRWM (Sustainable Resource and Waste Management), Hamburg University of Technology, 21079 Hamburg, Germany; kuchta@tuhh.de
2. DIATI (Department of Engineering for Environment, Land and Infrastructures), Politecnico di Torino, 10129 Torino, Italy; martina.bruno@polito.it
3. Bundesanstalt für Materialforschung und -Prüfung (BAM), 12200 Berlin, Germany; franz-georg.simon@bam.de
4. Sysav Utveckling AB, 20025 Malmö, Sweden; Raul.Gronholm@sysav.se
5. Heidemann Recycling GmbH, 28277 Bremen, Germany; m.hoppe@heidemann-recycling.de
* Correspondence: marco.abis@tuhh.de (M.A.); silvia.fiore@polito.it (S.F.)

Citation: Abis, M.; Bruno, M.; Simon, F.-G.; Grönholm, R.; Hoppe, M.; Kuchta, K.; Fiore, S. A Novel Dry Treatment for Municipal Solid Waste Incineration Bottom Ash for the Reduction of Salts and Potential Toxic Elements. *Materials* **2021**, *14*, 3133. https://doi.org/10.3390/ma14113133

Academic Editor: Antonio Gil Bravo

Received: 14 April 2021
Accepted: 21 May 2021
Published: 7 June 2021

Publisher's Note: MDPI stays neutral with regard to jurisdictional claims in published maps and institutional affiliations.

Copyright: © 2021 by the authors. Licensee MDPI, Basel, Switzerland. This article is an open access article distributed under the terms and conditions of the Creative Commons Attribution (CC BY) license (https:// creativecommons.org/licenses/by/ 4.0/).

Abstract: The main obstacle to bottom ash (BA) being used as a recycling aggregate is the content of salts and potential toxic elements (PTEs), concentrated in a layer that coats BA particles. This work presents a dry treatment for the removal of salts and PTEs from BA particles. Two pilot-scale abrasion units (with/without the removal of the fine particles) were fed with different BA samples. The performance of the abrasion tests was assessed through the analyses of particle size and moisture, and that of the column leaching tests at solid-to-liquid ratios between 0.3 and 4. The results were: the particle-size distribution of the treated materials was homogeneous (25 wt % had dimensions <6.3 mm) and their moisture halved, as well as the electrical conductivity of the leachates. A significant decrease was observed in the leachates of the treated BA for sulphates (44%), chlorides (26%), and PTEs (53% Cr, 60% Cu and 8% Mo). The statistical analysis revealed good correlations between chloride and sulphate concentrations in the leachates with Ba, Cu, Mo, and Sr, illustrating the consistent behavior of the major and minor components of the layer surrounding BA particles. In conclusion, the tested process could be considered as promising for the improvement of BA valorization.

Keywords: bottom ash; dry treatment; incineration; municipal solid waste; potential toxic elements; salts

1. Introduction

The mining of mineral aggregates is the largest extractive sector in the EU, which, on its own, exceeds the amount of all the minerals produced [1]. Nevertheless, its End-of-Life (EoL) input rate was estimated to be only 8 wt %. Typical EoL materials used as aggregates are construction and demolition waste (CDW) and bottom ash (BA) from municipal solid waste incineration (MSWI). However, the full recycling potential of these waste flows have not been tapped yet due to the existing gap between market prices and extraction, processing, and transportation costs [2].

BA is the main by-product of MSWI and counts approximately 25 wt % of MSW input for thermal valorization [3,4]; 71 Mt of MSW incinerated in Europe in 2018 produced about 18 Mt of BA. Current full-scale material recovery technologies applied to BA mainly focus on the separation of metals, with the most valuable components being aluminum and copper [5,6]; this leaves the mineral fraction unexploited, and which is usually landfilled. The mineral fraction has been estimated at 85–90 wt % of BA [4], resulting in 15–16 Mt/y of materials in Europe that could potentially be recycled as aggregates. Several studies have recently explored recycling alternatives for the BA mineral fraction to be used as construction material, e.g., as the sub-base layer for asphalt roads [7], as the source of construction sand [8] and as substitute material for concrete production [9,10]. The recycling of the BA mineral fraction as a secondary aggregate holds the potential of enhancing the profitability of BA management and is fully consistent with EU policy on Circular Economy. Material recovery from BA could entail a significant improvement in the circularity of the management of resources, not only limiting the request for primary aggregates but more particularly reducing the amount of waste sent to the landfill. Despite the low commercial value of secondary mineral aggregates, potential savings from landfill fees could entail significant economic benefits and ensure the profitability of BA mineral-fraction management. The production of primary aggregates in Europe in 2018 was 2431 Mt, which was composed of sand, gravel, and crushed rocks [11]. Considering this mass, the mineral fraction of BA could potentially replace 0.7 wt % of primary aggregates produced in Europe. This value was consistent with a recent study [12], which estimated a 0.6% potential substitution rate. Nonetheless, since about 806 Mt of non-hazardous waste generated in Europe are currently disposed in landfills, the recycling of the BA mineral fraction as a secondary aggregate might divert circa 2 wt % of the waste stream directed to landfills.

However, the potential reuse of the BA mineral fraction as a secondary aggregate is hindered by the presence of Potential Toxic Elements (PTEs) [13], which have negative environmental impacts [14,15]. Among the PTEs, cadmium, chromium, and molybdenum were the most found in MSWI BA [16,17]. Most European countries set threshold limits for recycled aggregates due to PTEs and chloride and sulphate leaching. A detailed analysis of the different leaching tests applied to BA and related limits in Europe were presented in a recent study [12]. In this framework, special attention should be devoted to studying BA composition, not only to detect the mineral phases for further geochemical dissolution modelling, but also to map their occurrence and locations in the solids. X-ray diffraction analyses (XRD), energy dispersive X-ray spectroscopy (EDX), and Scanning Electron Microscopy (SEM) were some analytical techniques adopted for such investigations. The key result of those characterization studies [18] was that the presence of chloride and sulphate salts is mostly limited to the surface layer coating the coarser BA particles. This observation justified the strong drops in chloride concentration observed by different authors [5,14,19] while performing percolation leaching tests on BA after washout and dissolution treatments. Similarly, the presence of other sparingly soluble salts, such as calcium sulphate and their similar releasing mechanisms, also suggests their accumulation in a layer coating the coarser BA particles. Finally, the existence of several PTEs was linked with the presence of calcite ($CaCO_3$), melilite, and iron oxides [16], and weathering products such as gypsum, ettringite, and zeolite have been proven to contain high amounts of PTEs.

Several efforts have been devoted to reducing the amounts of salts [20,21] and of the metals Zn, Cu, and Ni [22] from BA through the application of intensive washing processes, which were proven to be rather effective in the removal; however, run-off waters resulted as contaminated from the presence of heavy metals, chlorides, and sulphates [23]. Particularly, concern arose from the leaching of copper [24] and antimony [25], exceeding wastewater discharge limits. Therefore, wet processes aimed at reducing the leaching of salts and PTEs from the mineral fraction of BA, even if effective, still presented several critical downstream issues (e.g., wastewater treatment, sludge thickening, and disposal) in need of optimization. To our knowledge, no literature is available on dry treatment processes applied to the mineral fraction of BA; a dry process, if effective in reducing the release of salts and PTEs, could avoid any wastewater in need of further treatment. In this study, a novel dry treatment process was explored, based on the findings that most of the salts and PTEs released from BA are located on a thin, superficial layer coating coarser BA particles, which can be selectively removed by controlling the mutual abrasion of the particles in a tumbling mill. More precisely, while processing aggregates in a tumbling mill, the grinding of the material occurs. Grinding is driven by three main mechanisms [26]: impact (compression), chipping (attrition), and abrasion [27] when applied with normal, oblique, and parallel forces, respectively (Figure 1). The main objective of this study was to avoid and minimize the compression and chipping forces, and to maintain BA coarseness, while promoting the abrasion on the particles' surfaces.

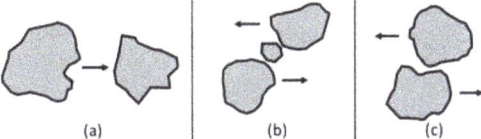

Figure 1. Grinding mechanisms: (a) compression, (b) chipping, (c) abrasion (adapted from Wills and Finch 2015 [27]).

The forces acting in a tumbling mill can be controlled by varying the operating rotational speed. High rotational speeds are associated with impact and compression forces, generated by particles that are cataracting and cascading (Figure 2).

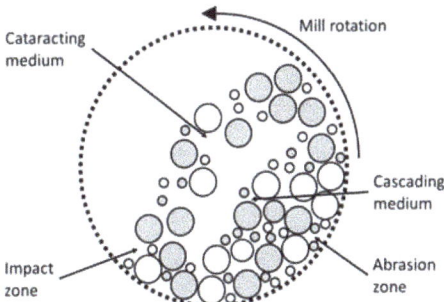

Figure 2. Aggregate motion in a tumbling mill (adapted from Ali et al., 2019 [28]; Wills 2016 [29]; Wills and Finch 2015 [27]).

Alternatively, abrasion is promoted, operating at rotational speeds below 30% of the critical speed Cs, above which particles are only centrifuged without being ground [30]. For low rotational speeds, particles slide on the drum surface remaining in the abrasion zone (Table 1).

Table 1. Empirical description of the grinding actions at different percentages of C_S. Numbers from 1 to 3 indicate the ascending degree of action (adapted from Gupta and Yan 2016 [30]).

% C_S	Sliding	Cascading	Centrifuging
10%	3	-	-
20%	3	-	-
30%	3	1	-
40%	2	1	-
50%	2	1	1
60%	2	2	1
70%	1	3	3
80%	1	3	2
90%	-	2	3

Consequently, the dry treatment process presented in this work did not rely simply on dissolution and mass transfer, which divert the contaminants to wastewater in need of further treatment (as in wet processes); instead, it took advantage of the natural abrasive behavior of BA particles, which scrub off each other's external contamination. Moreover, abrasion allows BA particles to preserve their original size, and therefore remain suitable for recycling as secondary aggregates. The output streams of the treatment investigated in this study were a cleansed coarse fraction of inert material and a fine fraction easily removable by sieving. In particular, the intensity of the abrasion process was controlled to limit its effect on the superficial abrasion of the particles' external layers, avoiding comminution and subsequently, particle-size reduction. This work was aimed at exploring the feasibility of a dry abrasion treatment for the reduction of the BA leaching potential for salts and contaminants, in the light of BA reuse as building materials. Furthermore, the effect of the applied treatment was assessed, comparing the leaching behaviors of treated and untreated BA samples. This approach was an advancement with regard to the current industrial trends, where there is no industrial alternative to the wet treatment of BA. To our knowledge, there are still no literature studies investigating abrasion processes applied to BA at the laboratory scale. The work in the present study was performed with batch experiments only within a laboratory setting. However, tumbling mills were applied in the comminution process of ores in continuous operation mode with throughputs in the range of thousands t/h [27]. In industrial processing plants, it is possible to increase the residence time in abrasion units, as well as install liners on the inner surface of rotating trommels in order to decrease the dimension of the processing unit and optimize its filling ratio.

2. Materials and Methods

2.1. Origin of the Samples

This research is part of the activities of the BASH-TREAT "Bottom ash treatment for an improved recovery of valuable fractions" project (ID-157), funded by ERA-NET Cofund in the 2017 call ERA-MIN2 "Research and Innovation Programme on Raw Materials to foster Circular Economy". Specifically, this research considered the valorization of the mineral fraction derived from state-of-the-art BA treatment. The samples of BA mineral fractions were collected from two BA treatment plants located in Germany (Plant A) and Sweden (Plant B). The primary focus of both plants was the recovery of ferrous and non-ferrous metals. In these plants, BA is usually stored for 12 weeks before being processed, in order to decrease the moisture content to values compatible with the adopted technologies. Plant A discharges the fines (<2 mm) and produces two coarse mineral fractions (2–8 mm and 8–40 mm), which, for this study, were mixed in homogeneous proportions (38 wt % and 62 wt %, respectively) based on what happened in the plant. Plant B produces a mineral fraction with dimensions in the range 4–26 mm. For this study, 15 incremental samples were collected in each plant, directly at the conveyor belt discharge (sampling the entire section of the belt). The 15 samples were collected at regular intervals within one working day, and later merged into a single composite sample for each plant.

For the abrasion experiments, the mixture 2–40 mm for Plant A and the 4–26 mm fraction for Plant B were further investigated. The fines (0–2 mm for Plant A, 0–4 mm for Plant B) were not included in the abrasion tests since these fractions were not relevant for their potential reuse as mineral aggregates, and due to the presence of contaminants in high concentrations.

2.2. Abrasion Tests

The samples 2–40 mm (from Plant A) and 4–26 mm (from Plant B) underwent the abrasion tests. Two different pilot-scale rotating units were operated with increasing processing times to establish intense abrasion forces. Firstly, an initial abrasion test was performed for 240 min using a concrete mixer with an inner radius of 50 cm and a height of 60 cm. The use of a concrete mixer did not allow the removal of fine particles, while minimal losses of fugitive dust from the main opening occurred. Hence, a second abrasion unit was developed, including a sieving device able to remove the fine materials gradually produced. A cylindrical sieving drum (Scheppach RS 400), was coated with a 2 mm mesh stainless steel grate. The fines produced were collected directly at the bottom of the cylindrical drum for further analysis and characterization. The feedstock (12–14 kg) and the rotational speed (set at 42 rpm) were chosen to avoid the cascading of particles. The abrasion process was investigated by varying the abrasion time (60 and 120 min) (Table 2). For all the experiments, the mass and moisture of the samples before and after abrasion were evaluated in order to estimate the production of fines, loss of materials (fugitive dust), and water content. No grinding aids (e.g., steel balls or rods) were added in any of the experiments in order to achieve an autogenous grinding regime [31].

Table 2. Parameters of the abrasion tests performed on the samples collected at plants A and B.

Sample ID	Processing Unit	Duration (min)	Plant A (2–40 mm)	Plant B (4–26 mm)
Raw	Untreated sample	-	✓	✓
CM	Concrete Mixer	240	✗	✓
Abr. 60	Sieving drum	60	✓	✓
Abr. 120	Sieving drum	120	✓	✓

2.3. Leaching Tests

Column leaching tests were performed to evaluate the cumulated release of salts and PTEs before and after abrasion, at liquid to solid (L/S) ratio values equal to 0.3, 1, 2, 3, and 4.0 L/kg. For each sample, the tests were performed in triplicates. The columns (40 cm height, 10 mm internal diameter) were packed with undried samples, and quartz sand was placed on the top and bottom layers of the column (2 mm thickness, grain size: 0.7–1.2 mm). After the saturation of the column, percolation speed was increased according to DIN 19528 [32]. Approximately 3.5–4 kg of the sample was used for each column. The leachates were collected in closed glass bottles for further analyses. For each element, the released amount E_i at any L/S value was calculated (Equation (1)):

$$E_i = (c_i \times V_i)/m_d \text{ (mg/kg)} \tag{1}$$

where i is the index of the eluted fraction (0.3, 1, 2, 3, and 4 L/kg); c_i is the concentration (in mg/L) of the respective element in the leachate volume V_i (in litres); m_d (in kg) is the dry mass of the sample in the column. The cumulative leached amount U (with $U_{L/S} = \Sigma E_i$, in mg/kg) for each element was given at L/S values of 2 and 4 L/kg (Equations (2) and (3)).

$$U_{L/S = 2} = E_{0.3} + E_1 + E_2 \tag{2}$$

$$U_{L/S = 4} = E_{0.3} + E_1 + E_2 + E_3 + E_4 \tag{3}$$

2.4. Analytical Procedures

The moisture content was evaluated on each BA mineral fraction sample by drying batches of materials in porcelain containers at 105 °C, according to DIN EN ISO 17829-1 [33]. The particle size distribution analysis was performed on ash samples before and after the abrasion experiments using a vibratory sieve shaker (Retsch AS 300 control), according to DIN EN 933-1 [34], without any wet removal of the fines.

The procedures adopted to analyze the leachates from the percolation leaching tests were as follows: Immediately after collecting each sample, the electrical conductivity (EC) was measured through a WTW Multi 3320 portable probe. Chlorides were analyzed by titration, following the DIN 38405-1 [35]. Dissolved organic carbon (DOC) was measured through a Multi N/C 2000 analyzer from Analytik Jena AG. Sulphates were analyzed through a Hach DR3900 spectrophotometer. For major elements and PTEs analyses (Ca, K, Na, Al, B, Ba, Co, Cr, Cu, Fe, Li, Mg, Mn, Mo, Ni, Sr, Ti, V, Zr), a part of the leachates was acidified with a small addition (1:100) of 65% nitric acid and analyzed through an ICP-OES Agilent 5100 spectrometer. All analyses were performed in triplicates.

2.5. Data Analysis

The experimental results derived from the analyses of the leachates that came from the column leaching tests were analyzed though a Pearson correlation analysis (two-tailed, 95% confidence) by means of the SPSS Statistics software. The Pearson correlation considered a significance level of $\alpha \leq 0.05$.

3. Results

3.1. Physical Characterization

The particle-size distributions of the BA mineral fraction samples (Figure 3a) were rather different; the samples collected at Plant A presented lower amounts of coarse particles compared to samples from Plant B, although obtaining a greater standard deviation. Considering the particle-size distribution after the abrasion tests (Figure 3b,c), it is possible to observe that, despite the differences between the raw samples, the particle size distribution of the treated samples assumed a coherent behavior, following a narrow bundle where about 25% of the particles had a dimension of below 6.3 mm. For Plant A the additional fine fraction produced was 25% and 38% of the input after 60 min and 120 min abrasion, respectively. For Plant B, the fines produced were 12% and 16% after 60 min and 120 min abrasion, respectively. The significant difference between the processes performed in the two plants was explained by the presence of near-size particles in the Plant A sample, in the fraction proximate to 2 mm (Plant A range: 2–40 mm); however, these were absent in Plant B (range: 4–26 mm). The future optimization of the abrasion process should consider the use of a smaller screen mesh to prevent the loss of particles slightly bigger than 2 mm, together with the abrasion products. Alternatively, the feedstock must be screened to avoid the presence of near-size particles (as occurred naturally for the Plant B material).

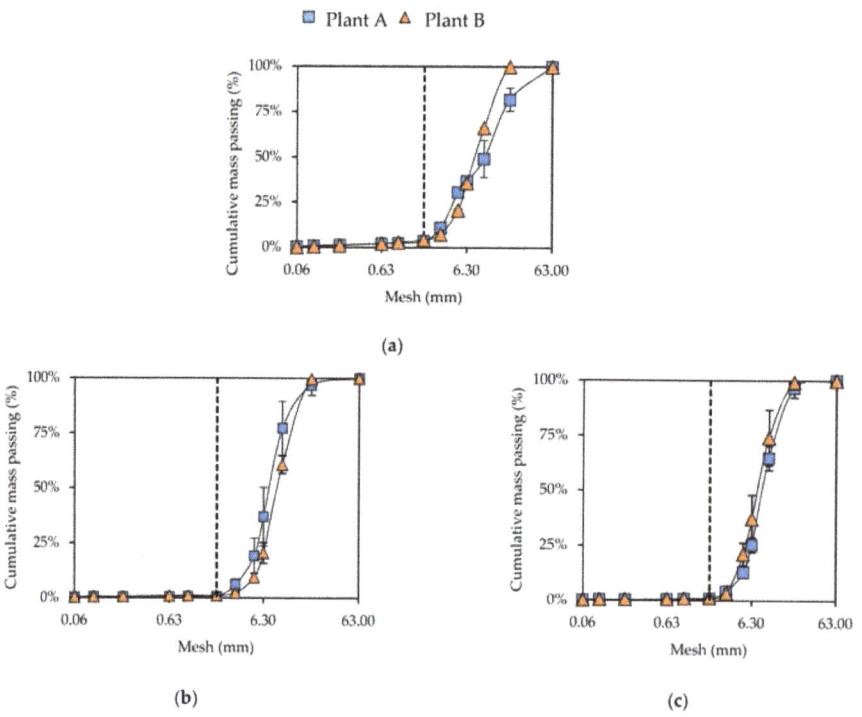

Figure 3. Particle size distribution of (**a**) the raw samples from Plant A and Plant B, of (**b**) samples treated for 60 min and of (**c**) samples treated for 120 min. Dotted line is mesh = 2 mm.

The moisture content decreased drastically in the samples processed with the abrasion unit with removal of the fine particles. Considering the samples from Plant A, the initial moisture was 7.8%, decreased to 4.2% after 60 min of abrasion, and further diminished to 2.8% after 120 min of abrasion. Similarly, for the samples from Plant B, the initial moisture was 4.8%, while the material abraded with the concrete mixer showed a moisture of 4.1%. The moisture was further decreased to 3.3% and 2.1% after 60 min and 120 min abrasion times, respectively (Figure 4). In both samples a limited decrease in the d_{50} after abrasion can be observed: in plant A, d_{50} decreased from 10.25 mm to 8.62 mm after 120 min of abrasion (−16%), while in plant B, d_{50} decreased from 8.03 mm to 7.58 mm after the same abrasion time (−6%).

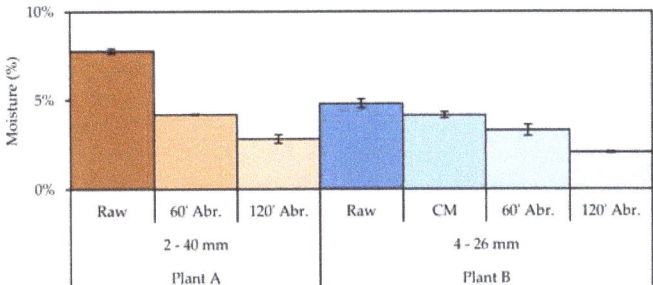

Figure 4. Moisture content of the raw samples from Plant A and Plant B, of the samples treated for 60 and 120 min abrasion (Abr.) time, and those treated with the concrete mixer (only Plant B).

3.2. Leaching Tests

The performance of the investigated abrasion process was estimated by comparing the results of column leaching tests prior to and after treatment, and by analyzing the release of major salts and PTEs.

3.2.1. Release of Major Salts

The purpose of the developed process was to promote the mutual friction between the ash particles. The abrasion, combined with the prompt removal of progressively generated fines and dust, led to a qualitative improvement of the particles' aspect and leaching potential. The intense abrasive forces in the abrasion units were capable of mechanically removing the outer shells of the particles by both smoothing and sanding the ash surfaces. Furthermore, it was suspected that a further contribution to the improvement of the leaching potential could be obtained by the detachment of unbound fine particles weakly adhering to coarser ash fractions, and therefore not screened in standard screening operations in BA processing plants.

The greyish coating layer observed for wet-quenched BA after ageing was removed, displaying the original colors of melt products and inert particles. Specifically, melt products were turned to their original dark brown color and in several cases, their vesicular structure was revealed. It was presumable that repeated collisions cause particle vibrations, in turn capable of detaching mineral phases contained in the ash pores, which are later removed from the process by the continuous turning of the particles. At the same time, refractory materials such as glass, ceramics, and the few metal particles still embedded in the mineral fraction were polished and freed from secondary weathering phases. Therefore, the reduction of salts and PTEs was finally achieved by the sieving out of the progressively generated products of abrasion from the treated coarser particles.

From a qualitative point of view, the removal of the coating layer from BA particles was echoed in an important change in their physical–chemical properties. In all cases, the abraded samples showed a significant decrease in the EC of the leachates (Table 3), which improved with growing abrasion times. Plant A leachates at L/S equal to 0.3 L/kg were characterized by EC values at around 20 mS/cm; the EC value halved at L/S ratio equal to

1 L/kg, and decreased to values below 2 mS/cm for L/S ratios over 3 L/kg. Regarding 60 min and 120 min abrasion times, it was possible to observe a reduction of 30% and 66% on the initial EC, decreased down to values below 1 mS/cm for L/S ratio equal to 4 L/kg. Similar results were achieved for the leachates of the samples from Plant B; the EC of the untreated samples was lower compared to that of the samples from Plant A (7 mS/cm), with the EC value almost halved after 120 min of abrasion time. Even in this case, the final EC values were below 1 mS/cm for L/S equal to 4 L/kg. The analysis of the EC trends clearly demonstrated a significant depletion of the soluble species in the treated BA samples. Linear correlations between EC values and the concentrations of the main salts in the leachates were proven (Figure 5). This, in turn, could be directly correlated with a significant reduction of chloride (r (58) = 0.98, $p < 0.01$) and sulphate (r (58) = 0.94, $p < 0.01$) concentrations. Thus, the EC could be used as a quick in-field control parameter for the evaluation of the effectiveness of an abrasion process.

Table 3. Electrical conductivity values measured in leachates, obtained at different L/S ratios. (Raw: untreated sample; 240 CM: samples treated in the concrete mixer for 240 min; 60–120 Abr.: samples treated in the sieving drum for 60 and 120 min abrasion (Abr.) time respectively).

L/S (L/kg)	Raw	240 CM	Plant A Abr. 60 (mS/cm)	Abr. 120
0.3	19.53 ± 0.31	Not investigated	13.31 ± 0.54	6.64 ± 0.30
1.0	9.63 ± 1.87		5.27 ± 0.57	3.64 ± 0.12
2.0	3.35 ± 0.13		2.47 ± 0.37	1.72 ± 0.10
3.0	2.03 ± 0.07		1.38 ± 0.10	1.03 ± 0.03
4.0	1.53 ± 0.07		0.93 ± 0.03	0.73 ± 0.01

L/S (L/kg)	Raw	240 CM	Plant B Abr. 60 (mS/cm)	Abr. 120
0.3	7.08 ± 1.24	7.40 ± 1.59	4.50 ± 0.35	3.88 ± 0.22
1.0	3.19 ± 0.35	3.73 ± 0.29	2.97 ± 0.16	2.77 ± 0.27
2.0	1.85 ± 0.12	2.00 ± 0.17	1.56 ± 0.07	1.37 ± 0.04
3.0	1.35 ± 0.08	1.24 ± 0.12	1.02 ± 0.02	1.10 ± 0.04
4.0	1.18 ± 0.06	1.07 ± 0.12	0.86 ± 0.19	0.63 ± 0.03

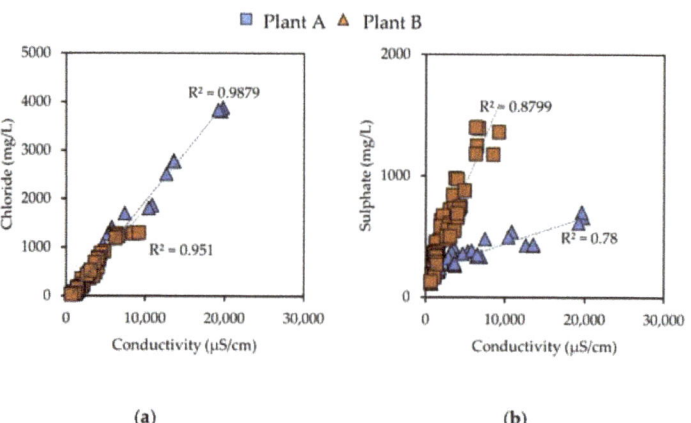

Figure 5. Correlations between electrical conductivity and (a) chloride and (b) sulphate concentrations in the leachates.

The cumulative releases of chloride and sulphate ions in the leachates before and after the abrasion tests are shown in Figure 6 and in Table 4. Chloride salts, which are highly soluble, exhibited leaching profiles characterized by a relevant drop in concentrations that are already at low L/S ratios. With regard to the samples from Plant A, the unprocessed BA presented initial chloride concentrations close to 3800 mg/L at an L/S ratio of 0.3 L/kg

(Figure 6a). These values were halved at an L/S ratio equal to 1 due to the high solubility of chlorides. Between L/S ratios equal to 1 and 2 L/kg, it was possible to observe the shift from the diffusion-controlled dissolution to the reaction-controlled dissolution of low-solubility mineral chloride phases. This was clearly visible in the cumulative concentration curve of chlorides, which tends to flatten for L/S ratios > 2 L/kg. Regarding the samples from Plant B, the processed samples showed decreasing initial chloride concentrations for growing abrasion times (Figure 6c). The effect of the abrasion was reflected mostly at low L/S ratios, where concentrations of 2700 mg Cl^-/L and 1800 mg Cl^-/L were observed at an L/S ratio of 0.3 L/kg, for 60 min and 120 min of abrasion time, respectively. This results in the flattening of the cumulative release profile at an L/S ratio equal to 1 L/kg, and the cumulative release limited to 2300 mg/kg and 1600 mg/kg for 60 min and 120 min of abrasion time, respectively. Similar behaviour was observed for the samples from Plant B. In this case, the unprocessed ash and the samples abraded in the concrete mixer showed comparable results. This could be attributed to the establishment of adhesive electrostatic forces between the fine material not promptly removed from the processing unit and the coarse aggregates. However, the concrete mixer abrasion unit was not further investigated. On the other hand, the cumulative chloride release at an L/S equal to 4 L/kg showed a reduction in the total released chlorides for the treated samples. This resulted in a depletion from circa 1050 mg Cl^-/kg for raw BA from Plant B, to 870 and 770 mg Cl^-/kg after 60 min and 120 min of abrasion time, respectively.

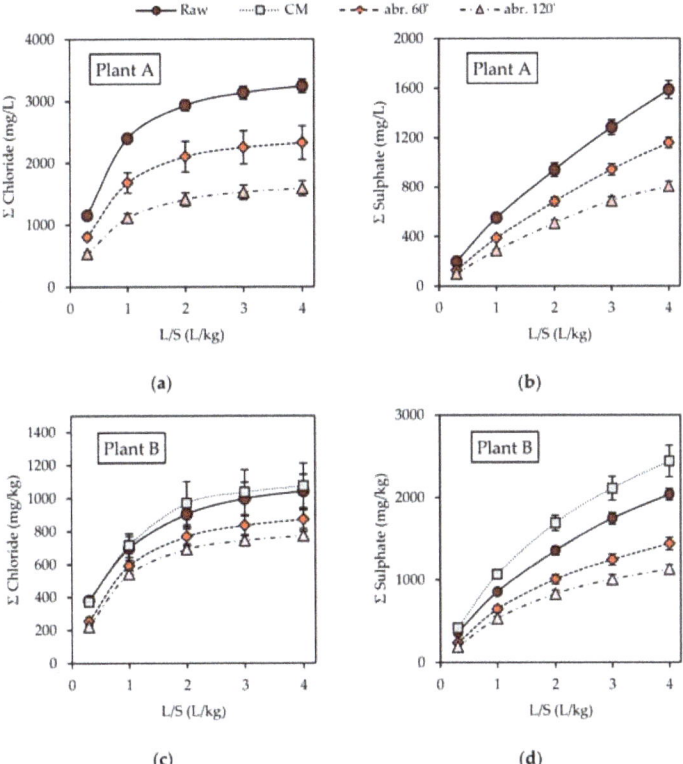

Figure 6. Trends of (a) chlorides and (b) sulphates' cumulative release from the raw BA samples, and of (c) chlorides and (d) sulphates' cumulative release after the abrasion tests (raw: untreated samples; CM: samples treated in the concrete mixer for 240 min; abr. 60 and abr. 120: samples treated in the sieving drum unit for 60 min and 120 min).

Table 4. Cumulative release for the elements considered at an L/S equal to 4 L/kg. (The results obtained from the concrete mixer are not shown). Reduction rates are calculated on the raw sample for increasing abrasion (Abr.) times.

Parameter	Plant A					Plant B				
	Raw	Concentration Abr. 60'	Abr. 120'	Reduction Rate Abr. 60'	Abr. 120'	Raw	Concentration Abr. 60'	Abr. 120'	Reduction Rate Abr. 60'	Abr. 120'
				Major Elements (mg/kg)						
Cl^-	3244 ± 111	2329 ± 275	1589 ± 123	−28%	−51%	1043 ± 102	872 ± 56	774 ± 27	−16%	−26%
SO_4^{2-}	1587 ± 73	1160 ± 42	806 ± 39	−27%	−49%	2037 ± 70	1441 ± 76	1132 ± 54	−29%	−44%
DOC	264 ± 16	180 ± 11	120 ± 9	−32%	−55%	157 ± 24	117 ± 3	96 ± 5	−26%	−39%
Ca	1360 ± 134	887 ± 33	668 ± 23	−35%	−51%	529 ± 29	404 ± 16	325 ± 8	−24%	−39%
K	627 ± 19	454 ± 36	345 ± 20	−27%	−45%	249 ± 21	203 ± 11	176 ± 7	−19%	−29%
Na	1224 ± 33	912 ± 78	711 ± 46	−25%	−42%	834 ± 80	674 ± 45	587 ± 33	−19%	−30%
				Minor Elements (μg/kg)						
Al	3913 ± 769	9822 ± 1267	17290 ± 4085	151%	342%	7222 ± 521	6192 ± 167	5964 ± 195	−14%	−17%
B	495 ± 126	373 ± 13	698 ± 167	−25%	41%	1891 ± 124	1593 ± 51	1815 ± 228	−16%	−4%
Ba	639 ± 62	557 ± 15	476 ± 28	−13%	−25%	220 ± 9	219 ± 12	206 ± 5	−1%	−7%
Co	11 ± 4	16 ± 2	18 ± 3	44%	55%	37 ± 4	49 ± 11	57 ± 3	33%	52%
Cr	276 ± 45	152 ± 17	141 ± 15	−45%	−49%	469 ± 76	290 ± 67	221 ± 43	−38%	−53%
Cu	1613 ± 76	868 ± 61	619 ± 135	−46%	−62%	448 ± 118	301 ± 172	181 ± 31	−33%	−60%
Fe	181 ± 65	206 ± 41	216 ± 42	14%	19%	323 ± 83	245 ± 19	316 ± 49	−24%	−2%
Li	500 ± 42	285 ± 17	184 ± 4	−43%	−63%	94 ± 8	56 ± 6	40 ± 6	−41%	−58%
Mg	351 ± 54	337 ± 8	460 ± 44	−4%	31%	347 ± 36	284 ± 53	337 ± 21	−18%	−3%
Mn	9 ± 1	13 ± 2	10 ± 2	40%	6%	20 ± 3	18 ± 3	22 ± 2	−10%	10%
Mo	586 ± 14	498 ± 19	440 ± 14	−15%	−25%	400 ± 18	375 ± 46	368 ± 39	−6%	−8%
Ni	38 ± 16	27 ± 17	30 ± 4	−28%	−21%	67 ± 14	96 ± 17	104 ± 15	43%	54%
Sr	8528 ± 700	5552 ± 263	3589 ± 39	−35%	−58%	1955 ± 89	1555 ± 118	1208 ± 38	−20%	−38%
Ti	45 ± 1	43 ± 3	43 ± 1	−5%	−6%	99 ± 1	100 ± 3	100 ± 1	1%	0%
V	203 ± 19	144 ± 6	126 ± 9	−29%	−38%	145 ± 8	144 ± 4	143 ± 5	0%	−1%
Zn	166 ± 37	143 ± 65	110 ± 33	−14%	−34%	80 ± 31	265 ± 138	20 ± 1	233%	−75%

The sulphates' leaching trends (Figure 6) were slightly different from those of chlorides, exhibiting a slower concentration decay and resulting in an almost linear cumulative release profile. This could be explained by the sparing solubility of sulphate species such as ettringite, which are governed by reaction-controlled dissolution. Hence, considering the samples collected from Plant A, concentrations of sulphates in the leachate derived from the raw samples (Figure 6b) decreased from 650 mg SO_4^{2-}/L at L/S equal to 0.3 L/kg, down to approximately 300 mg SO_4^{2-}/L at L/S equal to 4 L/kg. However, with regard to the samples from Plant B, the untreated ones were characterized by higher concentrations of sulphate in leachates (Figure 6d); more specifically, it was possible to observe a leaching profile that shared the features of both the reaction-controlled and diffusion-controlled dissolution of salts. Significant reductions in the concentration of sulphates were observed between L/S 0.3 and 1 L/kg, probably due to the presence of large amount of sulphate salts. Their depletion occurred especially until L/S equal to 1, after which a constant decrease was observed (Figure 6d). For the treated samples, a reduction in sulphate concentrations with growing abrasion times was observed in all samples. In contrast to the case of chlorides, the flattening of the cumulative release profiles for sulphate was partially appreciable after 120 min of abrasion time for L/S = 4 L/kg. Regarding the samples from Plant A, the cumulative release decreased from 1600 mg SO_4^{2-}/L down to around 1150 mg SO_4^{2-}/L (after 60 min abrasion) and 800 mg SO_4^{2-}/L (after 20 min abrasion). For samples derived from Plant B, the cumulative release of untreated material was over 2000 mg SO_4^{2-}/L and reduced to 1450 mg SO_4^{2-}/L and 1150 mg SO_4^{2-}/L after 60 min and 120 min of abrasion times, respectively.

3.2.2. Release of Other Compounds

The values for the cumulative release of major and minor elements in the leachates are displayed in Table 4. Chloride and sulphate trends have been previously discussed. The superficial abrasion was demonstrated to be capable of reducing DOC cumulative release to 39%, which is, together with sulphate, the main reason for copper release, as suggested by several studies [36,37] and also confirmed by this work. Despite the high standard deviations obtained by measuring Cu concentrations in the leachates, the decrease of DOC could be the cause of a drastic reduction in Cu concentrations. Furthermore, a reduction of both Cu and DOC after the treatment with the concrete mixer was observed (data not shown). This trend was not observed for other elements, suggesting an auxiliary mechanism acting during the abrasion. However, the reduction was expected to be less effective with the extension of the process duration. As mentioned in the introduction, salts and PTEs during ageing acted as a coating layer on BA particles. By extending the abrasion time enough to completely wear out the outer particle shell, surfaces of minerals formed within the incineration chamber and refractory materials started to be abraded. This behavior explained the higher measure of iron and nickel concentrations in the leachates after 120 min of abrasion time (Table 4). Iron in BA is in the form of non-soluble minerals, and despite its concentration in BA (31–150 g Fe/kg BA) (Astrup et al., 2016), its concentrations in the leachates were relatively low. Hence, what could be hypothesized is that by prolonging the abrasion time, iron minerals start to wear out, leading to increasing iron concentrations in the leachates. As for the main soluble species, the release of specific PTEs as Cu, Cr, and Mo can be tracked by following the progressive washout of the main salts. Molybdenum present as oxyanion MoO_4^{2-} is reported to exhibit high solubility, although the formation of minerals such as powellite ($CaMoO_4$) might have an influence on the leaching behavior [38]. However, in this study, notwithstanding a significant relationship with sulphate concentrations, Mo showed a slightly higher correlation with chloride concentrations ($r(58) = 0.97$, $p < 0.001$).

Another important piece of information obtained from exploring the correlations among the different elements is the speciation of the soluble salts. The main chloride forms in BA are halite (NaCl) and sylvite (KCl) [18]. This was also proven in this work, where the chloride correlations with K and Na concentrations in the leachates (detailed in the Supplementary Materials) suggested that halite was the primary cause of chloride release (Cl–Na: $r(58) = 0.99$, $p < 0.001$; Cl–K: $r(58) = 0.99$, $p < 0.001$). Similarly, the correlation between calcium and sulphate concentrations ($r(58) = 0.98$, $p < 0.001$) suggested the dissolution of gypsum, a typical product resulting from BA weathering processes [16].

The efficiency of the proposed dry treatment was finally assessed by studying the correlations between the major elements and the PTEs. The statistical analysis performed on all the concentration values measured in the leachates (Tables S1 and S2, Supplementary Materials) allowed us to obtain some interesting results (Figure 7). Chloride and sulphate concentrations in the leachates showed a good correlation with Ba, Cu, Mo, and Sr. It can be concluded that the dissolution trends of the PTEs were coherent with the behavior followed by the bearing salts (e.g., mainly chlorides and sulphates), present in the outer layers coating the BA particles.

The variables portrayed in Figure 7 show a large positive relationship; the values of the coefficient R^2 determine the quality of the fitting of the linear correlation model applied to the considered variables. Nonetheless, the strength of the correlation between the variables was expressed by the Pearson coefficient R (see Supplementary Materials Tables S1 and S2). All the portrayed relationships displayed a significance level below the threshold $\alpha \leq 0.05$.

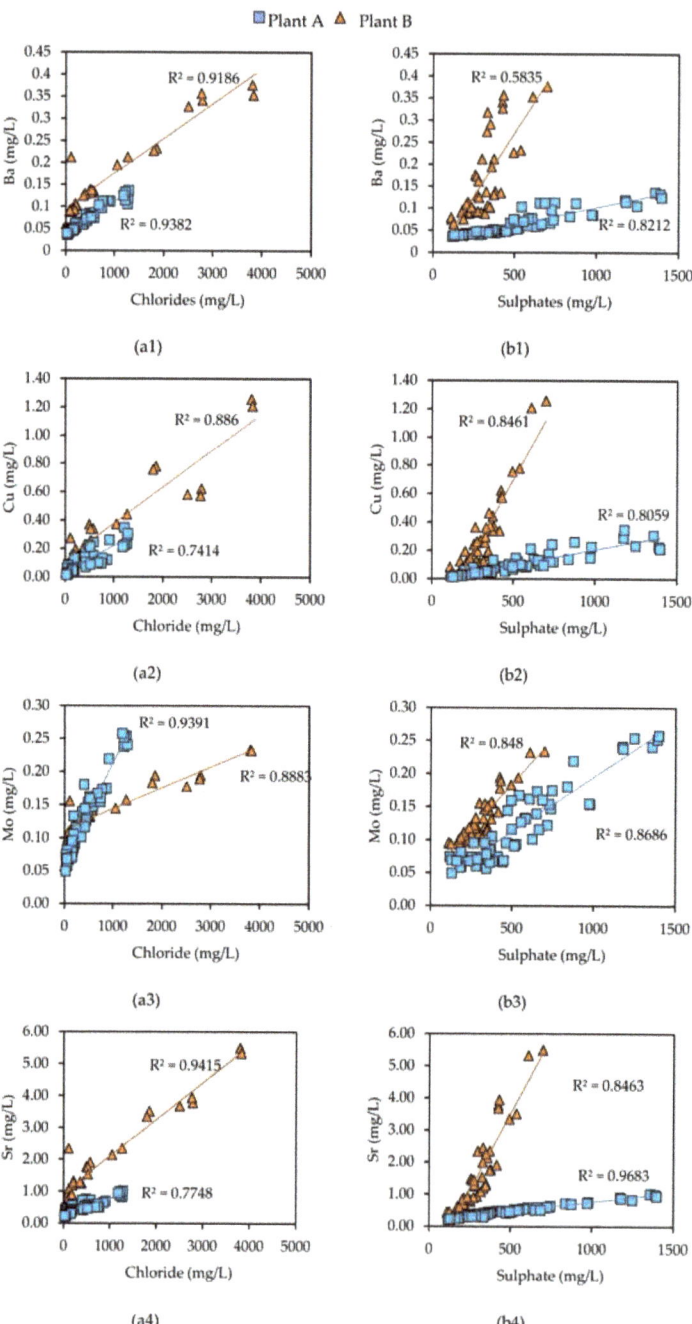

Figure 7. Correlations between chlorides (**a1–a4**) and sulphates (**b1–b4**) and Ba, Cu, Mo, and Sr in the leachates derived from the column leaching tests.

4. Conclusions

This work investigated at the laboratory scale a dry treatment process based on abrasion, used for the removal of salts and PTEs from BA mineral fraction particles. Two different abrasion units were developed and operated at different times. The best results were obtained using an attrition unit equipped with a screen for the removal of the progressively generated fines. In this case, a relevant drop in the electrical conductivity was associated to a reduction of chlorides (up to 26%) and sulphates (up to 44%), in turn coherent with lower amounts of PTEs released from BA in percolation tests (up to 53% for chromium, 60% for copper, and 8% for molybdenum). On the other hand, prolonged abrasion times showed slightly higher concentrations of Fe, Co, and Ni in leachates, suggesting that the process reached deeper layers of iron-containing incineration products. Further tests are required to assess the release of these metals in aggregate mixtures and its correlation with the duration of abrasion. The statistical analysis revealed good correlations between chloride and sulphate concentrations in the leachates, and the concentrations of barium, copper, molybdenum, and strontium, thus revealing consistent behavior for the major salts and the cited minor components present in the layer surrounding the BA particles. Overall, the studied approach is a valid alternative to wet processes for the reduction of the leaching potential of salts and PTEs in coarse bottom ash samples. However, two key issues need to be further investigated: the specific effect of abrasion on BA components (inerts, refractory materials, and incineration products), and the influence of abrasion on the release of PTEs in the leachates.

The results of our research cannot be compared with previous studies on the topic due to the originality of the process, making it impossible to contextualize our findings. As a general conclusion, any BA valorization strategy should consider that the parent material could affect the overall impact of the results of the abrasion because of two key issues. Firstly, the chemical composition: salts and PTEs, if included in BA composition (e.g., sulphates in residues from renovation activities and/or gypsum), would be released as fine particles. Therefore, it was necessary to equip the abrasion unit with a screen for the removal of fines to limit their recirculation and adhesion to the abraded particles. Secondly, the material hardness: the lower it was, the more effective the abrasion was. Refractory materials (glass, ceramics) and incineration products (silicates/pyroxenes, oxides/spinels, and hematite) commonly found in BA are harder than the weathering phases, making possible their further removal through abrasion. In the absence of these harder fractions, it was predictable that salts and PTEs would land directly with the fines that had already been separated in the BA processing plants.

Supplementary Materials: The following are available online at https://www.mdpi.com/article/10.3390/ma14113133/s1, Table S1: Results of statistical analysis of the correlations among the major and minor compounds analysed in the leachates (samples of bottom ash mineral fraction from PLANT A); Table S2: Results of statistical analysis of the correlations among the major and minor compounds analysed in the leachates (samples of bottom ash mineral fraction from PLANT B).

Author Contributions: Conceptualization, methodology, experimental investigation, data curation, writing—original draft preparation, M.A. and M.B.; conceptualization, methodology, supervision, S.F.; writing—review and editing, M.A., M.B., F.-G.S., R.G., M.H., K.K. and S.F.; project administration and funding acquisition, K.K., F.-G.S., R.G., M.H. and S.F. All authors have read and agreed to the published version of the manuscript.

Funding: The authors would like to acknowledge the financial support to the project BASH-Treat (ID 157) provided by the ERA-MIN2 Research & Innovation Programme on Raw Materials to Foster Circular Economy, by the German Federal Ministry of Education and Research (BMBF, FKZ 033RU005 A-C) and by the Italian Ministry of University and Research (MIUR). The authors are thankful to project partners Heidemann Recycling GmbH and Sysav Utveckling AB.

Institutional Review Board Statement: Not applicable.

Informed Consent Statement: Not applicable.

Data Availability Statement: The full data related to this research are not publicly available due to the confidentiality agreement signed by the Consortium of BASH-TREAT project.

Acknowledgments: The authors gratefully acknowledge the support of the "Open access publishing" program of Hamburg University of Technology for part of the article processing fee.

Conflicts of Interest: The authors declare no conflict of interest.

References

1. British Geological Survey. *BGS, British Geological Survey World Mineral Production 2013–2017*; British Geological Survey: Nottingham, UK, 2019. Available online: http://www.bgs.ac.uk/mineralsuk/statistics/worldStatistics.html (accessed on 2 June 2020).
2. Publications Office of the European Union. European Commission Study on the Review of the list of Critical Raw Materials. In *Non-Critical Raw Materials Factsheets*; Publications Office of the European Union: Luxembourg, 2017.
3. Thomé-Kozmiensky, E. *Abfallverbrennungsanlagen—Deutschland—2016 | 2017*; Thomé-Kozmiensky GmbH (Anlagendokumentation, 2): Nietwerder, Germany, 2018.
4. Neuwahl, F.; Cusano, G.; Benadives, J.G.; Holbrook, S.; Serge, R. *Best Available Techniques (BAT) Reference Document for Waste Treatment Industries*; Publications Office of the European Union: Luxembourg, 2019. [CrossRef]
5. Astrup, T.; Muntoni, A.; Polettini, A.; Pomi, R.; van Gerven, T.; van Zomeren, A. Chapter 24—Treatment and Reuse of Incineration Bottom Ash. In *Environmental Materials and Waste//Environmental Materials and Waste. Resource Recovery and Pollution Prevention*; Prasad, M.N.V., Shih, K., Eds.; Academic Press: Cambridge, MA, USA, 2016; pp. 607–645. [CrossRef]
6. Šyc, M.; Simon, F.G.; Hykš, J.; Braga, R.; Biganzoli, L.; Costa, G.; Funari, V.; Grosso, M. Metal recovery from incineration bottom ash: State-of-the-art and recent developments. *J. Hazard. Mater.* **2020**, *393*. [CrossRef] [PubMed]
7. van Praagh, M.; Johansson, M.; Fagerqvist, J.; Grönholm, R.; Hansson, N.; Svensson, H. Recycling of MSWI-bottom ash in paved constructions in Sweden—A risk assessment. *Waste Manag.* **2018**, *79*, 428–434. [CrossRef] [PubMed]
8. Minane, J.R.; Becquart, F.; Abriak, N.E.; Deboffe, C. Upgraded Mineral Sand Fraction from MSWI Bottom Ash: An Alternative Solution for the Substitution of Natural Aggregates in Concrete Applications. *Procedia Eng.* **2017**, *180*, 1213–1220. [CrossRef]
9. Ginés, O.; Chimenos, J.M.; Vizcarro, A.; Formosa, J.; Rosell, J.R. Combined use of MSWI bottom ash and fly ash as aggregate in concrete formulation: Environmental and mechanical considerations. *J. Hazard. Mater.* **2009**, *169*, 643–650. [CrossRef] [PubMed]
10. Keulen, A.; van Zomeren, A.; Harpe, P.; Aarnink, W.; Simons, H.A.E.; Brouwers, H.J.H. High performance of treated and washed MSWI bottom ash granulates as natural aggregate replacement within earth-moist concrete. *Waste Manag.* **2016**, *49*, 83–95. [CrossRef] [PubMed]
11. Brown, T.J.; Idoine, N.E.; Wrighton, C.E.; Raycraft, E.R.; Hobbs, S.F.; Shaw, R.A.; Everett, P.; Kresse, C.; Deady, Y.E.A.; Bide, T. *World Mineral Production 2014–2018*; British Geological Survey: Nottingham, UK, 2016.
12. Blasenbauer, D.; Huber, F.; Lederer, J.; Quina, M.J.; Blanc-Biscarat, D.; Bogush, A.; Bontempi, E.; Blondeau, J.; Chimenos, J.M.; Dahlbo, H.; et al. Legal situation and current practice of waste incineration bottom ash utilisation in Europe. *Waste Manag.* **2020**, *102*, 868–883. [CrossRef] [PubMed]
13. Pourret, O.; Bollinger, J.C.; Hursthouse, A. Heavy Metal: A misused term? *Acta Geochim.* **2021**. [CrossRef]
14. Di Gianfilippo, M.; Hyks, J.; Verginelli, I.; Costa, G.; Hjelmar, O.; Lombardi, F. Leaching behaviour of incineration bottom ash in a reuse scenario: 12years-field data vs. lab test results. *Waste Manag.* **2018**, *73*, 367–380. [CrossRef]
15. Dijkstra, J.J.; van Zomeren, A.; Meeussen, J.C.L.; Comans, R.N.J. Effect of accelerated aging of MSWI bottom ash on the leaching mechanisms of copper and molybdenum. *Environ. Sci. Technol.* **2018**, *40*, 4481–4487. [CrossRef]
16. Alam, Q.; Schollbach, K.; van Hoek, C.; van der Laan, S.; Wolf, T.; de Brouwers, H.J.H. In-depth mineralogical quantification of MSWI bottom ash phases and their association with potentially toxic elements. *Waste Manag.* **2019**, *87*, 1–12. [CrossRef]
17. Chimenos, J.M.; Segarra, M.; Fernández, M.A.; Espiell, F. Characterization of the bottom ash in municipal solid waste incinerator. *J. Hazard. Mater.* **1999**, *64*, 211–222. [CrossRef]
18. Alam, Q.; Lazaro, A.; Schollbach, K.; Brouwers, H.J.H. Chemical speciation, distribution and leaching behavior of chlorides from municipal solid waste incineration bottom ash. *Chemosphere* **2020**, *241*, 124985. [CrossRef]
19. Alam, Q.; Florea, M.V.A.; Schollbach, K.; Brouwers, H.J.H. A two-stage treatment for Municipal Solid Waste Incineration (MSWI) bottom ash to remove agglomerated fine particles and leachable contaminants. *Waste Manag.* **2017**, *67*, 181–192. [CrossRef]
20. Holm, O.; Simon, F.G. Innovative treatment trains of bottom ash (BA) from municipal solid waste incineration (MSWI) in Germany. *Waste Manag.* **2017**, *59*, 229–236. [CrossRef]
21. Steketee, J.J.; Duzijn, R.F.; Born, J.G.P. Quality Improvement of MSWI Bottom Ash by Enhanced Aging, Washing and Combination Processes. In *Studies in Environmental Science: Waste Materials in Construction*; Goumans, J.J.J.M., Senden, G.J., van der Sloot, H.A., Eds.; Elsevier: Amsterdam, The Netherlands, 1997; Volume 71, pp. 13–23. Available online: http://www.sciencedirect.com/science/article/pii/S0166111697801847 (accessed on 10 April 2021).
22. Sun, X.; Yi, Y. Acid washing of incineration bottom ash of municipal solid waste: Effects of pH on removal and leaching of heavy metals. *Waste Manag.* **2021**, *120*, 183–192. [CrossRef]
23. Hu, Y.; Zhao, L.; Zhu, Y.; Zhang, B.; Hu, G.; Xu, B.; He, C.; Di Maio, F. The fate of heavy metals and salts during the wet treatment of municipal solid waste incineration bottom ash. *Waste Manag.* **2021**, *121*, 33–41. [CrossRef]

24. Quek, A.; Xu, W.; Guo, L.; Wu, D. Heavy Metal Removal from Incineration Bottom Ash through Washing with Rainwater and Seawater. *Int. J. Waste Resour.* **2016**, *6*, 203. [CrossRef]
25. Alam, Q.; Schollbach, K.; Florea, M.V.A.; Brouwers, H.J.H. Investigating washing treatment to minimize leaching of chlorides and heavy metals from MSWI bottom ash. In Proceedings of the 4th International Conference on Sustainable Solid Waste Management, Limassol, Cyprus, 23 June 2016.
26. Swapan Kumar, H. (Ed.) Mineral Exploration. In *Principles and Applications*; Elsevier Science: San Diego, CA, USA, 2018.
27. Wills, B.A.; Finch, J.A. Chapter 7—Grinding Mills. In *Wills' Mineral Processing Technology. An Introduction to the Practical Aspects of Ore Treatment and Mineral Recovery*, 8th ed.; Wills, B.A., Finch, J., Eds.; Butterworth-Heinemann: Amsterdam, The Netherlands, 2015; pp. 147–179.
28. Ali, Y.; Garcia-Mendoza, C.D.; Gates, J.D. Effects of 'impact' and abrasive particle size on the performance of white cast irons relative to low-alloy steels in laboratory ball mills. *Wear* **2019**, *426*, 83–100. [CrossRef]
29. Barry, A.W. (Ed.) *Wills' Mineral Processing Technology*, 8th ed.; Butterworth-Heinemann: Boston, CA, USA, 2016.
30. Gupta, A.; Yan, D.S. Chapter 7—Tubular Ball Mills. In *Mineral Processing Design and Operations*, 7th ed.; Gupta, A., Yan, D.S., Eds.; Elsevier: Amsterdam, The Netherlands, 2016; pp. 189–240. [CrossRef]
31. Gupta, A.; Yan, D.S. Chapter 9—Autogenous and Semi-Autogenous Mills. In *Mineral Processing Design and Operations*, 6th ed.; Elsevier: Amsterdam, The Netherlands, 2016; pp. 234–254.
32. Deutsche Institut für Normung e.V. DIN 19528:2009-01, Elution von Feststoffen_- Perkolationsverfahren zur gemeinsamen Untersuchung des Elutionsverhaltens von anorganischen und organischen Stoffen. Available online: https://www.beuth.de/en/standard/din-19528/104285985 (accessed on 25 May 2021).
33. Deutsche Institut für Normung e.V. Geotechnische Erkundung und Untersuchung_- Laborversuche an Bodenproben_- Teil_1: Bestimmung des Wassergehalts (ISO_17892-1:2014); Deutsche Fassung EN_ISO_17892-1:2014. Available online: https://www.beuth.de/en/standard/din-en-iso-17892-1/208096261 (accessed on 25 May 2021).
34. Deutsche Institut für Normung e.V. DIN EN 933-1:2012-03, Prüfverfahren für geometrische Eigenschaften von Gesteinskörnungen_- Teil_1: Bestimmung der Korngrößenverteilung_- Siebverfahren; Deutsche Fassung EN_933-1:2012. Available online: https://www.beuth.de/en/standard/din-en-933-1/148605855 (accessed on 25 May 2021).
35. Deutsche Institut für Normung e.V. DIN 38405-1:1985-12, Deutsche Einheitsverfahren zur Wasser-, Abwasser- und Schlammuntersuchung; Anionen (Gruppe_D); Bestimmung der Chlorid-Ionen_(D_1). Available online: https://www.beuth.de/en/standard/din-38405-1/1263215 (accessed on 25 May 2021).
36. Arickx, S.; de Borger, V.; van Gerven, T.; Vandecasteele, C. Effect of carbonation on the leaching of organic carbon and of copper from MSWI bottom ash. *Waste Manag.* **2010**, *30*, 1296–1302. [CrossRef]
37. Hyks, J.; Astrup, T.; Christensen, T.H. Leaching from MSWI bottom ash: Evaluation of non-equilibrium in column percolation experiments. *Waste Manag.* **2009**, *29*, 522–529. [CrossRef]
38. Nilsson, M.; Andreas, L.; Lagerkvist, A. Effect of accelerated carbonation and zero valent iron on metal leaching from bottom ash. *Waste Manag.* **2016**, *51*, 97–104. [CrossRef]

Article

Effect of Metal Lathe Waste Addition on the Mechanical and Thermal Properties of Concrete

Marcin Małek [1], Marta Kadela [2,*], Michał Terpiłowski [1], Tomasz Szewczyk [1], Waldemar Łasica [1] and Paweł Muzolf [1]

1. Faculty of Civil Engineering and Geodesy, Military University of Technology in Warsaw, ul. Gen. Sylwestra Kaliskiego 2, 00-908 Warsaw, Poland; marcin.malek@wat.edu.pl (M.M.); mic.terpilowski@gmail.com (M.T.); tomasz.szewczyk6@gmail.com (T.S.); waldemar.lasica@wat.edu.pl (W.Ł.); pawel.muzolf@wat.edu.pl (P.M.)
2. Building Research Institute (ITB), ul. Filtrowa 1, 00-611 Warsaw, Poland
* Correspondence: m.kadela@itb.pl; Tel.: +48-603-60-12-48

Citation: Małek, M.; Kadela, M.; Terpiłowski, M.; Szewczyk, T.; Łasica, W.; Muzolf, P. Effect of Metal Lathe Waste Addition on the Mechanical and Thermal Properties of Concrete. *Materials* 2021, 14, 2760. https://doi.org/10.3390/ma14112760

Academic Editor: Franco Medici

Received: 11 February 2021
Accepted: 21 May 2021
Published: 23 May 2021

Publisher's Note: MDPI stays neutral with regard to jurisdictional claims in published maps and institutional affiliations.

Copyright: © 2021 by the authors. Licensee MDPI, Basel, Switzerland. This article is an open access article distributed under the terms and conditions of the Creative Commons Attribution (CC BY) license (https://creativecommons.org/licenses/by/4.0/).

Abstract: The amount of steel chips generated by lathes and CNC machines is 1200 million tons per year, and they are difficult to recycle. The effect of adding steel chips without pre-cleaning (covered with production lubricants and cooling oils) on the properties of concrete was investigated. Steel waste was added as a replacement for fine aggregate in the amounts of 5%, 10% and 15% of the cement weight, which correspond with 1.1%, 2.2% and 3.3% mass of all ingredients and 0.33%, 0.66% and 0.99% volume of concrete mix, respectively. The slump cone, air content, pH value, density, compressive strength, tensile strength, tensile splitting strength, elastic modulus, Poisson's ratio and thermal parameters were tested. It was observed that with the addition of lathe waste, the density decreased, but mechanical properties increased. With the addition of 5%, 10% and 15% metal chips, compressive strength increased by 13.9%, 20.8% and 36.3% respectively compared to plain concrete; flexural strength by 7.1%, 12.7% and 18.2%; and tensile splitting strength by 4.2%, 33.2% and 38.4%. Moreover, it was determined that with addition of steel chips, thermal diffusivity was reduced and specific heat capacity increased. With the addition of 15% metal chips, thermal diffusivity was 25.2% lower than in the reference sample, while specific heat was 23.0% higher. No effect was observed on thermal conductivity.

Keywords: recycling; lathe waste; CNC machining; sustainable development; mix modification; workability; mechanical properties; thermal properties

1. Introduction

In 2019, global steel production was estimated at 1869.9 Mt, an increase of 3.4% compared to 2018 [1,2]. The construction sector generates a huge demand for steel [3,4]. Due to the development of the construction industry, a further increase in steel production is forecasted. The widespread use of steel products in the industry results in energy consumption [5], CO_2 emission [6,7] and production of steel waste in various amounts and sizes [8–11]. Some steel waste is recycled [12–14]. Metal chips are generated during the cutting, milling and turning process as a side effect of manufacturing elements with specific geometric dimensions and surface finish [15,16]. These wastes represent about 3–5% by weight of metal casting. Moreover, it was estimated that industrial lathes produce about 3–4 kg of chips per working day [17,18], and according to an ICI report, the amount of waste generated by lathes and CNC machines is up to 1200 Mt per year [19–21]. Due to contamination of the chip surface with oils or other coolants during the machining process, the storage of this waste has a negative impact on the environment, and its cleaning generates additional costs [22]. Because of this, as well as due to their elongated spiral shape, small size and surface contamination, the recycling of metal chips is difficult [23]. In addition, the generated chips can have different properties due to different types of materials being processed.

It is possible to use a variety of waste as concrete components [24,25]. In literature it can be found that metal chips used in concrete mix are classified by other scientists as a replacement for aggregate or steel fibers.

With the addition of steel chips as a replacement for fine and coarse aggregate at 22%, 33% and 44% of cement mass, Maanvit et al. [26] obtained 19%, 27% and 19% respective increases in compressive strength compared to plain concrete and 50, 100 and 75% respective increases in flexural strength. Ismail and Al.-Hashmi [27] obtained 13% and 17% increases in compressive strength with the addition of iron filings at 28% and 37% of cement mass as a replacement for fine aggregate compared to the base sample, a but slight decrease (approximately 2%) for a 19% addition. The tensile strength increased by 23%, 24% and 28% respectively. With an increase in the amount of the addition to 56%, 75% and 94% of cement mass, 5%, 8% and 22% respective increases in compressive strength and 9%, 28% and 41% respective increases in flexural strength compared to the base sample were determined [28]. Alwaeli and Nadziakiewicz [29] demonstrated that with the addition of steel waste at 68%, 136%, 203% and 271% of cement mass (25%, 50%, 75% and 100% of fine aggregate), compressive strength respectively increased by 24%, 30%, 43% and 50% in relation with plain concrete. An inverse result was demonstrated by Hemanth Tunga et al. [30], who for addition of steel waste as replacement fine aggregate at 11%, 17% and 22% of cement weight obtained decreases in compressive strength. However, they used a mix of metal and plastic chips.

The addition of steel waste as fibers has been tested by other researchers [31–34]. Kumaran et al. [31] reported 5%, 11% and 9% increases in compressive strength and 9%, 19% and 10% increases in flexural strength for concrete with fiber addition at 7%, 10% and 13% of cement mass compared to plain concrete. Gawatre et al. [32] obtained an increase in compressive strength (11%) with the addition of fiber up to 25% of cement mass, and then a decrease. Similar results were demonstrated by Dharmaraj [33]. The compressive strength increased up to 10% of shredded scrap iron addition and then decreased. The maximum compressive strength was 53%; however, they used fly ash. Equally high increases in compressive strength (40%, 51% and 62%) were determined by Seetharam et al. [19] for concrete with fiber additions of 10, 20 and 30% of cement weight. For similar amounts of fibers (31%, 48% and 65% of cement mass), Mohammed et al. [34] obtained only 7%, 12% and 15% higher compressive strength compared to the reference sample.

Based on the above, it can be concluded that the results for concretes with the addition of metal chips are very different, and in order to commonly use them, further research is required. Moreover, different scientists use varied classifications of steel chips (some as aggregate, some as dispersed fibers). Therefore, this study aimed to assess the effect of steel chips as a replacement for fine aggregate on the mechanical and thermal properties of concrete. Compared to other research, lathe chips were used without pre-cleaning (covered with production lubricants and cooling oils), which is new and aims to increase the use of waste materials from the production process in construction. Tests of mixed (consistence, air content and pH value) and hardened concrete with 5, 10 and 15 wt.% of steel chips (compressive, flexural and split tensile strength, modulus of elasticity and Poisson's coefficient, thermal conductivity and diffusivity and specific conductivity) were carried out. Steel chips in the amounts of 5, 10 and 15 wt.% of the cement weight, which correspond with 1.1%, 2.2% and 3.3% of mass of all ingredients and 0.33%, 0.66% and 0.99% of volume of concrete mix, respectively, were used. The obtained results show that it is possible to efficiently and ecologically manage lathe waste in concrete while improving its mechanical parameters.

2. Materials

2.1. Specimen Preparation

The mix was prepared using the following materials: Portland cement, aggregate, tap water, admixture and steel waste.

2.1.1. Cement

CEM I 42.5R Portland cement according to the EN 197:1:2011 [35] standard was used (Górażdże Cement Works, Opole, Poland). The properties [36] and chemical composition of the cement are shown in Tables 1 and 2, respectively.

Table 1. Physical properties of cement [36].

Specific Surface Area (m²/kg)	Specific Gravity (kg/m³)	Compressive Strength after Days (MPa)	
		2 Days	28 Days
3.874	3050–3140	27.8	59.3

Table 2. Chemical composition of cement [37].

Compositions	SiO$_2$	Al$_2$O$_3$	Fe$_2$O$_3$	CaO	MgO	SO$_3$	Na$_2$O	K$_2$O	Cl
Unit (vol. %)	19.5	4.9	2.9	63.3	1.3	2.8	0.1	0.9	0.05

2.1.2. Aggregate

Crushed basalt sand with fractions of 0–4 mm was used as aggregate in the mixture, see Figure 1.

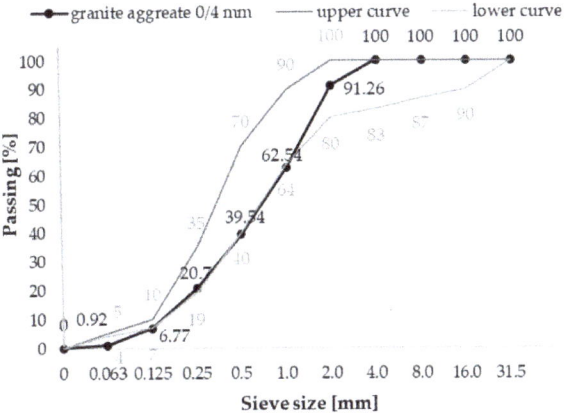

Figure 1. Particle size distribution curve for the used aggregate.

2.1.3. Admixture

Superplasticizer (Atlas Duruflow PE-531, Bydgoszcz, Poland) was used to reduce the amount of water required to liquefy the mixture. The product complies with the EN 934-2:2009+A1:2012 [38] standard. Figure 2 shows the chemical composition of the superplasticizer.

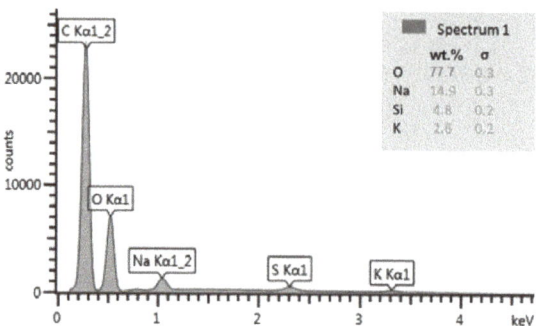

Figure 2. Chemical composition of admixture (%) [37].

2.1.4. Addition of Steel Waste

Curved and slightly twisted steel chips (without large, stringy drill chips) of post-production origin generated by CNC machine tools were used (Figure 3). The chips were made from steel grade 18CrNiMo7-6 compliant with EN ISO 683-3:2019 [39]. This steel is used in the production of toothed wheels, gears and shafts. The waste is heterogeneous in terms of its material, shape, dimensions (Table 3) and degree of contamination. Its chemical composition, determined using XRF spectrometry, is shown in Table 4.

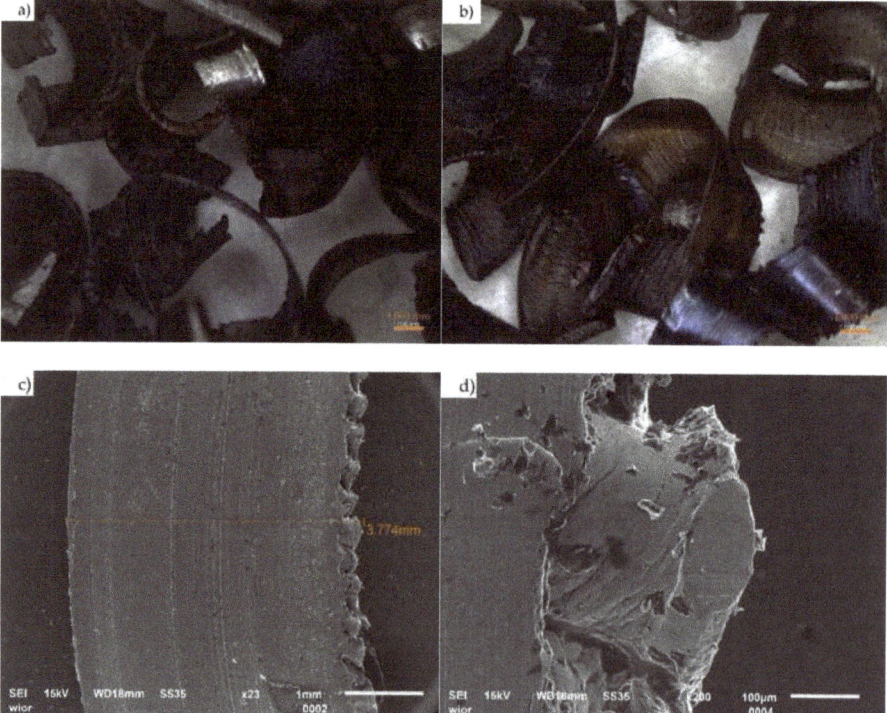

Figure 3. (**a**,**b**) Optical microscope images of steel chips used; (**c**,**d**) scanning electron microscope images of the steel chips' surface (**c**) and edge (**d**).

Table 3. Chip dimensions.

Description	Length (mm)	Width (mm)	Thickness (mm)
Dimension range	8.8–16.8	3.8–4.6	0.237–0.244
Mean value	12.3 ± 0.5	4.1 ± 0.5	0.240 ± 0.02

Table 4. Chemical composition, XRF results.

Compositions	Fe	C	Mn	Si	P	S	Cr	Ni	Mo	W	V	Cu
Unit (vol.%)	95.38	0.18	0.5	0.32	0.016	0.020	1.5	1.5	0.26	0.02	0.1	0.2

In this study, steel chips without pre-cleaning (covered with production lubricants and cooling oils) were used. Tests of organic substance content on the chip surface carried out at 450 °C showed the presence of organic substances of post-production origin in an amount of about 6%. By exposing the material to the temperature of 900 °C, the amount of organic substances increased to about 10% (Table 5). The increased loss on ignition (LOI) was probably due to the softening of the chips and the evaporation of their compounds.

Table 5. Results of loss on the ignition methods at 450 °C and 900 °C.

No.	Initial Weight (g)	Final Weight (g)	LOI (%)
1 (450 °C)	17.385	16.287	6.32
2 (450 °C)	19.976	18.596	6.91
3 (450 °C)	15.541	14.533	6.49
Average	-	-	6.57
1 (900 °C)	20.219	17.947	11.24
2 (900 °C)	18.935	16.913	10.68
3 (900 °C)	19.541	17.499	10.45
Average	-	-	10.79

2.2. Mixture Composition

Four types of mixtures were tested: a reference (without an additive of steel waste) and three mixes with different amounts of steel waste. Mixture composition is shown in Table 6. The composition of the reference mixture (M0) was established according to the method of designing the composition of the concrete recipe with an increased sand point, or sand concrete with a sand point value above 90%. The main idea behind the design of this type of concrete is the lack of coarse aggregate above 4.0 mm. The high proportion of fine-grained aggregate in the pile makes it necessary to cover the grains of the fine fractions with cement slurry, i.e., to use more cement (Table 5). Moreover, the strength of the concrete is mainly determined by the percentage of coarse-grained aggregate in the mixture composition. The designed (high) strength for concrete without coarse aggregate was obtained by adding a high cement content.

In the other mixtures (M1–M3, Table 6), the addition of steel waste was used as a fine aggregate replacement. The water–cement ratio for each mixture was 0.49.

Table 6. Mix proportions (1 m³).

Mix Symbol	Cement (kg)	Water (kg)	Admixture (kg)	Aggregate (kg)	Lathe Waste (wt.% of Cement)	Lathe Waste (kg)
M0	511	250	0.51	1535.00	0	0.00
M1				1526.50	5	25.55
M2				1518.01	10	51.10
M3				1509.52	15	76.65

2.3. Mixture Production

First, dry ingredients (the aggregate and steel chips) were mixed for 3 min. Then water mixed with a superplasticizer in the amount of 1% of cement weight was added. All components were mixed for 6 min, and then the mixture was poured into molds.

Molds in the shapes of cubes with dimensions of 150 mm × 150 mm × 150 mm; cylinders with dimensions of 150 mm × 300 mm and beams with dimensions of 100 mm × 100 mm × 500 mm and 40 mm × 40 mm × 160 mm were used. The mixture in the mold was compacted using a vibrating table. The top of the mold was prevented from water evaporation for 24 h after sample preparation. All samples were made in laboratory conditions (at 21 °C and 50% humidity). Samples were removed from molds after 36 h. The samples were stored in water until testing on day 28 in accordance with EN 12390-2:2019 [40].

3. Research Methodology

3.1. Testing the Concrete Mixtures

The consistency, air content and pH value of each concrete mixture were tested.

3.1.1. Slump Test

The concrete mix consistency was determined using an Abrams cone (Merazet, Poznań, Poland) according to EN 12350-2:2019 [41] (Figure 4a). The slump cone test (SC) was carried out for five samples for each mixture.

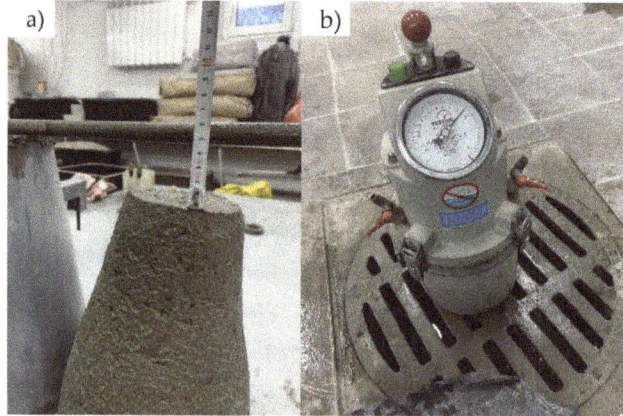

Figure 4. (a) Slump test, (b) measurement of air content (own photos).

3.1.2. Air Content

Air content was tested using the pressure method according to EN 12350-7:2019 [42]. The sample was placed in an 8-liter container of porosimeter (Merazet, Poznan, Poland),

and the vessel was filled with water under pressure. The measurement was recorded after pressure equalization (Figure 4b). The test was carried out for five samples for each mixture.

3.1.3. PH Value

Acidity and alkalinity (pH) values were determined for the liquid phase extracted from fresh mixtures in accordance with PN-EN 1015-3:2001+A2:2007 [43]. The test was carried out using a Testo 206 ph2 device (Testo, Pruszków, Poland). The measured pH value was recorded after 180 s. The test was carried out for five samples for each mixture.

3.2. Test of Hardened Samples

The tests were carried out after 28 days of concrete hardening. Density and mechanical and thermal properties were determined for samples made from the analyzed mixtures (M0, M1, M2 and M3).

3.2.1. Density

Density of hardened samples was determined for 150 mm × 150 mm × 150 mm cubes according to EN 12390-7:2019 [44]. The test was carried out for five samples for each mixture.

3.2.2. Mechanical Properties

Tests were performed using a Zwick machine (Zwick, Ulm, Germany) with a force range of 0–5000 kN. Compressive strength was tested on 150 mm × 150 mm × 150 mm samples (cubes) according to EN 12390-3:2019 [45], see Figure 5a. Flexural strength was tested on 100 mm × 100 mm × 500 mm beams using a three-point bending system with support spacing of 300 mm (Figure 5b) according to EN 12390-5:2019 [46]. Splitting tensile strength was performed on 150 mm × 300 mm cylindrical samples according to EN 12390-6:2010 [47]—Figure 6a. Elastic modulus and Poisson's ratio were tested on 150 mm × 300 mm cylindrical samples in accordance with EN 12390-13:2014 [48] by using 100 mm long strain gauges on two opposite sides of the specimens at half their height (Figure 6b). The edges of the compressive stressed samples were ground to avoid the effects of asymmetrical forces.

Each test was carried out for ten samples for each mixture.

Figure 5. (**a**) Compressive and (**b**) flexural strength tests (own photos).

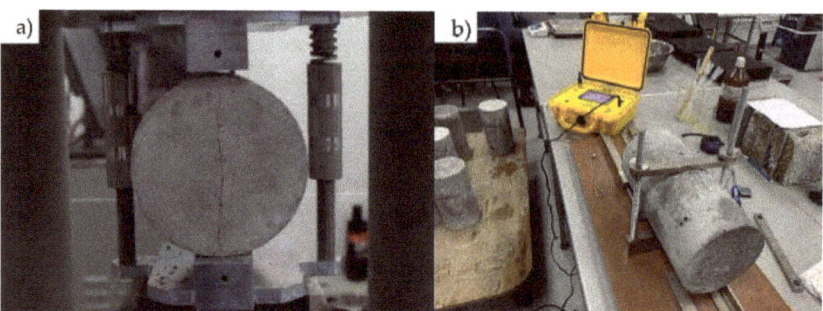

Figure 6. (**a**) Tensile splitting strength test and (**b**) measurement of Young's modulus (own photos).

3.2.3. Thermal Properties

Thermal conductivity, thermal diffusivity and specific heat tests were carried out using the ISOMET2114 analyzer (Applied Precision Ltd., Bratislava, Slovakia). The test consisted in assessing the temperature change of the tested material per heat flow impulses. The heat was generated by electric resistor heaters on the probe with a diameter of 60 mm, which was in direct contact with the concrete sample with a minimum thickness of 25 mm. It was assumed that heat propagation takes place in an unrestricted body. The temperature was recorded as a function of time. Five measurements were taken at different locations for each sample. The average of the resulting readings was taken as the final result. The test was carried out on ten samples for each mixture.

4. Results

4.1. Concrete Mixture

The obtained results of the concrete mixture tests are shown in Table 7. The presented values are averages of five samples for each mixture. No particle agglomeration during mixing and pouring was demonstrated for each mixture.

Table 7. Results of the slump test, air content and pH value tests.

Mix Symbol	Lathe Waste			Slump Cone (mm)	Air Content (%)	pH Value (-)
	(wt.% of Cement)	(wt.% of Component)	(vol.% of Mix)			
M0	0	0	0	40 ± 1	2.3 ± 0.1	12.02 ± 0.03
M1	5	1.1	0.33	50 ± 2	2.8 ± 0.1	12.09 ± 0.03
M2	10	2.2	0.66	45 ± 1	3.1 ± 0.1	12.16 ± 0.04
M3	15	3.3	0.99	40 ± 2	3.2 ± 0.1	12.20 ± 0.03

The mixture workability increased for mixtures M1 and M2 compared to the reference mixture (Table 7). This may be due to the use of a superplasticizer in this study, because slump can be increased by addition of chemical admixtures (e.g., superplasticizer) without changing the water–cement ratio [49].

The tested mixtures were S1 class (M0 and M3) and S2 class (M1 and M2). Shewalul [50] obtained the same class of workability for 42.5 concrete class with the addition of steel scraps in the amount of 15% of cement content (0.5% vol. of concrete), which is analogous to the content of steel waste in this study.

The air content of tested mixtures with the addition of steel chips (Table 7) increased (from 2.8% ± 0.1% to 3.2% ± 0.1%) in comparison to plain concrete (2.3% ± 0.1%). This is probably a result of the irregular shapes of the chips. Furthermore, a slight effect of adding steel chips on the pH value of the mixture was observed (Table 7). Based on this, it can be concluded that the addition of chips using steel bars will not affect chip corrosion and the potential reinforcement of concrete.

4.2. Hardened Concrete

The obtained results are presented in Tables 8 and 9. These values are the averages of five samples for each mixture for density and ten samples for other properties. On the basis of the obtained results, the relationships between the mechanical/thermal properties and the content of the used addition of lathe steel chips were prepared (see Section 5).

Table 8. Experimental results of material and mechanical properties.

Mix Symbol	Lathe Waste (wt.% of Cement)	Lathe Waste (wt.% of Component)	Lathe Waste (vol.% of Mix)	Density (kg/m^3)	Compressive Strength (MPa)	Flexural Strength (MPa)	Splitting Tensile Strength (MPa)	Young's Modulus (GPa)	Poisson's Ratio (-)
M0	0	0	0	2172 ± 2	50.4 ± 0.3	10.8 ± 0.1	2.89 ± 0.03	32.0 ± 0.4	0.120 ± 0.03
M1	5	1.1	0.33	2109 ± 2	57.4 ± 0.7	11.6 ± 0.1	3.01 ± 0.05	32.2 ± 0.3	0.121 ± 0.04
M2	10	2.2	0.66	2021 ± 3	60.9 ± 0.5	12.2 ± 0.1	2.85 ± 0.03	33.2 ± 0.4	0.123 ± 0.03
M3	15	3.3	0.99	1986 ± 2	68.7 ± 0.7	12.8 ± 0.1	4.00 ± 0.04	34.0 ± 0.3	0.123 ± 0.04

Table 9. Experimental results of thermal properties.

Mix Symbol	Lathe Waste (wt.% of Cement)	Lathe Waste (wt.% of Component)	Lathe Waste (vol.% of Mix)	Thermal Conductivity (W/m·K)	Thermal Diffusivity ×10^6 (m^2/s)	Specific Heat (J/m^3·K)
M0	0	0	0	1.6 ± 0.1	1.09 ± 0.07	696 ± 4
M1	5	1.1	0.33	1.6 ± 0.1	1.05 ± 0.07	768 ± 4
M2	10	2.2	0.66	1.4 ± 0.2	0.90 ± 0.07	799 ± 4
M3	15	3.3	0.99	1.4 ± 0.3	0.82 ± 0.07	856 ± 4

5. Discussion

Lathe scraps are added to concrete as replacement for aggregate or as fiber. Depending on this, its addition can be calculated in relation either to the aggregate or to the cement (see Section 1). In order to compare the results, in this article, the content of lathe chips was related to cement content. This is a proprietary approach to the issue of lathe chips as a component of concrete mix. Discussion of the individual properties of the mixtures and hardened concrete with the addition of steel chips is presented below.

5.1. Slump Cone

With the addition of steel chips in the amount from 5% to 15% of cement mass, the workability of the mixture decreased (Figure 7). This trend is consistent with the observations of other scientists, regardless of use as aggregate replacement [26–28,51] or as dispersed fibers [17,19,21,32]. Prasad et al. [51] determined decreases in slump cone by 21%, 37% and 58% for steel waste up to 4.75 mm in length and up to 2 mm thick in the amount of 22%, 33% and 44% of the cement mass, respectively. The same results were obtained by Maanvit et al. [26] for steel waste with length of 10–20 mm, thickness of 0.25 mm and in the same amount. Ismail and Al.-Hashmi [28] also observed a 23% decrease in the slump cone for concrete with the addition of steel chips as a substitute for fine aggregate at 56% of cement weight, where 53% of chips with a size of 0.6–1.18 mm were used. While the same researchers obtained an 8% decrease for chips added at 38% of cement mass, in that case, 92% of the chips used were 1.18–2.36 mm in length [27]. A similar decrease in the slump cone with increasing fiber content compared to plain concrete was obtained by Gewatre et al. [32], who used steel chips as fiber in the range of 8–42% of cement weight. Seetharam et al. [19] reported more decreases in the slump cone for a concrete mixture with addition of lathe chips at 10%, 20% and 30% of cement weight. For 10% addition, the slump cone decreased by 12% compared to the reference mix. Abbas [17] determined a 25% decrease for steel fiber addition at 15.3% of cement weight and a 50% decrease for addition at 31.2% of cement weight due to poor anchoring of lathe chips. Purohit et al. [21] obtained a decrease in slump cone of 6%, 12%, 18% and 18% respectively for spiral steel chips 25–40 mm long and 0.3–0.75 mm thick at 3%, 6%, 9% and 12% of cement weight.

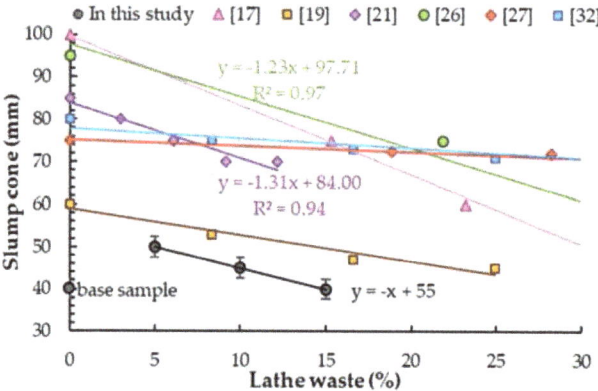

Figure 7. Slump cone test results.

The less liquid consistency obtained in this study is due to the composition of the mixture. Moreover, it can be observed that for longer chips presented in this study and in [21,26], similar decreases in slump cone for the same steel chips content were determined (the slopes of the curves are similar (Figure 7)).

5.2. Density

With the increase in addition of steel waste (5%, 10% and 15% of cement mass), concrete density decreased by 2.9%, 7.0% and 8.6% compared to the reference sample, and this correlation is linear (Table 8). This was surprising because steel waste is heavier than granite aggregate. In order to explain this phenomenon, photographs of the concrete samples' microstructure with different contents of steel chips were taken (Figure 8). It can be observed that there is no aggregation of steel chips and no air pockets. The chips are distributed uniformly and the sample is not segregated, which indicates that the test samples were made correctly. This phenomenon is probably related to the bulk density and that the aggregate did not fit in the places of the wound lathe chips. This will be the subject of further research.

Moreover, similar observations were determined by Mohammed et al. [34], who obtained a 2.7% decrease in the density of the sample with addition of steel chips as fibers in the amount of 65% of cement weight compared to plain concrete (Figure 9). An inverse correlation was found by other researchers; however, the increase was insignificant (mostly less than 5% compared to plain concrete). Ismail and Al.-Hashmi [27] used steel chips, most of which (92%) were between 1.18 and 2.36 mm, as replacement fine aggregate, and they obtained a 3.2% increase in density for a sample with an addition of 37% of cement weight. Ismail and Al.-Hashmi [28] used chips, the majority (53%) of which were between 0.6 and 1.18 mm in length, and reported an 8% increase in density for concrete with an addition of steel waste at 94% of cement weight compared to a reference sample. Slight increases in density were determined by Abbas [17] and Qureshi and Ahmed [52], who used steel chips as fibers. Abbas [17], using steel chips (both straight and spiral, 50 mm in length, 1 mm in thick, 2 mm in wide), reported a 2% increase in density for concrete with an addition of chips at 15% of cement weight and 3.6% for concrete with an addition at

31% of cement weight. The same increase in density (1.9%) was obtained by Qureshi and Ahmed [52] for concrete with steel chips (up to 80 mm in size) at 52% of cement weight.

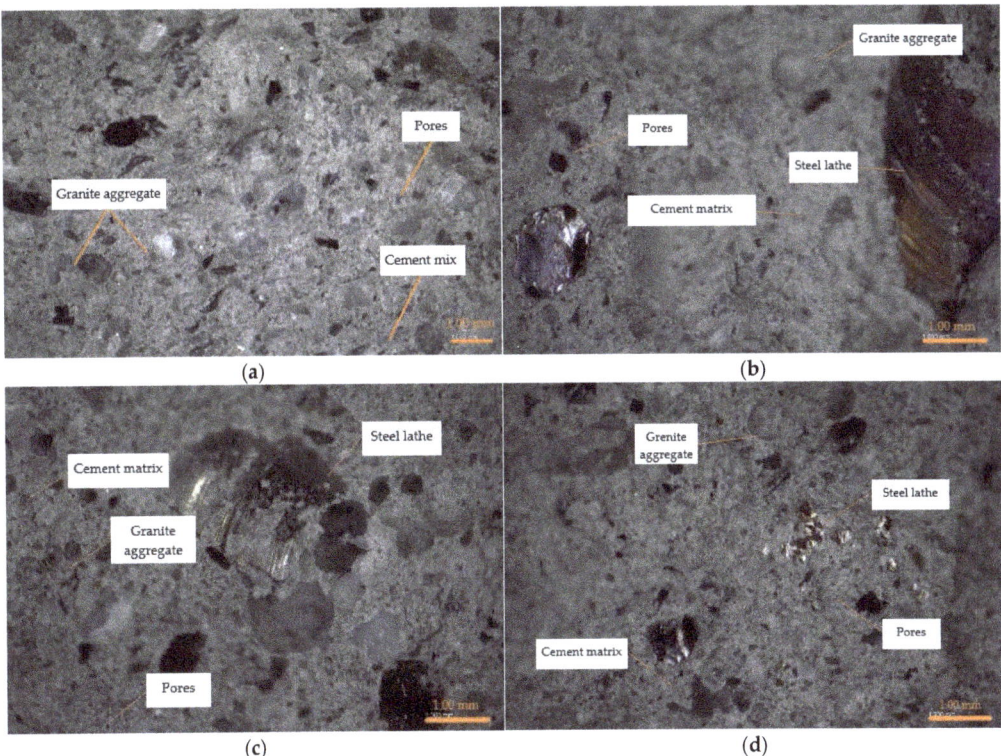

Figure 8. Concrete microstructure (OPTA-TECH, Warsaw, Poland): (**a**) M0, (**b**) M1, (**c**) M2, (**d**) M3.

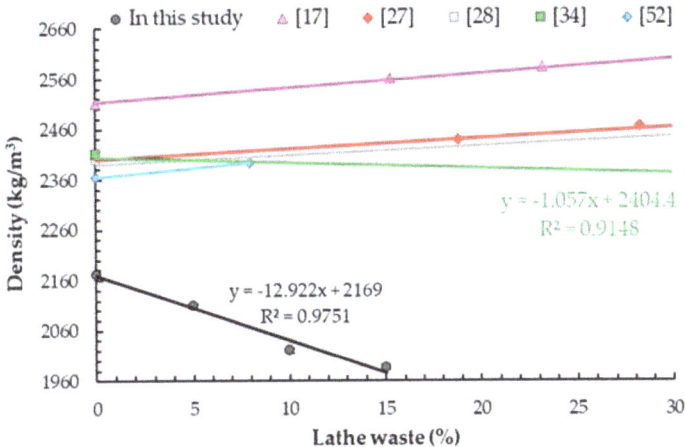

Figure 9. Density test results.

5.3. Compressive Strength

Table 8 shows the results of the compressive strength tests of the samples after 28 days of hardening. Compressive strength increased by 13.9% for M1, 20.8% for M2 and 36.3% for M3 in relation to the reference sample (50.4 ± 0.3 MPa). The same results were determined by Arunakanthi and Ch. Kumar [53] and Shewalul [50]. Arunakanthi and Ch. Kumar [53] obtained 15%, 22% and 25% increases in compressive strength for concrete with chips added at 6%, 12% and 19% of cement mass. Prabu et al. [54] demonstrated about a 20% improvement for 5% and 10% of cement mass used as fine aggregate replacement. Shewalul [50] achieved a 26.8% increase for steel chips addition at 13% of cement mass, which is similar to results obtained in this study. Moreover, increases in compressive strength for concrete with steel chips as a substitute for fine aggregate have been observed by other scientists. Shukla [20] obtained lower compressive strength increases (5 and 14% for the addition of chips in the amount at 6% and 12% of cement mass) than in this study. The same effect was demonstrated by Ismail and Al.-Hashmi [27,28], while a higher compressive strength increase (23%) was demonstrated by Hemanth Tunga et al. [30] for addition at 6% of cement mass. Sheikh and Reza [55] obtained a 35.9% improvement, but they used a lathe waste as coarse aggregate replacement. Alwaeli and Nadziakiewicz [29] determined higher increases in compressive strength (24%, 30%, 43% and 50%), but for much larger content of steel chips (68%, 136%, 203% and 271% of cement weight and 25%, 50%, 75% and 100% of fine aggregate).

The same trend was observed for lathe chips used as steel fibers [26,32,51]. Kumaran et al. [31] reported 5%, 11% and 9% increase in compressive strength for concrete with fiber addition in amounts analogous to this study (7%, 10% and 13% of cement mass) compared to plain concrete. Ghumare [56] observed 16.0%, 19.5% and 21.3% improvement for addition at 4%, %6 and 8% of cement mass and a w/c ratio equal to 0.4 (the calculated increase for 5% of cement mass was equal to 17.7%). Althoey and Hosen [57] obtained 5% and 13% increases for the same cement class as in this study and for metal chips added at 10.8% and 21.6% of cement mass, respectively.

A higher compressive strength increase (40%) was obtained by Seetharam et al. [19] for addition at 10% of cement mass, while Purohit et al. [21] demonstrated increases of 3%, 4%, 20% and 8% for steel chips added at 3%, 6%, 9% and 12% of cement weight respectively.

A similar increase in compressive strength in relation to the reference sample (15% and 13%) as in this study was observed by Mansi et al. [18] for addition of steel fibers at 1% and 2% of the sample weight. 11–13% increases were also observed by other scientists [58–60]. Ashok et al. [61] and Shrivastavaa and Joshib [62] obtained a maximum strength increase of about 3% for concrete with lower content of steel chips added (at 0.5–2% of cement weight).

The samples with metal lathe waste addition were destroyed in the same way as the reference sample. The relation between compressive strength and the addition of chips is linear. The obtained results are in line with the observations of other scientists (Figure 10). For addition lathe chips as fibers at 7%, 10% and 13% of cement weight, Kumaran et al. [31] determined 5%, 11% and 9% increases in compressive strength in relation to the reference sample, but those values are almost half of those obtained in this study. Abbas [17] obtained a slight 2% increase in strength for the addition of lathe chips at 15%, 23% and 31% of cement weight in relation to the reference sample. For additives of 31%, 48% and 65% by weight of cement, Mohammed et al. [34] obtained compressive strength increases of 7%, 12% and 15% respectively compared to the reference sample. It can be observed that the obtained correlation between compressive strength and lathe waste content is similar to relationships determined by other scientists [17,31,53]. In this case the slope of the curve is an approximate (Figure 10).

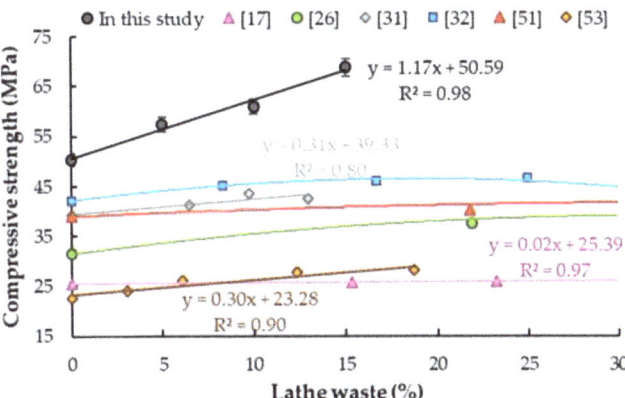

Figure 10. Results of compressive strength tests of concrete samples with different steel chip contents.

Concrete with analogous added content to that in this study (6%, 11%, 17% and 22% of cement weight) was tested by Hemanth Tunga et al. [30], and they obtained a 23% increase in compressive strength for a 6% addition of chips, but then a decrease. The same correlation was determined by Gawatre et al. [32] for concrete with lathe chips added at 8–42% of cement weight, but the maximum increase in compressive strength (11%) was obtained for an addition of 25% of cement weight. The increase in compressive strength and then decrease was also observed by Maanvit et al. [26] and Prasad et al. [51]. Dharmaraj [33] also obtained an increase in compressive strength up to 53% for 10% of shredded scrap iron addition, and then a decrease.

An almost incredible increase in compressive strength with the increase in steel waste addition can be observed here, while the experimental studies of other scientists indicate a rather slight or even negligible increase. This may be related to the adopted mixture composition (see Section 2).

5.4. Flexural Strength

The results of the flexural strength of the samples are shown in Table 8. It was observed that with the addition of steel lathe waste at 5%, 10% and 15% of cement weight, the increase in flexural strength was 7.1%, 12.7% and 18.2% respectively compared to the reference sample (10.8 ± 0.1 MPa). Figure 11 shows a linear relationship between the tensile flexural strength of the sample and the content of steel chip addition. The same conclusion was determined by Arunakanthi and Ch. Kumar [53]. They obtained higher increases in flexural strength (18%, 27% and 38%) compared to the base sample for the same content of steel chip addition (3%, 6%, 9% and 12%), but in this case, steel chips were used as fiber. However, if the amount of lathe chips is taken in relation to the cement content, it can be observed that correlation between flexural strength and lathe waste content is similar (the slope of the curves is an approximate (Figure 11)). The same correlation was determined by Kumaran et al. [31], who reported 9%, 19% and 10% increases in flexural strength for concrete with fiber addition at 7%, 10% and 13% of cement mass compared to plain concrete. This is related to the prevention of crack formation and the bridging of cracks analogously to the case of using traditional fibers [63–65] or recycled fibers [66–68]. This phenomenon for concrete with lathe chips as fiber was presented by Shrivastavaa and Joshi [62]. In this study, the steel chips were used as replacement of fine aggregate, so these tests were not analyzed.

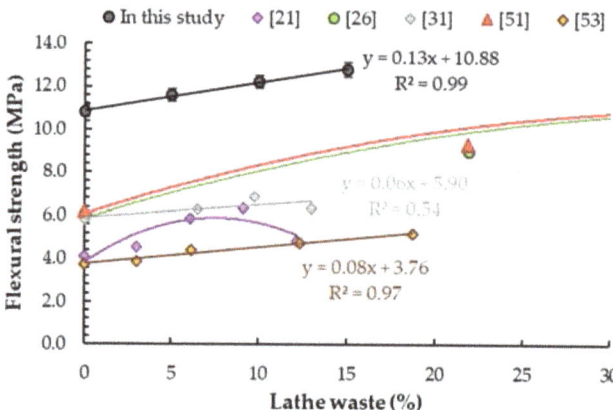

Figure 11. Results of tensile flexural strength tests of concrete samples with different steel chip contents.

The increase in flexural strength with increasing steel chip content is in line with the observations of other scientists (Figure 11). Similar improvement (13.15%) was observed by Ghumare [56] for concrete with lathe chip content at 8% of cement mass and a w/c ratio equal to 0.45, but the value was lower than the results obtained in this study. Prabu et al. [54] obtained similar values of flexural strength for M30 concrete with 5% and 10% addition of steel chips and w/c = 0.4, but the increase was much higher than in this study. Ismail and Al.-Hashmi [27] obtained a similar increase in flexural strength (22%) in relation to the base sample for steel chips added as replacement for fine aggregate at 19% of cement weight. In another test, Ismail and Al.-Hashmi [28] obtained a slightly lower increase in tensile strength (9%) for the same content of chips, but this was probably due to the use of a mix of metal and plastic chips in this case. Additions at 22%, 33% and 44% of cement weight were tested by Maanwit et al. [26], who obtained 50%, 100% and 75% increases in flexural strength in relation to the reference sample. Similar increases (50%, 94% and 69%) were obtained by Prasad et al. [51] for the same amounts of addition of lathe chips but at a greater size (as replacement for fine and coarse aggregate). Dharmaraj [33] studied concrete with the addition of fly ash and shredded scrap iron as steel fiber at 10–25% of cement weight and obtained an increase in compressive strength of up to 3% for 15% scrap iron. Ashok et al. [60] used a slight amount of steel chip addition in the range of 0.5–2% of cement weight and obtained a maximum increase in strength of 42% compared to the reference sample. A decrease in strength was presented by Seetharam et al. [19] for the addition of steel chips as fiber at 10%, 20% and 30% of cement weight.

5.5. Splitting Tensile Strength

Results of splitting tensile strength tests of samples with different steel chip contents are shown in Table 8. It was observed that with the addition of steel lathe waste at 5%, 10% and 15% of cement weight, the increase in splitting tensile strength was 4.2%, 33.2% and 38.4% respectively compared to the reference sample (2.89 ± 0.03 MPa). Prabu et al. [54] obtained the same values of flexural strength and the same increase for M30 concrete with 5% and 10% addition of steel chips and w/c = 0.4. Shewalul [50] achieved similar values (4.37 MPa) for M25 concrete with steel chips used as fine aggregate replacement at 13% of cement mass, but the increase was much lower (11.2%). Increases in splitting tensile strength compared to base samples were also observed by other scientists (Figure 12). Kumaran et al. [31] reported lower increases in splitting tensile strength (11%, 15% and 13%) compared to the reference sample for the same addition as in this study (7%, 10% and 13% of cement weight). Hemanth Tunga et al. [30] obtained the highest increase in splitting tensile strength (14%) for a sample with 6% steel chips addition of cement mass, and then a decrease. Maanvit et al. [26] obtained increases in strength of 38%, 79% and 51% in relation to the reference sample for samples with addition of steel chips at 22%, 33% and 44% of cement weight respectively. Moreover, Prasad et al. [51], for the same amounts of addition, but for a mix of steel and plastic chips, determined much lower increases in tensile split strength at 3%, 7% and 3%. The same correlation was observed by Mohammed et al. [34] and Kumaran et al. [31]. Mohammed et al. [34] obtained splitting tensile strength increases of 42%, 11% and 20% for concrete with additions of 31%, 48% and 65% by weight of cement, respectively. Ashok et al. [61] tested a lower amount of additive in the range of 0.5–2.0% of cement weight, and they obtained the maximum increase in splitting tensile strength at 20%.

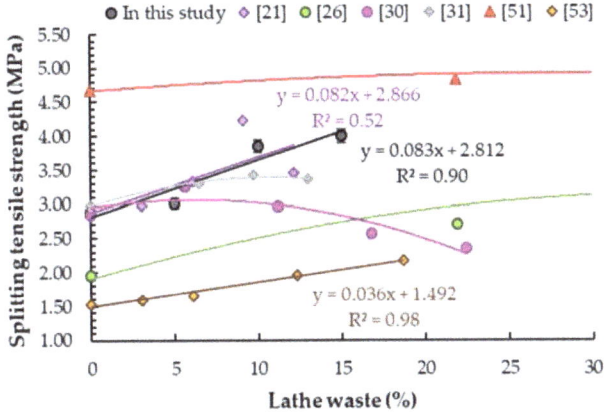

Figure 12. Results of tensile splitting strength tests of concrete samples with different steel chip contents.

Analogously to other mechanical properties, it can be observed that the slope of the obtained curve for correlation between splitting tensile strength and lathe waste content is very similar to relationships determined by other scientists [21,53], especially to the results obtained by Purohit et al. [21] (Figure 12).

5.6. Elastic Modulus and Poisson's Ratio

With the addition of steel waste, the elasticity modulus increased from 0.6% to 6.2% compared to plain concrete (32.0 ± 0.4 GPa) (Table 8). This phenomenon is consistent with results obtained by Shewalul [50], but values demonstrated in this study were higher.

Poisson's ratio for the reference sample was 0.120 ± 0.03. It can be observed that the results for samples M1–M3 were within the measurement error limit (Table 8).

5.7. Thermal Properties

Thermal parameters for concrete samples are presented in Table 9. The addition of lathe waste (between 5% and 15% of cement mass) did not affect thermal conductivity.

For a 5%, 10% and 15% addition of lathe waste, a 3.7%, 17.5% and 25.2% decrease in thermal diffusivity was obtained in relation to the reference sample. A linear correlation between steel chip content and thermal diffusivity was demonstrated (Figure 13). In addition, Figure 13 shows specific heat results for concrete with different amounts of steel chips. With the addition of 5%, 10% and 15% of lathe waste, an increase in specific heat by 10.4%, 14.8% and 23.0% respectively was obtained compared to the base sample. The obtained correlation is similar to concrete with steel fibers [14].

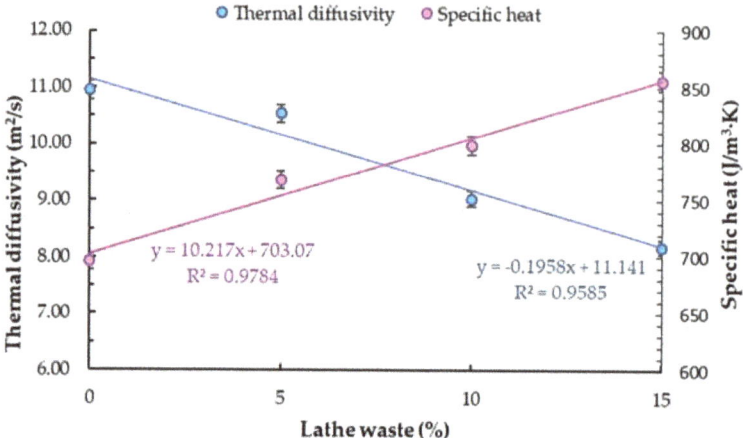

Figure 13. Specific heat and thermal diffusivity of concrete.

6. Conclusions

The aim of the study was to assess the possibility of using post-production waste materials in the form of lathe chips without pre-cleaning (covered with production lubricants and cooling oils) in concrete. In this study, three different contents of addition (5%, 10% and 15% of cement weight) were used as substitutes for fine aggregate. The following conclusions can be drawn on the basis of this experimental study's results:

1. For a waste addition of 5% and 10% of cement weight, the workability of the concrete mix increased and was qualified to the S2 class. The reference samples and the samples with additions of 15% waste lathe were S1 class.
2. The air content of tested mixtures with the addition of steel chips (Table 7) increased (from 2.8% ± 0.1% to 3.2% ± 0.1%) in comparison to plain concrete (2.3% ± 0.1%). This is probably a result of the irregular shape of chips.
3. The compressive strength of concrete after 28 days increased linearly from 50.4 MPa to 68.7 MPa in relation to the reference sample. For an addition of steel chips at 15% of cement weight, a 36.3% increase in compressive strength was obtained.
4. The flexural strength increased linearly from the reference value of 10.83 MPa to 12.8 MPa. This corresponds to an 18.2% increase in flexural strength for the 15 wt.% addition of steel chips.
5. The splitting tensile strength for the 15 wt.% additive increased by 38.4% compared to the reference sample.
6. A slight increase in the elastic modulus from about 1% to 6% was observed for additions of metal lathe waste from 5% to 15% of cement weight.
7. It was demonstrated that the addition of steel chips as a substitute for fine aggregate does not affect the thermal conductivity of concrete.
8. A 3.7%, 17.5% and 25.2% decrease in thermal diffusivity was obtained for the addition of steel chips at 5%, 10% and 15% of cement weight.
9. The specific heat for concrete with addition of 15% steel chips as a replacement for fine aggregate was higher by 23.0% compared to the reference sample.

Moreover, in this study it was determined that thanks the adoption of lathe chip content in relation to the cement content, correlations between properties and lathe chip content for different concrete mixes are similar. The slopes of the curves are approximate.

Based on the obtained results and correlations between properties and lathe chip content, differences can be observed between these results and results obtained by other scientists due to the use of different cement classes, different aggregate type and fraction, type of plasticizers and the type, quality and level of contamination of the added steel chips.

Moreover, it was determined that use of lathe chips without pre-cleaning (covered with production lubricants and cooling oils) as replacement for fine aggregate in the amount of 5% to 15% can improve the mechanical properties of concrete. It was particularly surprising that better results were obtained than with chips processed using the classical method entailing their melting.

This paper is part of a wider research project aimed at developing environmentally friendly concrete in accordance with the principles of sustainable development and closed-loop economy by using recycled construction materials in concrete. Taking into account the improvements in the mechanical properties of concrete through the addition of chips, further testing of this material, especially fatigue testing, will be scheduled.

Author Contributions: Conceptualization and methodology, M.M. and M.K.; investigation, M.M., M.K., W.Ł. and P.M.; data curation, M.K. and M.M.; formal analysis, M.K.; funding acquisition, M.M.; project administration, M.M.; resources, M.M., M.T., T.S., W.Ł. and P.M.; supervision—M.K.; validation, M.K.; visualization, M.M. and M.K.; writing—original draft preparation, M.M., M.K., M.T. and T.S.; writing—review and editing, M.K. All authors have read and agreed to the published version of the manuscript.

Funding: This work was financially supported by the Dean of Faculty of Civil Engineering and Geodesy of the Military University of Technology as part of a scholarship no. 1/DHP/2020 and by the Military University of Technology under research project UGB no. 22-870.

Institutional Review Board Statement: Not applicable.

Informed Consent Statement: Not applicable.

Data Availability Statement: Not applicable.

Conflicts of Interest: The authors declare no conflict of interest.

References

1. World Steel Association. Global Crude Steel Output. 2020. Available online: https://www.worldsteel.org/media-centre/press-releases/2020/Global-crude-steel-output-increases-by-3.4--in-2019.html (accessed on 14 December 2020).
2. IEA. *Global Status Report for Buildings and Construction 2019*; IEA: Paris, France, 2019. Available online: https://www.iea.org/reports/global-status-report-for-buildings-and-construction-2019 (accessed on 14 December 2020).
3. Gajdzik, B. Retrospect changes in manufacture of crude steel by processes in Polish steel industry. *Pr. Instyt. Metal. Żel.* **2015**, *67*, 54–59.
4. Duff, R.; Lenox, M.J. Path to 2060: Decarbonizing the Industrial Sector. 2018. Available online: https://www.researchgate.net/publication/329810198_Path_to_2060_Decarbonizing_the_Industrial_Sector (accessed on 14 August 2020).
5. Gajdzik, B. Energy intensity of steel production—Retrospective and prognostic analysis. *Pr. Instyt. Metal. Żel.* **2016**, *68*, 34–41.
6. World Steel Association. Steel's Contribution to a Low Carbon Future and Climate Resilient Societies. 2017. Available online: https://www.steel.org.au/resources/elibrary/resource-items/steel-s-contribution-to-a-low-carbon-future-and-cl/download-pdf.pdf/ (accessed on 14 August 2020).
7. IPCC Intergovernmental Panel on Climate Change. *Guidelines for National Greenhouse Gas Inventories—Chapter 4—Metal Industry Emissions; Japan*. 2006. Available online: https://www.ipcc-nggip.iges.or.jp/public/2006gl/pdf/3_Volume3/V3_4_Ch4_Metal_Industry.pdf (accessed on 14 August 2020).
8. Directive 2008/98/EC of the European Parliament and of the Council of 19 November 2008 on Waste and Repealing Certain Directives. Available online: http://data.europa.eu/eli/dir/2008/98/oj (accessed on 14 August 2020).
9. Adamczyk, J.; Dylewski, R. Recycling of construction waste in terms of sustainable building. *Probl. Ekorozw.* **2010**, *5*, 125–131.
10. Directive 2018/851 of the European Parliament and of the Council of 30 May 2008 Amending Directive 2008/98/EC on Waste. Available online: http://data.europa.eu/eli/dir/2018/851/oj (accessed on 14 August 2020).
11. Barros, J.A.O.; Etse, G.; Ferrara, L.; Folino, P.C.; Koenders, E.A.B.; Toledo Filho, R.D. The EnCoRe project: An international network for sustainable concrete. *Rev. Rout. Transp.* **2014**, *43*. Available online: https://aqtr.com/association/actualites/encore-project-international-network-sustainable-concrete (accessed on 20 May 2021).
12. El-Sayed, T.A. Flexural behavior of RC beams containing recycled industrial wastes as steel fibers. *Constr. Build. Mater.* **2019**, *212*, 27–38. [CrossRef]
13. Caggiano, A.; Folino, P.; Lima, C.; Martinelli, E.; Pepe, M. On the mechanical response of hybrid fiber reinforced concrete with recycled and industrial steel fibers. *Constr. Build. Mater.* **2017**, *147*, 286–295. [CrossRef]
14. Małek, M.; Jackowski, M.; Łasica, W.; Kadela, M. Influence of polypropylene, glass and steel fiber on the thermal properties of concrete. *Materials* **2021**, *14*, 1888. [CrossRef] [PubMed]

15. Manaswini, C.; Vasu, D. Fibre reinforced concrete from industrial waste—A review. *Int. J. Innovat. Res. Sci. Eng. Technol.* **2015**, *4*, 11751–11758.
16. Rodríguez, J.M.; Carbonell, J.M.; Cante, J.C.; Oliver, J. Continuous chip formation in metal cutting processes using the Particle Finite Element Method (PFEM). *Int. J. Sol. Struct.* **2017**, *120*, 81–102. [CrossRef]
17. Abbas, A.H. Management of steel solid waste generated from lathes as fiber reinforced concrete. *Eur. J. Sci. Res.* **2011**, *4*, 481–485.
18. Mansi, A.H.; Galal, O.H.; Lafi, M. The utilisation of lathe steel waste fibers to improve plain concrete. In Proceedings of the Ninth International Conference on Advances in Civil, Structural and Mechanical Engineering, Rome, Italy, 7–8 December 2019.
19. Seetharam, P.G.; Bhuvaneswari, C.; Vidhya, S.; Vishnu Priya, M. Studies on properties of concrete replacing lathe scrap. *Int. J. Eng. Res. Technol.* **2017**, *6*, 382–386. [CrossRef]
20. Shukla, A.K. Application of CNC waste with recycled aggregate in concrete mix. *Int. J. Eng. Res. Appl.* **2013**, *3*, 1026–1031.
21. Purohit, R.; Dulawat, S.; Ahmad, E. To enhance mechanical properties of concrete by using lathe steel scarp as reinforced material. *J. Eng. Sci.* **2020**, *11*, 206–214.
22. Bendikiene, R.; Čiuplys, A.; Kavaliauskiene, L. Circular economy practice: From industrial metal waste to production of high wear resistant coatings. *J. Clean. Prod.* **2019**, *229*, 1225–1232. [CrossRef]
23. Jassim, A.K. Sustainable Solid Waste Recycling. Available online: https://www.intechopen.com/books/skills-development-for-sustainable-manufacturing/sustainable-solid-waste-recycling (accessed on 14 August 2020).
24. Vasoya, N.K.; Varia, H.R. Utilization of Various Waste Materials in Concrete a Literature Review. *Int. J. Eng. Res. Technol.* **2015**, *4*, 1122–1126.
25. Małek, M.; Łasica, W.; Jackowski, M.; Kadela, M. Effect of waste glass addition as a replacement for fine aggregate on properties of mortar. *Materials* **2020**, *13*, 3189. [CrossRef] [PubMed]
26. Maanvit, P.S.; Prasad, B.P.; Vardhan, M.H.; Jagarapu, D.C.K.; Eluru, A. Experimental examination of fiber reinforced concrete incorporation with lathe steel scrap. *IJITEE* **2019**, *9*, 3729–3732.
27. Ismail, Z.Z.; Al-Hashmi, E.A. Reuse of waste iron as partial replacement of sand in concrete. *Waste Manag.* **2008**, *28*, 2048–2053. [CrossRef] [PubMed]
28. Ismail, Z.Z.; Al-Hashmi, E.A. Validation of using mixed iron and plastic wastes in concrete. In Proceedings of the Second International Conference on Sustainable Construction Materials and Technologies, Ancona, Italy, 28–30 June 2010.
29. Alwaeli, M.; Nadziakiewicz, J. Recycling of scale and steel chips waste as a partial replacement of sand in concrete. *Constr. Build. Mater.* **2012**, *28*, 157–163. [CrossRef]
30. Hemanth Tunga, G.N.; Rasina, K.V.; Akshatha, S.P. Experimental investigation on concrete with replacement of fine aggregate by lathe waste. *Int. J. Eng. Technol.* **2018**, *10*, 834–837.
31. Kumaran, M.; Nithi, M.; Reshma, K.R. Effect of lathe waste in concrete as reinforcement. *Int. J. Res. Adv. Technol.* **2015**, *6*, 78–83.
32. Gawatre, D.W.; Haldkar, P.; Nanaware, S.; Salunke, A.; Shaikh, M.; Patil, A. Study on addition of lathe scrap to improve the mechanical properties of concrete. *Int. J. Innovat. Res. Sci. Eng. Technol.* **2016**, *5*, 8573–8578.
33. Dharmaraj, R. Experimental study on strength and durability properties of iron scrap with flyash based concrete. *Mater. Today Proc.* **2020**, *37*, 1041–1045. [CrossRef]
34. Mohammed, H.J.; Abbas, H.A.; Husain, M.A. Using of recycled rubber tires and steel lathes waste as fibers to reinforcing concrete. *Iraq. J. Civ. Eng.* **2011**, *1*, 27–38.
35. EN 197-1:2011. *Cement—Part 1: Composition, Specifications and Conformity Criteria for Common Cements*; European Committee for Standardization: Brussels, Belgium, 2011.
36. Górażdże Group. Cement, Concrete, Aggregate. Technical Data Sheet CEM I 42.5 R. Available online: https://www.gorazdze.pl/pl/cement-portlandzki-pn-en-197-1-cem-i-425r (accessed on 14 August 2020).
37. Małek, M.; Jackowski, M.; Łasica, W.; Kadela, M.; Wachowski, M. Mechanical and material properties of mortar reinforced with glass fiber: An experimental study. *Materials* **2021**, *14*, 698. [CrossRef] [PubMed]
38. EN 934-2:2009+A1:2012. *Cement—Admixtures for Concrete, Mortar and Grout—Part 2: Concrete Admixtures—Definitions, Requirements, Conformity, Marking and Labelling*; European Committee for Standardization: Brussels, Belgium, 2012.
39. EN ISO 683-3:2019. *Heat-Treatable Steels, Alloy Steels and Free-Cutting Steels—Part 3: Case—Hardening Steels*; International Organization for Standardization: Geneva, Switzerland, 2019.
40. EN 12390-2:2019. *Testing Hardened Concrete—Part 2: Making and Curing Specimens for Strength Tests*; European Committee for Standardization: Brussels, Belgium, 2019.
41. EN 12350-2:2019. *Testing Fresh Concrete—Part 2: Slump Test*; European Committee for Standardization: Brussels, Belgium, 2019.
42. EN 12350-7:2019. *Testing Fresh Concrete—Part 7: Air Content—Pressure Methods*; European Committee for Standardization: Brussels, Belgium, 2019.
43. EN 1015-3:2001+A2:2007. *Methods of Test for Mortar for Masonry—Part 3: Determination of Consistence of Fresh Mortar (by Flow Table)*; European Committee for Standardization: Brussels, Belgium, 2007.
44. EN 12390-7:2019. *Testing Hardened Concrete—Part 7: Density of Hardened Concrete*; European Committee for Standardization: Brussels, Belgium, 2019.
45. EN 12390-3:2019. *Testing Hardened Concrete—Part 3: Compressive Strength of Test Specimens*; European Committee for Standardization: Brussels, Belgium, 2019.

46. EN 12390-5:2019. *Testing Hardened Concrete—Part 5: Flexural Strength of Test Specimens*; European Committee for Standardization: Brussels, Belgium, 2019.
47. EN 12390-6:2010. *Testing Hardened Concrete—Part 6: Tensile Splitting Strength of Test Specimens*; European Committee for Standardization: Brussels, Belgium, 2010.
48. EN 12390-13:2014. *Testing Hardened Concrete—Part 13: Determination of Secant Modulus of Elasticity in Compression*; European Committee for Standardization: Brussels, Belgium, 2014.
49. Ferrari, L.; Kaufmann, J.; Winnefeld, F.; Plank, J. Multi-method approach to study influence of superplasticizers on cement suspensions. *Cem. Concr. Res.* **2011**, *41*, 1058. [CrossRef]
50. Shewalul, Y.W. Experimental study of the effect of waste steel scrap as reinforcing material on the mechanical properties of concrete. *Case Stud. Constr. Mater.* **2021**, *14*, e00490.
51. Prasad, P.B.; Sai Maanvit, P.; Jagarapu, D.C.K.; Eluru, A. Flexural behavior of fiber reinforced concrete incorporation with lathe steel scrap. *Mater. Today Proc.* **2020**, *33*, 196–200. [CrossRef]
52. Qureshi, T.; Ahmed, M. Waste metal for improving concrete performance and utilisation as an alternative of reinforcement bar. *Int. J. Res. Appl.* **2015**, *2*, 97–103.
53. Arunakanthi, E.; Chaitanya Kumar, J.D. Experimental studies on fiber reinforced concrete (FRC). *Int. J. Civ. Eng. Technol.* **2016**, *7*, 329–336.
54. Prabu, M.; Vignesh, K.; Saii Prasanna, N.; Praveen, C.; Mohammed Nafeez, A. Experimental study on concrete in partial replacement of fine aggregate with lathe waste. *Int. J. Sci. Eng. Res.* **2020**, *11*, 68–72. Available online: https://www.ijser.org/researchpaper/Experimental-study-on-concrete-in-partial-replacement-of-fine-aggregate-with-lathe-waste.pdf (accessed on 17 May 2021).
55. Sheikh, M.A.; Reza, M. Strengthening of concrete using lathe scrap waste. *Int. Res. J. Eng. Technol.* **2020**, *7*, 464–470. Available online: https://issuu.com/irjet/docs/irjet-v7i1077/2 (accessed on 17 May 2021).
56. Ghumare, S.M. Strength Properties of Concrete Using Lathe Waste Steel Fibers. Available online: https://www.slideshare.net/smghumare/lathe-waste-steel-fibrous-concrete-85751842 (accessed on 17 May 2021).
57. Althoey, F.; Hosen, M.A. Physical and mechanical characteristics of sustainable concrete comprising industrial waste materials as a replacement of conventional aggregate. *Sustainability* **2021**, *13*, 4306. [CrossRef]
58. Alfeehan, A.; Mohammed, M.; Jasim, M.; Fadehl, U.; Habeeb, F. Utilizing industrial metal wastes in one-way ribbed reinforced concrete panels. *Rev. Ing. Constr.* **2020**, *35*, 246–256. Available online: https://ricuc.cl/index.php/ric/article/view/1020 (accessed on 17 May 2021). [CrossRef]
59. Anusha Bharathi, M.; Samyukta, R.; Subashini, K.; Balakumar, P. Experimental study on strength parameters of lathe waste fiber reinforced concrete. *Int. J. Eng. Technol. Sci. Res.* **2018**, *5*, 948–953. Available online: http://www.ijetsr.com/images/short_pdf/1525021840_948-953-site184_ijetsr.pdf (accessed on 11 January 2021).
60. Haldkar, P.; Salunke, A. Analysis of Effect of Addition of Lathe Scrap on the Mechanical Properties of Concrete. *Int. J. Sci. Res.* **2016**, *5*, 2321–2325. Available online: https://www.ijsr.net/get_abstract.php?paper_id=NOV163133 (accessed on 17 May 2021).
61. Ashok, S.P.; Suman, S.; Chincholkar, N. Reuse of steel scarp from lathe machine as reinforced material to enhance properties of concrete. *Glob. J. Eng. Appl. Sci.* **2012**, *2*, 164–167.
62. Shrivastavaa, P.; Joshib, Y.P. Reuse of lathe waste steel scrap in concrete pavements. *Int. J. Eng. Res. Appl.* **2014**, *4*, 45–54. Available online: https://issuu.com/www.ijera.com/docs/j0412045559 (accessed on 17 May 2021).
63. Chalioris, C.E.; Panagiotopoulos, T.A. Flexural analysis of steel fibre reinforced concrete members. *Comp. Concr.* **2018**, *22*, 11–25.
64. Jakubovskis, R.; Jankutė, A.; Urbonavičius, J.; Gribniak, V. Analysis of mechanical performance and durability of self-healing biological concrete. *Constr. Build. Mat.* **2020**, *260*, 119822. [CrossRef]
65. Kytinou, V.K.; Chalioris, C.E.; Karayannis, C.G.G.; Elenas, A. Effect of steel fibers on the hysteretic performance of concrete beams with steel reinforcement—Tests and analysis. *Materials* **2020**, *13*, 2923. [CrossRef] [PubMed]
66. Małek, M.; Jackowski, M.; Łasica, W.; Kadela, M. Characteristics of recycled polypropylene fibers as an addition to concrete fabrication based on portland cement. *Materials* **2020**, *13*, 1827. [CrossRef] [PubMed]
67. Mohammed, A.A.; Rahim, A.A.F. Experimental behavior and analysis of high strength concrete beams reinforced with PET waste fiber. *Constr. Build. Mat.* **2020**, *244*, 118350. [CrossRef]
68. Małek, M.; Łasica, W.; Kadela, M.; Kluczyński, J.; Dudek, D. Physical and mechanical properties of polypropylene fibre-reinforced cement-glass composite. *Materials* **2021**, *14*, 637. [CrossRef] [PubMed]

Article

Neutralization of Acidic Wastewater from a Steel Plant by Using CaO-Containing Waste Materials from Pulp and Paper Industries

Tova Jarnerud *, Andrey V. Karasev and Pär G. Jönsson

KTH Royal Institute of Technology, SE-100 44 Stockholm, Sweden; karasev@kth.se (A.V.K.); parj@kth.se (P.G.J.)
* Correspondence: jarnerud@kth.se; Tel.: +46-705-465-011

Abstract: In this study, CaO-containing wastes from pulp and paper industries such as fly ash (FA) and calcined lime mud (LM) were utilized to neutralize and purify acidic wastewaters from the pickling processes in steel mills. The investigations were conducted by laboratory scale trials using four different batches of wastewaters and additions of two types of CaO-containing waste materials. Primary lime (PL), which is usually used for the neutralization, was also tested in the same experimental set up in the sake of comparison. The results show that these secondary lime sources can effectively increase the pH of the acidic wastewaters as good as the commonly used primary lime. Therefore, these secondary lime sources could be potential candidates for application in neutralization processes of industrial acidic wastewater treatment. Moreover, concentrations of metals (such as Cr, Fe, Ni, Mo and Zn) can decrease dramatically after neutralization by using secondary lime. The LM has a purification effect from the given metals, similar to the PL. Application of fly ash and calcined lime mud as neutralizing agents can reduce the amount of waste from pulp and paper mills sent to landfill and decrease the need for nature lime materials in the steel industry.

Keywords: reusing of wastes; secondary lime; neutralization; reduce landfill; acidic wastewater treatment; sustainable production

Citation: Jarnerud, T.; Karasev, A.V.; Jönsson, P.G. Neutralization of Acidic Wastewater from a Steel Plant by Using CaO-Containing Waste Materials from Pulp and Paper Industries. *Materials* **2021**, *14*, 2653. https://doi.org/10.3390/ma14102653

Academic Editor: Franco Medici

Received: 6 April 2021
Accepted: 14 May 2021
Published: 18 May 2021

Publisher's Note: MDPI stays neutral with regard to jurisdictional claims in published maps and institutional affiliations.

Copyright: © 2021 by the authors. Licensee MDPI, Basel, Switzerland. This article is an open access article distributed under the terms and conditions of the Creative Commons Attribution (CC BY) license (https://creativecommons.org/licenses/by/4.0/).

1. Introduction

The production chain of stainless steel is made up of a series of steps, where the steel undergoes various treatments to reach the desired material properties and surface qualities. During treatments at high temperatures, such as hot rolling and annealing, alloying metals from the steel matrix diffuses to the surface, react with the surrounding air oxygen and form oxide layers on steel surface. Underneath the oxide layers, some chromium depleted zones exist since the chromium diffused to the surface. These zones have lower corrosion resistances and strengths [1]. In order to remove these chromium depleted zones, a pickling process can be applied. Pickling is a chemical cleaning process that is commonly used in various steelmaking processes in a steel mill for removal of impurities (such as contaminants, corrosion products or scale) from a steel surface. The treatment aims to make the material more receptive to further processing or use. The pickle liquor usually contains strong acids such as sulfuric acid (H_2SO_4) and hydrochloric acid (HCl) [2]. After treating the steel in a pickling bath, the steel is rinsed with water. After rinsing, the wastewater contains high amounts of acids, and thereby has a low pH value. Before disposing or reusing of the wastewater, it has to be neutralized by addition of base reagents to avoid harming the recipient. The neutralization process is a type of a replacement reaction: Acid + Base → Salt + Water. An acid-base reaction is a chemical reaction that involves the exchange of one or more hydrogen ions. When an acid is dissolved in water, the solution has a greater hydrogen ion activity than that of pure water. When a base is dissolved in water it can accept hydrogen ions. The neutralization process is influenced by the choice of chemicals, the dosage of chemicals, the pH value and the mixing rate [1]. The pH is a major factor in neutralization and metal removal through precipitation. The metals form insoluble hydroxides at higher pH, which may be removed from the wastewater to the sediments [3]. A pH range of 8.0–11.0 minimizes the solubilities of metal hydroxides [4]. In industrial practice, the acidic wastewater is usually neutralized with natural slaked lime (denoted as primary lime in this study) to raise the pH value up to normal neutral levels.

However, by replacing primary lime with secondary lime-containing waste materials, natural resources can be saved, waste generation can be decreased and the resource efficiency can be improved. Moreover, the usage of secondary lime will also decrease the greenhouse gas emissions, since calcination of natural limestone (mainly containing $CaCO_3$) generates large amounts of CO_2 gas (780 ton CO_2 per ton of burnt CaO produced [5]). Furthermore, it is obvious that we, the modern society, need to redirect linear material flows towards more circular flows. For instance, the possibilities to utilize secondary lime materials (fly ash and calcined lime mud) from pulp and paper production as slag formers in electric arc furnace and argon oxygen decarburization stainless steelmaking processes have been successfully tested as is described in previous publications [6,7].

In several publications ([4,8–11]) during the last decade it was reported that steelmaking slags, waste limestone materials from the marble industry [12], and cement kiln dust [13] and other secondary lime sources can be successfully applied for neutralization of acidic wastewaters and purification of industrial waters. Lime mud and recovery boiler ash can be used to remove heavy metal contaminations from metal finishing wastewater. The residual metal concentration of chromium, copper, led and zinc decreases, as the pH level of the solution increases due to direct precipitation of heavy metals by the carbonate precipitation agents calcite and burkeite. Sediments such as metal carbonates forms when carbonite reacts with heavy metals [14].

Today, some steel mills reuse the sediments from the neutralization process as flux and some steel mills send the sediments to landfills, due to that the chemical composition makes it unsuitable for reuse [15].

In 2019, pulp and paper mills only in Sweden sent 136,000 tons of various wastes to landfills [16]. According to the European council directive 1999/31/EC on the landfill of waste, prevention, recycling and recovery of waste should be encouraged as should the use of recovered materials and energy so as to safeguard natural resources and to obviate a wasteful use of land [17]. The content of CaO-compounds in the wastes from pulp and paper industries can reach 60–90%, which is comparable or significantly larger compared to that in steelmaking slags. Therefore, it was assumed that those waste materials can also be applied for neutralization of acidic wastewater in steelmaking industries. A simplified flow scheme of the pickling wastewater is shown in Figure 1.

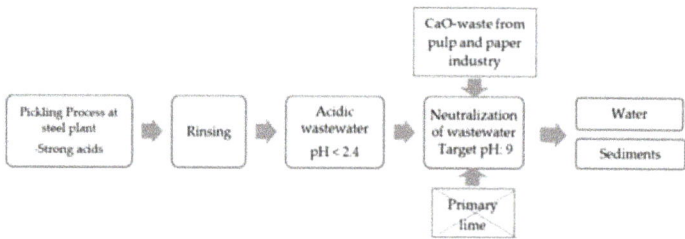

Figure 1. Flow from pickling to disposal of water.

This study focused on investigating the possibilities to use some different CaO-containing wastes obtained from pulp and paper industries to replace primary lime in the neutralization process. Furthermore, the efficiency of neutralization of the acidic wastewater was evaluated as a function of the chemical compositions of the neutralized waters and added wastes.

2. Experimental

The main aim of the experimental trials was to determine the proper amount of different secondary lime materials and method of addition to raise the "potential of hydrogen" (pH) of the acidic wastewater to a value of 9 during a maximum process time of 30 min. The initial pH value of the investigated acidic wastewater was in the interval of 1.3 to 2.4. Though the pH of regular water has a value of 7, in this study, the aim was to raise the pH value to 9 since the absorption of some elements (such as Ni) from the reactants are maximized at high pH levels [8]. Furthermore, flocculants added in industrial applications will significantly lower the pH of the neutralized water.

2.1. Materials

In this study, two types of CaO compounds containing wastes from pulp and paper industries (so-called secondary lime sources) were used in the experimental trials, namely fly ash (FA) (Stora Enso, Hyltebruk, Sweden) and calcined lime mud (LM) (SCA, Obbola, Sweden). The FA is formed by combustion of internal and external fuels (sludge from the recycled paper and wood fuels) and LM consists of the excess chemicals from the recovery boiler in a pulp and paper mill. Moreover, an industrial primary lime (PL), which is usually applied for neutralization process in steelmaking plants, was also used in the experimental trials for sake of comparison. The main components of the lime-containing materials are given in Table 1. The LM has almost similar CaO content (90.6 wt%) as the PL (95.2 wt%). The FA has lower CaO compound contents (61.5%). Moreover, not all of the detected CaO is free CaO, since calcium silicate (Ca_2SiO_4) and gehlenite ($Ca_2Al_2SiO_7$) are also present in the FA [7]. These secondary lime materials and the primary lime were used in form of powders, as delivered from the pulp and paper mills and from a steel mill. Typical photographs of the powder materials used for experimental trials are shown in Figure 2. Moreover, for some trials, the powders were additionally calcined at 1050 °C for 60 min to enable a comparison of their efficiency as neutralizing agents. Four batches of different acidic wastewaters (AWW, BWW, CWW and DWW) from two various steelmaking plants were used in the trials of this study. It should be pointed out that the acidic wastewaters from steelmaking plants are mixtures of different acids and technological waters in various combinations, depending on pickling processes by production of different steel grades. As a result, the chemical composition and pH level of each wastewater can significantly vary from batch to batch. In this study, the pH of the given acidic wastewaters equals on average to 1.92 ± 0.14 for AWW, 2.08 ± 0.14 for BWW, 1.53 ± 0.15 for CWW and 2.21 ± 0.11 for DWW.

Table 1. Contents of main components in the lime-containing materials, and primary lime as reference (wt%).

Materials	CaO	SiO_2	Al_2O_3	Fe_2O_3	K_2O	Na_2O	MgO	P_2O_5	S	Others
Fly ash (FA)	61.5	15.8	8.90	0.60	0.07	0.08	3.10	0.26	0.40	9.29
Calcined lime mud (LM)	90.6	0.18	0.07	0.04	0.10	0.48	1.05	0.75	0.11	6.62
Primary lime (PL)	95.2	1.00	0.50	0.20	0.04	0.04	1.70	0.01	0.08	1.23

2.2. Method

Acidic wastewater (210 mL) was poured into a glass beaker and stirred during 5 min by magnetic stirring to obtain a homogenization of the liquid. Then, the stirring was stopped, and the pH value of the wastewater was measured. The chosen amount of powder of secondary or primary lime was added into the beaker containing wastewater. Thereafter, the mixture was stirred for 5 min between the stops for measurements of the pH value of the solution. The pH value did not increase during the stirring stops for measurement. Therefore, only the stirring time was taken into account. The last two steps of the procedure were repeated until reaching almost constant values of pH, as illustrated in Figure 3, or until a pH value of 9 was reached.

Figure 2. Typical photos of (**a**) fly ash powder, and (**b**) calcined lime mud powder.

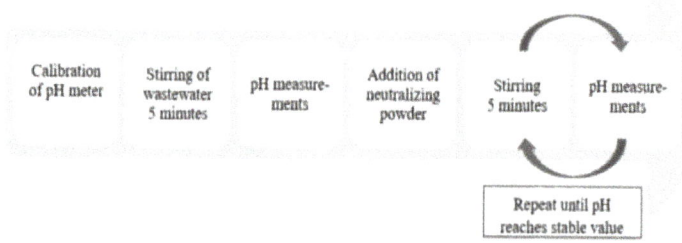

Figure 3. Workflow for neutralization experiments.

The stirring device was run at approximately a constant speed of 500 RPM (in the span 470–530 RPM) for most trials, but stirring at speeds of ~200 and ~1000 RPM were also tested in some trials. The pH measurements were conducted by using a VWR pHenomenal IS 2100L pH-meter (VWR International, Germany). Every time the pH value was checked, three measurements were completed before a mean value was calculated. The pH-meter was calibrated by using standard technical buffers (pH 4.00, 7.00 and 10.00) every day before the experiments were started, and after three experiments had been carried out. It was also checked after a completed set of experiments. In total, 88 neutralizing experiments were made of which 40 were with FA, 26 with LM and 22 with PL.

3. Results and Discussion

3.1. Repeatability of pH Measurements

It was found that the average standard deviations (σ) for three values of pH measured in solution at the same stage of the experiment can vary between 0.1 and 0.3, which corresponds approximately to 1–3% of measured pH values in most trials. Therefore, the repeatability of the pH measurements in the given experimental trials are considered to be reliable in this study.

Moreover, the repeatability of the pH measurements was evaluated in several experimental trials, which were carried out with the same materials and at the same given conditions. The results obtained in trials with an addition of 9.5 g of fly ash (FA) and 4.3 g of lime mud (LM) per liter of wastewater C are shown in Figure 4. It can be seen that the pH levels of the solutions consisting of the same type of wastewater and same type and ratio of neutralizing powder develop consistently, if the solution is stirred with same intensity.

3.2. Effect of Stirring Rate on Efficiency of the Neutralization Process

In order to keep the cost, and amount of sediments, at the lowest levels possible and still meet the required neutralizing rate, a ratio of neutralizer and wastewater for the prevailing circumstances needs to be established. It was found that the efficiency of the neutralization process depends significantly on the stirring intensity, as shown in Figure 5. In these trials, the same amount of lime mud powder (4.3 g/L) was added into wastewater C.

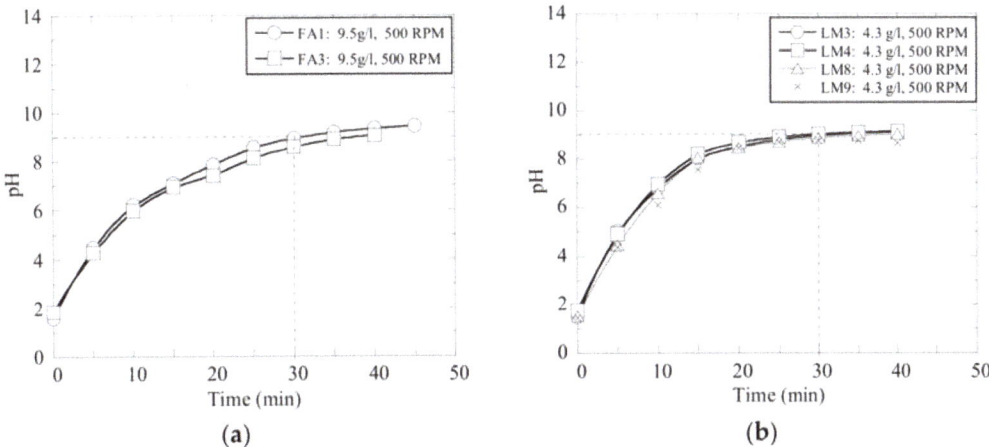

Figure 4. Repeatability of several experiments with (**a**) fly ash and (**b**) calcined lime mud.

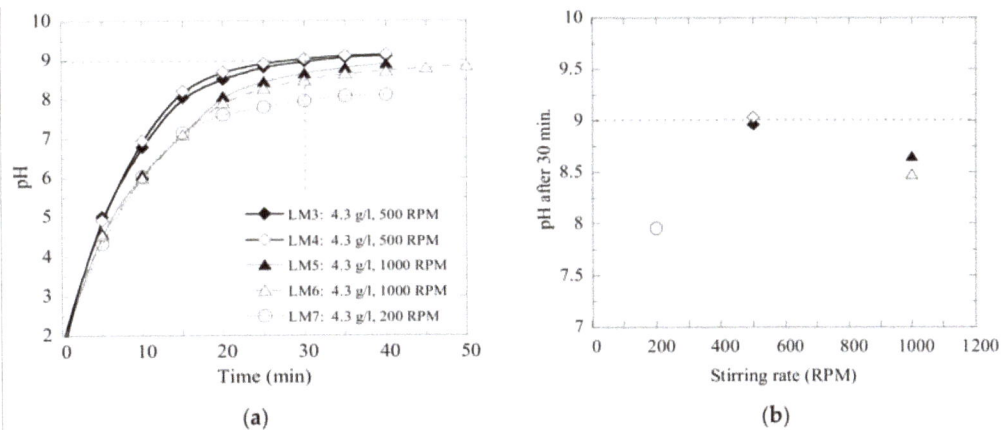

Figure 5. Effect of different stirring intensities (200, 500 and 1000 RPM) on the pH values. (**a**) the neutralization process and (**b**) the reached pH value after 30 min of stirring at different stirring intensities.

It was found that the most efficient results were obtained when using the stirring rate of 500 RPM. In this case, the pH value 9 was reached after 25 min of stirring. The least efficient process was obtained when using the lowest stirring intensity (200 RPM), as can be seen in Figure 5. Finally, the batch stirred with a 1000 RPM speed reached a pH value of 8.8 after 45 min of stirring. However, the pH value did not reach the required pH level 9 during 30 min. A similar tendency was found by Zinck and Aube [18] by evaluation of the effect of a mixing process on neutralization and cleanliness of wastewaters. They reported that the conditions of formation and growth of precipitated particles are less productive with an increased mixing rate more than optimum value. Thus, all experimental trials discussed below were carried out using the same stirring rate of 500 RPM.

3.3. Ratio of Required Neutralizer and Wastewater

A process of increasing the pH value of acidic wastewaters by addition of CaO-containing materials is defined in this study as a neutralization process. As can be seen in Figure 6, various wastes from pulp and paper industries have different reactivities (defined as a possibility of added material to increase the pH value of wastewater) with acidic wastewater. It is obvious that the obtained pH value increases with an increased amount of added CaO-containing materials. According to the requirements of industrial technological processes, the neutralization of wastewaters should be mostly finished during 30 min after the addition of reagents. The final required value of pH should be equal to 9, because the following additions of industrial flocculants significantly decrease the final pH value of the water.

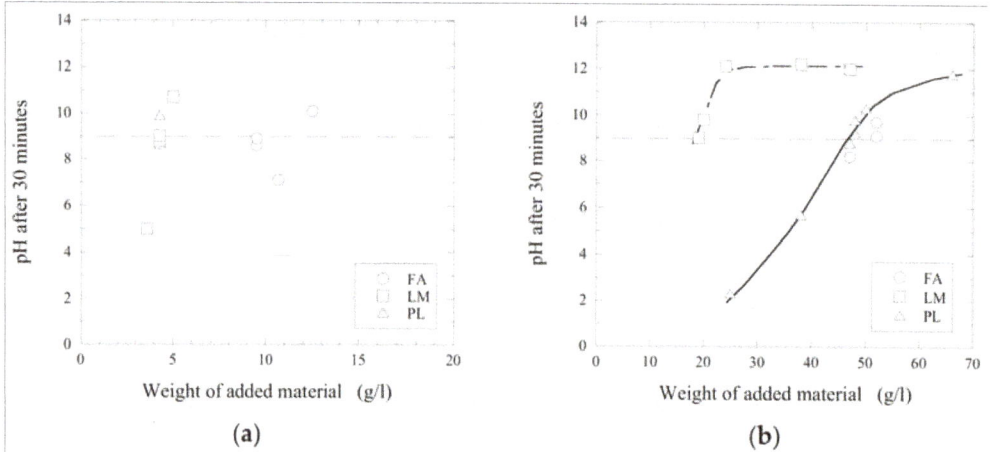

Figure 6. Neutralization effects obtained 30 min after the addition of fly ash (FA), lime mud (LM) and primary lime (PL) in to (**a**) C and (**b**) D acidic wastewaters.

Figure 6 shows a neutralization effect obtained in acidic wastewaters C and D after the additions of different amounts of fly ash (FA), calcined lime mud (LM) and primary lime (PL) after 30 min of stirring (at 500 RPM). It can be seen that the neutralization of wastewater C up to the given value of pH = 9 required approximately a 2.2 times larger amount of fly ash (9.5 g/L) compared to that of calcined lime mud (4.3 g/L). This can be explained by the significantly higher concentration of CaO in the LM compared to that in FA, as given in Table 1. Moreover, the original FA and PL powders consist of various complex CaO-containing components (such as $CaCO_3$ and CaO-SiO_2-Al_2O_3), which can be more stable and resistant in contact with acidic water compared to pure CaO. The content of such stable components can significantly decrease the reactivity of the investigated materials. A similar effect was observed for wastewater D: The required weight of FA is almost 2.7 times larger than that of LM. Even though FA contains less free CaO than LM and PL, it contains 1.8 times more MgO than PL, and 3 times more than LM. MgO reacts with water to form magnesium hydroxide ($Mg(OH)_2$), which is known for its acid neutralizing capability [13].

Moreover, it was found that a further increase of added material (~24 g/L of LM and ~66 g/L of PL in D-wastewater) larger than some specific value did not promote an increase of the obtained pH value larger than ~12–12.3, as shown in Figure 6b.

It should be pointed out that an effect of the same amount of added waste material on the pH values of different wastewaters can vary significantly. In Figure 7, some results are plotted to compare the amounts of FA and LM needed to neutralize the different batches of wastewaters. It can be seen that wastewater D requires significantly more neutralizers than the other batches of wastewater (for FA 52 g/L in the D-wastewater compared to 9–11 g/L in other wastewaters and for LM 19 g/L in the D-wastewater compared to 4 g/L in the A- and C-wastewaters). Moreover, it was found that the C-wastewater requires only 4 g/L of primary lime, while the D-wastewater requires about 48 g/L of PL to reach similar results. Specifically, this amount is 12 times higher.

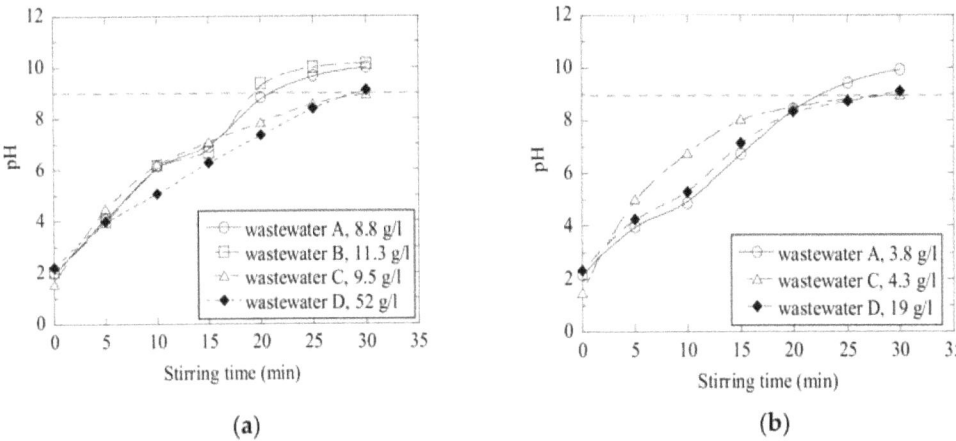

Figure 7. Neutralization of (a) wastewater A–D using fly ash and (b) wastewater A, C and D using calcined lime mud.

It is important to note that the neutralization effect of acidic wastewaters can significantly depend on both compositions and other characteristics of the added powder materials (such as size distribution of particles and moisture). For instance, small sizes of particles promote faster neutralization reactions. The particle size distributions (PSD) of FA and LM powders, which were achieved by a sieve shaker range analysis using a Retch AS200 Tap, are shown in Figure 8. It can be seen that the particles of LM (<56 μm) used in the present study are much smaller than those of FA powder (40~250 μm), which can also promote a higher efficiency of the LM compared to the FA due to the higher surface/volume ratio.

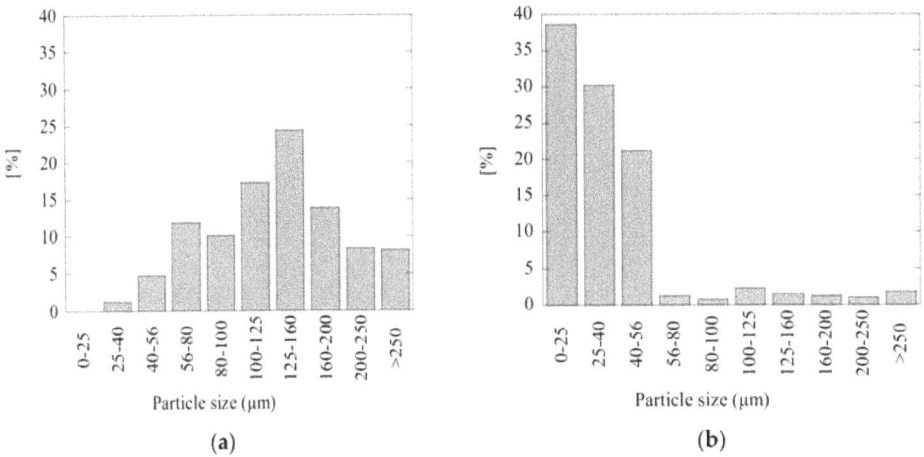

Figure 8. Particle-size distribution for (a) fly ash powder and (b) calcined lime mud powder used in this study.

In order to evaluate effect of moisture in powder materials on the neutralization process, the original powders of fly ash and primary lime were additionally calcined in a laboratory muffle furnace at 1050 °C for 60 min. Neutralization effects obtained 30 min after the addition of fly ash and primary lime powders without and with additional calcination (open and filled marks, respectively) into C and D acidic wastewaters are shown in Figure 9. By calcination of FA and PL, volatile matter and moisture are removed from the powders by thermal degradation and vaporization. This, in turn, leads to a higher concentration of substances being active in the neutralization process. It is clear that the amount of powder can be decreased significantly for both fly ash and primary lime when it is calcined. For instance, by using an additional calcination, the amount of added powder can be decreased by ~1.6 time for FA (from 52 up to 33 g/L) and 2.4 time for PL (from 48 to 20 g/L) to reach a similar pH value of approximately 9 in the D-wastewater.

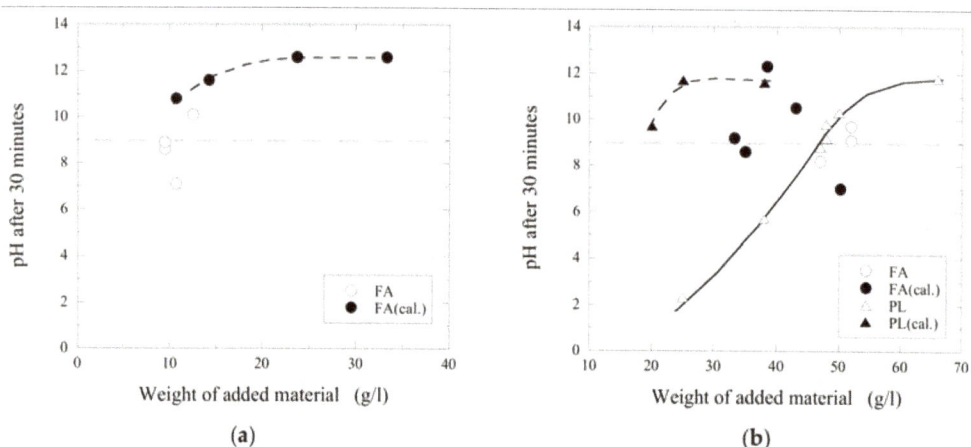

Figure 9. Neutralization effects obtained after 30 min after the addition of fly ash (FA) and primary lime (PL) powders without and with additional calcination in to (**a**) C and (**b**) D acidic wastewater.

In Figure 10, it can be seen that LM 19 g/L reaches a pH value of 9 in 30 min, and the original PL powder at 38 g/L only reaches a pH value of 5.7 during the same time. When the PL powder was calcined prior to the trials, the efficiency is improved drastically and the maximum pH value (~12 pH) is reached in less than 10 min by using the same amount of powder. However, the calcination of the materials means of course an additional technological process and cost. Thus, this needs to be considered in the future.

3.4. Purification of Acidic Wastewaters During Neutralization Process

The cleaning of the wastewater is important, and it goes hand in hand with the neutralization. A variety of chemical and physical means are generally accomplished to remove heavy metals from wastewaters. Hydroxide precipitation, ion exchange, adsorption and membrane processes are some examples of treatments. Chemical precipitation is by far the most widely used process to remove heavy metals from wastewaters in the industry [19]. Neutralization and precipitation are double displacement reactions. The principle of neutralization of acidic waters by using lime lies in the insolubility of heavy metals for alkaline conditions, resulting in precipitates of hydroxides of these metals. The first step of the neutralization process is the lime dissolution. Here, the lime reacts with water and is then dissolved to increase the pH value, which can be illustrated by the following reactions:

$$CaO + H_2O \rightarrow Ca(OH)_2 \qquad (1)$$

$$Ca(OH)_2 \rightarrow Ca^{2+} + 2OH^- \qquad (2)$$

When the pH value has increased, the hydroxide ions precipitate the metals. The precipitation reaction is shown with Zn as example [20]:

$$Zn^{2+} + 2OH^- \rightarrow Zn(OH)_2 \quad (3)$$

In this study, the concentrations of some metals (such as Cr, Fe, Ni, Zn and Mo) were determined in the C- and D-wastewaters (CWW and DWW, respectively) before and after neutralization by addition of the FA, LM and PL. The element determination was carried out by using conventional optical emission spectrophotometers inductively coupled plasma (OES-ICP) using a SPECTRO ARCO AES-ICP model: FHX in an industrial laboratory. The obtained results are given in Table 2.

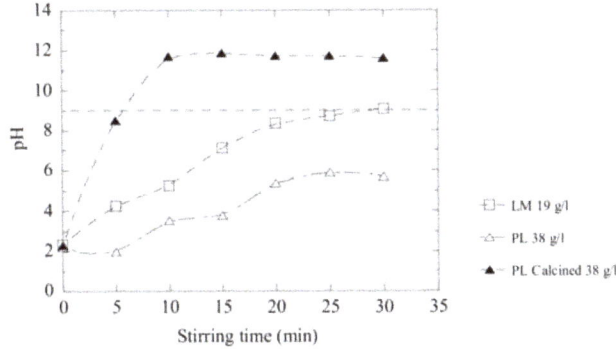

Figure 10. Neutralization effects obtained after the addition of LM, PL and calcined PL into D-acidic wastewater.

Table 2. Results from chemical composition determinations of some metals in wastewater C and D as received, and after the addition of FA, LM and PL.

Sample	Added Neutralizer (g/L)	pH After 30 min	Cr (mg/L)	Fe (mg/L)	Ni (mg/L)	Zn (mg/L)	Mo (mg/L)
CWW Before neutralization	0		202	609	179	<0.01	40.1
C + FA *	33	12.6 in 5 min	0.13	0.02	<0.01	0.04	3.48
C + LM	4.3	8.7	0.45	1.40	1.20	0.01	6.93
C + PL	4.3	9.9	0.43	0.53	0.29	<0.01	15.7
DWW Before neutralization	0		618	2350	1084	1.5	117
D + FA	52	9.1	0.08	0.10	0.04	<0.01	5.06
D + FA *	33	8.6	0.34	0.35	0.28	<0.01	4.66
D + LM	19	9.1	0.10	0.54	1.67	0.01	1.06
D + PL	47	8.8	0.11	0.45	156	0.47	0.17
D + PL	48	9.2	0.03	0.12	0.15	0.01	2.46

* powder used after additional calcination at 1050 °C during 60 min.

First of all, it can be seen that the DWW contained significantly larger concentrations of these metal elements compared to those in the CWW. More specifically, the concentrations of metals in D-wastewater are about 2.9–3.9 times larger with respect to the elements Cr, Fe and Mo, ~6.1 times for Ni and more than 100 times for Zn. Furthermore, it was found that the concentrations of these metals in the wastewater decreased dramatically (except for Zn) during the neutralization process in all trials when using of PL as well as LM and FA. For instance, the concentrations of Cr after neutralization decreased approximately 450–1550 times in the CWW and 1800–20,600 times in the DWW experiments. It should be pointed out that the reduction of concentrations of the given metals during neutralization of DWW is 4–60 times larger (depending on the metal) compared to the CWW results. It may be a reason why D-wastewater needs significantly larger amount of added materials for neutralization up to similar pH level.

Moreover, it can be seen that the LM has a purification effect in the acidic wastewaters from the given metals similar to the PL, while the FA has lower purification effect compared to the LM and PL.

Based on obtained results, it can be concluded that the fly ash and lime mud, which were obtained as CaO-containing wastes from pulp and paper industry, can be successfully applied instead of the common primary lime for neutralization of acidic wastewaters in steelmaking. It can promote to decrease a problem of landfill of wastes from pulp and paper industries and to reduce mining of natural limestone, which is used for production of primary lime by a calcination process generating large amount of CO_2 gas.

4. Conclusions

Two types of CaO-containing waste materials (calcined lime mud and fly ash) from pulp and paper mills were tested as secondary lime sources for the neutralization of four batches of industrial acidic wastewaters (AWW, BWW, CWW and DWW) after the pickling process in steel production of two steel plants. The results obtained in the laboratory scale trials can be summarized as follows:

1. The secondary lime (in the form of fly ash (FA) and calcined lime mud (LM) from pulp and paper industry) can be successfully used instead of natural primary lime (PL) for neutralization of industrial acidic wastewaters in steelmaking plants;
2. A stirring rate of acidic wastewater with added CaO-contained materials was selected as 500 RPM, which showed better kinetic of neutralization process in the given experiments;
3. The neutralization of industrial acidic wastewaters up to the given pH value of 9 during 30 min required about 2.2–2.7 times less LM compared to that of FA. It can be explained by the significantly higher concentration of CaO in the LM and much smaller particle sizes compared to that in FA. Similar amounts of fly ash and primary lime is needed to reach the same efficiency of the neutralization process;
4. Additional calcination of PL and FA powders (at 1050 °C for 60 min) can considerably increase the neutralization effect of acidic wastewaters. For instance, the amount of added FA and PL can be decreased approximately 1.6 and 2.4 times for FA and PL, respectively, to reach a pH value of 9 in the D-wastewater;
5. Concentrations of metals (such as Cr, Fe, Ni, Mo and Zn) in steelmaking acidic wastewaters decrease dramatically after neutralization by using secondary lime. The LM has a purification effect in the acidic wastewaters from the given metals similar to the PL, while the FA has lower purification effect compared to the LM and PL.

Author Contributions: Conceptualization, A.V.K. and T.J.; methodology, A.V.K. and T.J.; validation, T.J. and A.V.K.; formal analysis, T.J.; investigation, T.J.; writing—original draft preparation, T.J.; writing—review and editing, A.V.K. and P.G.J.; visualization, T.J.; supervision, A.V.K. and P.G.J. All authors have read and agreed to the published version of the manuscript.

Funding: This research was funded by Sweden's Innovation Agency (VINNOVA) within the frame of the OSMet S2 project (dnr: 2017-01327) and by a scholarship from Jernkontoret, Prytziska foundation no 2, and KTH ITM-foundation AH Göransson.

Institutional Review Board Statement: Not applicable.

Informed Consent Statement: Not applicable.

Data Availability Statement: Data sharing is not applicable to this article.

Acknowledgments: The authors would like to thank Outokumpu Stainless Avesta, Sandvik Materials Technology Sandviken, Stora Enso Hylte Mill and SCA Obbola for the materials used in experimental trials. The chemical analysis laboratory at Outokumpu Avesta are acknowledged for performing the chemical composition determinations of the water samples. Furthermore, Chuan Wang at SWERIM AB, Luleå is acknowledged for his contributions in the Osmet S2 project.

Conflicts of Interest: The authors declare no conflict of interest.

References

1. Dahlgren, L. Treatment of Spent Pickling Acid from Stainless Steel Production A Review of Regeneration Technologies with Focus on the Neutralisation Process for Implementation in Chinese Industry. Master's Thesis, KTH Royal Institute of Technology, Stockholm, Sweden, 2010.
2. Worldstainless.org. Available online: Passivating_Pickling_EN_2007 (accessed on 3 July 2020).
3. Forsido, T.; McCrindle, R.; Maree, J.; Mpenyana-Monyatsi, L. Neutralisation of acid effluent from steel manufacturing industry and removal of metals using an integrated electric arc furnace dust slag/lime process. *SN Appl. Sci.* **2019**, *1*, 1605. [CrossRef]
4. Gunatilake, S.K. Methods of Removing Heavy Metals from Industrial Wastewater. Department of Natural Resources Sabaragamuwa. *JMESS* **2015**, *1*, 12–18.
5. Jiang, B.; Xia, D.; Yu, B.; Xiong, R.; Ao, W.; Zhang, P.; Cong, L. An environment-friendly process for limestone calcination with CO_2 looping and recovery. *J. Cleaner Prod.* **2019**, *240*, 118147. [CrossRef]
6. Jarnerud, T.; Hu, X.; Karasev, A.V.; Wang, C.; Jönsson, P.G. Application of Fly Ash from Pulp and Paper Industries as Slag Formers in Electric Arc Furnace Stainless Steel Production. *Steel Res. Int.* **2020**, *91*, 2000050. [CrossRef]
7. Hu, X.; Jarnerud, T.; Karasev, A.V.; Jönsson, P.G.; Wang, C. Utilization of fly ash and waste lime from pulp and paper mills in the Argon Oxygen Decarburization process. *J. Clean. Prod.* **2020**, *261*, 121182. [CrossRef]
8. De Colle, M.; Jönsson, P.; Karasev, A.; Gauffin, A.; Renman, A.; Renman, G. The Use of High-Alloyed EAF Slag for the Neutralization of On-Site Produced Acidic Wastewater: The First Step towards a Zero-Waste Stainless-Steel Production Process. *Appl. Sci.* **2019**, *9*, 3974. [CrossRef]
9. Puthucode, R. Neutralization of acidic wastewaters with the use of landfilled Electric Arc Furnace (EAF) high-alloyed stainless-steel slag: An upscale trial of the NEUTRALSYRA Project. MSc Thesis, KTH, Royal Institute of Technology, Stockholm, Sweden, 2019.
10. Brännberg Fogelström, J.; Lundius, A.; Pousette, H. Neutralizing Acidic Wastewater from the Pickling Process Using Slag from the Steelmaking Process: A Pilot Study in Project "Neutralsyra". Master's Thesis, KTH Royal Institute of Technology, Stockholm, Sweden, 2019.
11. Zvimba, J.N.; Siyakatshana, N.; Mathye, M. Passive neutralization of acid mine drainage using basic oxygen furnace slag as neutralization material: Experimental and modelling. *Water Sci. Technol.* **2017**, *75*, 1014–1024. [CrossRef] [PubMed]
12. Petruzzelli, D.; Petrella, M.; Boghetich, G.; Calabrese, P.; Petruzzelli, V.; Petrella, A. Neutralization of Acidic Wastewater by the Use of Waste Limestone from the Marble Industry. Mechanistic Aspects and Mass Transfer Phenomena of the Acid−Base Reaction at the Liquid−Solid Interface. *Ind. Eng. Chem. Res.* **2009**, *48*, 399–405. [CrossRef]
13. Mackie, A.; Boilard, S.; Walsh, M.E.; Lake, C.B. Physicochemical characterization of cement kiln dust for potential reuse in acidic wastewater treatment. *J. Hazard.* **2009**, *173*, 283–291. [CrossRef] [PubMed]
14. Sthiannopkao, S.; Sreesai, S. Utilization of pulp and paper industrial wastes to remove heavy metals from metal finishing wastewater. *J. Environ. Manage.* **2009**, *90*, 3283–3289. [CrossRef] [PubMed]
15. Prochazka, J. Recirkulation av Metallhydroxidslam. Master's Thesis, KTH Royal Institute of Technology, Stockholm, Sweden, 2007.
16. Skogsindustrierna.org. Available online: https://miljodatabas.skogsindustrierna.org/simdb/Web/main/report.aspx?id=103 (accessed on 29 March 2021).
17. Official Journal of the European Communities. Available online: https://eur-lex.europa.eu/legal-content/EN/TXT/PDF/?uri=CELEX:31999L0031&from=EN (accessed on 29 March 2021).
18. Zinck, J.; Aubé, B. Optimization of lime treatment processes. *CIM Bull.* **2000**, *93*, 98–105.
19. Young, K.; In-Liang, J. Photocatalytic reduction of Cr(VI) in aqueous solutions by UV irradiation with the presence of titanium dioxide. *Water Res.* **2001**, *35*, 135–142. [CrossRef]
20. Aubé, B.; Zinck, J.; Eng, M. Lime Treatment of Acid Mine Drainage in Canada. In Proceedings of the Brazil-Canada Seminar on Mine Rehabilitation, Florianópolis, Brazil, 1–3 December 2003.

Article

Effect of Acid Leaching Pre-Treatment on Gold Extraction from Printed Circuit Boards of Spent Mobile Phones

Nicolò Maria Ippolito [1], Franco Medici [2,*], Loris Pietrelli [3] and Luigi Piga [2]

[1] Department of Industrial and Information Engineering and Economics, University of L'Aquila, Via Giovanni Gronchi 18, Zona industrial Pile, 67100 L'Aquila, Italy; nicolomaria.ippolito@univaq.it
[2] Department of Chemical Engineering, Materials and Environment, Sapienza University of Rome, Via Eudossiana 18, 00184 Rome, Italy; luigi.piga@uniroma1.it
[3] Department of Chemistry, Sapienza University of Rome, P.le Aldo Moro 5, 00185 Rome, Italy; lpietrelli@gmail.com
* Correspondence: franco.medici@uniroma1.it

Citation: Ippolito, N.M.; Medici, F.; Pietrelli, L.; Piga, L. Effect of Acid Leaching Pre-Treatment on Gold Extraction from Printed Circuit Boards of Spent Mobile Phones. *Materials* **2021**, *14*, 362. https://doi.org/10.3390/ma14020362

Received: 8 December 2020
Accepted: 11 January 2021
Published: 13 January 2021

Publisher's Note: MDPI stays neutral with regard to jurisdictional claims in published maps and institutional affiliations.

Copyright: © 2021 by the authors. Licensee MDPI, Basel, Switzerland. This article is an open access article distributed under the terms and conditions of the Creative Commons Attribution (CC BY) license (https://creativecommons.org/licenses/by/4.0/).

Abstract: The effect of a preliminary acid leaching for the recovery of gold by thiourea from printed circuit boards (PCBs) of spent mobile phones, was investigated. Preliminary leaching is aimed to recover copper in the leachate that would compete with gold in the successive leaching of the residue with thiourea, thus preventing the formation of the gold-thiourea complex. Two hydrometallurgical routes were tested for the recovery of copper first, and gold after. The first one was based on a two-step leaching that utilizes sulfuric acid and hydrogen peroxide in the preliminary leaching and then thiourea for the recovery of gold in the successive leaching: A copper and gold recovery of 81% and 79% were obtained, respectively. In the second route, nitric acid was used: 100% of copper was recovered in the leachate and 85% of gold in the thiourea successive leaching. The main operative parameters, namely thiourea and ferric sulphate concentrations, leach time, liquid-solid ratio, and temperature were studied according to a factorial plan strategy. A flowsheet of the processes was proposed, and a mass balance of both routes was obtained. Finally, qualitative considerations on the technical and economic feasibility of the different routes were made.

Keywords: printed circuit boards; spent mobile phones; thiourea; precious metals; hydrometallurgy; factorial plans

1. Introduction

Electrical and Electronic Equipment (EEE) were significantly growing all over the world before the corona virus crisis, owing to the continuous development of the technology. Therefore, the amount of waste of this equipment (WEEE) was expected to increase in the next few years [1]. Spent mobile phones belong to WEEE. Consumers are driven to change their mobile phone by the continuous production of new and fashionable models with higher functionality so that the average usage of these components is around 2 years.

The presence of recoverable gold, in printed circuit boards (PCBs) of spent mobile phones, as well as of silver and palladium, makes them among the most valuable components of WEEE, and their recycling would also reduce the environmental impact of the waste due to the presence of heavy metals like copper and nickel [2]. Moreover, the recovery of gold would improve the collection, treatment and recycling of spent mobile phones then contributing to the circular economy according to the EU directive [3]. PCBs of mobile phones are made of 63% metals, 24% ceramics and 13% polymers [4], copper being the main component among the metals. The concentration of gold is higher than the one present in typical ores [5–8] and has changed as a function of the different types and of the year of production of mobile phones; the tendency is to reduce gold in the PCB to lower their production costs. Li et al. [9] determined the concentration of metals contained in PCBs mobile phones: Cu 39.9%, Ag 540 mg/kg and Au 43 mg/kg. Camelino et al. [10]

detected Cu 65.7%, Ag 285 mg/kg, Au 168 mg/kg, and Pd 110 mg/kg. Petter et al. [11] analysed the composition of mobile phones from different companies and year of manufacture and found that the highest concentration of Au was 880 mg/kg. Despite the lower content of gold in PCBs with respect to the content of copper, its higher value makes that 90% of the economic potential of the PCBs phones is given by gold.

A number of hydro-metallurgical [12] and pyro-metallurgical [12] and bio-metallurgical [13–15] processes have been proposed for the recovery of gold [16]. Most processes use hydrometallurgical techniques for economic and environmental reasons that provide the dissolution of most metals as a preliminary step and then the dissolution of gold in alkaline or acid media. The process generally starts with a mechanical treatment that includes the shredding in order to increase the liberation grade of valuable metals. A thorough literature survey on the current status of leaching of precious metals was published 8 years ago by Zhang et al. [17]. Li et al. [9] studied the dissolution of gold and silver in thiourea solution taking into account the influence of particle size, thiourea and oxidant (Fe^{3+}) concentrations and temperature. The best conditions that permitted to obtain 90% of gold dissolution and 50% of silver dissolution were: particle size 150 µm, thiourea concentration 25 g/L, Fe^{3+} = 0.6% w/v, room temperature, and 2 h of leach time. Gurung et al. [18] found that the best conditions to dissolve 3200 mg/kg gold present in PCBs were 38 g/L of thiourea, 0.05 mol/L sulfuric acid at 45 °C and 53–75 µm particle size. Copper is present in PCBs in a concentration of around 30% and with the formation of the complex Cu-thiourea enhances the consumption of thiourea that was expected to form the complex gold-thiourea that is the necessary step to achieve the dissolution of the precious metal [10,19]. Birloaga et al. [20] reports the leaching of spent computer-printed circuit boards, which are different from the boards of spent mobile phones that are the material used in this paper.

Considering the scarcity of publications that deeply illustrate the influence of copper removal before the gold leaching with thiourea from PCBs spent mobile phones, the aim of this work was to study the effect of a preliminary treatment with H_2O_2/H_2SO_4 or HNO_3 to highlight the differences occurring on the successive gold extraction when the two different oxidant solutions are preliminarily applied on the same material. In order to optimize the whole process, experimental tests were carried out with the use of a few statistical factorial plans to investigate the effect of the following factors: thiourea and ferric sulphate concentration, leach time, liquid-solid ratio, and temperature. This also permitted to determine the influence of those experimental parameters on the gold extraction against the purely experimental error. Therefore, a qualitative technical-economic comparison between the two preliminary treatments was proposed. The experimental data were used to describe two possible alternative flowsheets and their respective mass balances according to the best operative conditions.

2. Fundamentals

Gold is a noble metal and is not oxidized neither in water nor in air. The dissolution of gold according to the reaction (1):

$$Au^+ + e^- \leftrightarrows Au^0 \qquad (1)$$

does not practically occur due to the high standard Gibbs free energy of formation of the ion Au^+ (ΔG_f^0 = 163,063 kJ/mol). The standard redox potential, E^0, of this reaction is 1.69 V, and is easily calculated by Equation (2):

$$\Delta G_r^0 = -nFE^0 \qquad (2)$$

where $\Delta G_r^0 = -\Delta G_f^0$, n is the number of electrons involved in the reaction and F is the Faraday constant 96,487 C/mol. So, gold cannot be oxidized and dissolved even by concentrated nitric acid ($NO_3^- + 4H^+ + 3e^- \to NO + 2H_2O$, E^0 = 0.96 V). However, the standard potential is related to a unitary activity (a = 1) and not to the effective activity of gold ions in the solution. Hence, the effective redox potential of reaction (1) is calculated by Nerst Equation (3):

$$E = 1.69 + 0.059 \log [Au^+] \text{ at } 25\,°C, \tag{3}$$

The reaction between the ion Au^+ and particular ligands forms very stable complexes, according to the following reaction (4):

$$Au^+ + nL^{y-} \leftrightharpoons AuLn^{1-ny} \tag{4}$$

and leads to a decrease of the activity of Au^+ ions in the solution, which are blocked in the complex, and the effective redox potential of reaction (1) decreases. Usually, CN^- is used as a ligand in gold treatment that is an environmentally undesirable reagent but thiourea ($CS(NH_2)_2$) can replace CN^- even though thiourea is in suspicion of being a carcinogen for long exposures. If thiourea is added to the solution where solid gold is present in equilibrium with its Au^+ ions, the following reaction can be written, which spontaneously proceeds towards the formation of the Au-thiourea complex [21]:

$$Au^+ + 2\,CS(NH_2)_2 \leftrightharpoons Au(CS(NH_2)_2)_2^+ \tag{5}$$

The stability constant of the Au-thiourea complex can be expressed as:

$$K_{st} = 2 \times 10^{23} = \frac{[Au(CS(NH_2)_2)_2]^+}{[Au]^+[CS(NH_2)_2]^2} \tag{6}$$

If reaction (1) is subtracted from reaction (5), the following reaction is obtained

$$Au(CS(NH_2)_2)_2^+ + e^- \leftrightharpoons Au^0 + 2\,CS(NH_2)_2 \tag{7}$$

whose effective reduction potential is calculated by substituting $[Au^+]$ of Equation (6) in Equation (3):

$$E = 0.315 + 0.059 \log \frac{[Au(CS(NH_2)_2)_2]^+}{[CS(NH_2)_2]^2} \tag{8}$$

This means that the system Au^+/Au^0, in the presence of thiourea, becomes reduced with respect to other systems like oxygen for example, ($2\,H_2O \leftrightharpoons 4\,H^+ + O_2 + 4\,e^-$, $E^0 = 1.23$ V). Nevertheless, the fugacity of oxygen in a water solution is not sufficient to achieve the proper potential to oxidize gold. On the contrary, the lower potential value of the reaction that forms the Au-cyanide complex ($E^0 = -0.57$ V) at alkaline pHs, allows the gold to be oxidized by the oxygen dissolved in water. Moreover, the reduction of oxygen is slow in thiourea solutions and a stronger oxidizing agent is needed. Very strong oxidants should be avoided when working with thiourea that is oxidized to formamidine disulphide, which causes loss of thiourea and then decomposes irreversibly into cyanamide and sulphur [22]. Sulphur can passivate the surface of gold and prevent it from being dissolved. Moreover, precipitation of sulphur can drag a part of the Au-thiourea complex, then decreasing the recovery of dissolved gold in the solution. In practice, a medium oxidant has to be used to dissolve Au, like $Fe_2(SO_4)_3$, being 0.77 V the standard potential of the equilibrium reaction $Fe^{3+} \leftrightharpoons Fe^{2+}$ [23]. This oxidant was used in the leaching tests shown in this paper. The total gold dissolution is shown by reaction (9):

$$2\,Au^0 + 4\,CS(NH_2)_2 + Fe_2(SO_4)_3 \leftrightharpoons [Au(CS(NH_2)_2)_2]_2\,SO_4 + 2\,FeSO_4 \tag{9}$$

The potentials of the electrochemical reactions cited in the text, as a function of pH, are reported in Figure 1, and show a potential-pH diagram that was constructed by the authors taking into account the relations between the equilibrium constants of those reactions and their standard Gibbs free energy, whose numerical values at 298 K were taken by Gaspar et al. [24]. The potentials reported in the figure were calculated with the actual concentration of gold present in the PCBs used in this study and the concentrations of thiourea and ferric sulphate used for the experimental tests, that were 5×10^{-4} mol/L (all gold dissolved), 0.33 mol/L and 0.11 mol/L (expressed as Fe^{3+}), respectively.

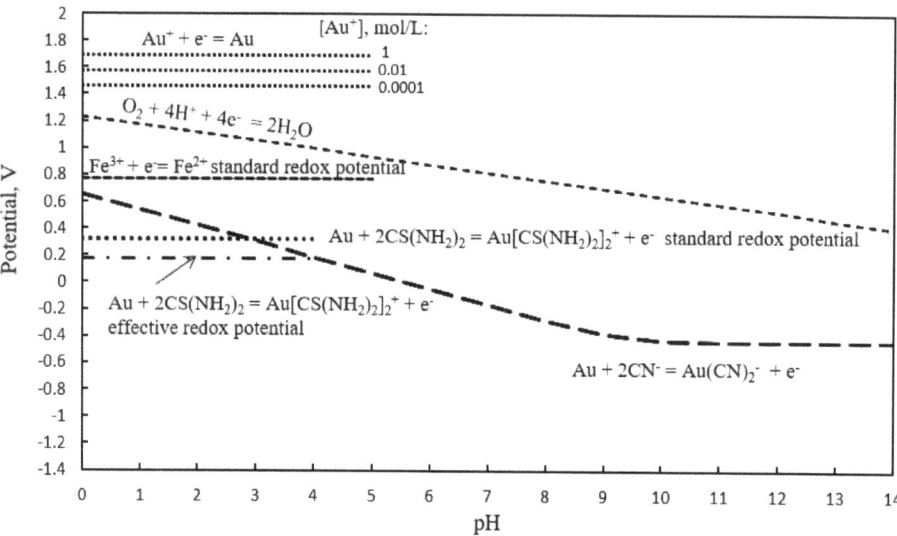

Figure 1. Potential-pH equilibrium diagram illustrating features of the Au-thiourea-H_2O and Au-cyanide-H_2O systems at 25 °C. $[CN^-]_{total} = 1 \times 10^{-4}$ mol/L, $Au(CN)_2^- = 1 \times 10^{-3}$ mol/L, $Au^+ = 5 \times 10^{-4}$ mol/L, $CS(NH_2)_2 = 0.33$ mol/L, $Fe^{3+} = 0.11$ mol/L.

The equilibrium concentration of the species involved can be calculated as reported in Table 1.

Table 1. Equilibrium concentration calculated.

	Au^+	+ $2CS(NH_2)_2$	⇌ $Au(CS(NH_2)_2)_2^+$
initial, mol/L	5×10^{-4}	0.33	-
reacted, mol/L	X	2X	-
equilibrium, mol/L	$5 \times 10^{-4} - X$	$(0.33 - 2X)$	X

Considering that X can be up to 5×10^{-4} mol/L, the thiourea concentration at the equilibrium may be approximated by 0.33 mol/L. After simple calculations, the concentrations of Au^+ and of the Au-thiourea complex at the equilibrium are 2.3×10^{-26} mol/L and 5×10^{-4} mol/L, respectively. The redox potential of reaction 7 decreases up to 0.18 V and is shown up to pH 4 as the stability of the complex decreases at higher pHs, depending on both the pH and on the gold and thiourea concentrations [24]. The redox potential of the equilibrium reaction $Fe^{3+} \leftrightarrows Fe^{2+}$ is 0.71 V and is able to oxidize and dissolve gold according to reaction (9).

3. Materials and Methods

3.1. Preparation and Characterization of Waste PCBs

Experimental studies were carried out with 10 kg of mobile phone PCBs supplied by a WEEE recycling plant that had previously removed all the electronic components like capacitors, cables, resistors, etc. The sample was firstly crushed by a Retsch SM 2000 cutting (Haan, Germany) mill up to −4 mm after various steps of comminution, reducing the size of the output grid after each step. This product was further ground with a Fritsch pulverisette 9 vibratory steel ring mill for 10 min. The resulting powder was sieved with a 0.5 mm screen and the +0.5 mm and −0.5 mm fractions were obtained. The +0.5 mm fraction was analysed with X-ray fluorescence by using a Bruker-Tracer IV SD (Billerica, MA, USA). The −0.5 mm fraction was subjected to an automatic sampling with PT 100—Retsch to obtain representative samples for chemical analyses and leaching tests. Chemical analyses of this fraction were carried out on six 1 g portions of the powder that were dissolved with 1:3 nitric acid and hydrochloric acid (aqua regia) at 90 °C for 3 h. The cooled digestion solution was filtered to remove plastics and ceramics and analysed for the main constituents and precious metals by flame atomic absorption spectroscopy (AAS).

3.2. Leaching Experiments and Test Planning

All chemical reagents utilized for leaching experiments were analytical grade: H_2SO_4 (96%), H_2O_2 (30% w/v), HNO_3 (65%), $CS(NH_2)_2$ (99%), and anhydrous $Fe_2(SO4)_3$. The leaching tests were performed in a 250 mL bottle flask with 10 g of sample powder and 100 mL of leaching solution under 250 rpm stirring and constant pH 1. Samples of the leaching solution were taken at different times and analysed by AAS for copper and gold, in order to study the kinetics of extraction. At the end of each test, the solution was separated from the residue by centrifugation; no paper filter was used to avoid loss of gold in the filter. The residue was washed, dried at 105 °C, weighed, and then attacked with aqua regia to determine the content of copper and gold not dissolved that was added to the content of the two metals in the solution. This procedure was used due to the heterogeneity of powders of PCBs and permitted to determine the reconstituted feed that was taken as the concentrations of the metals in the initial sample to calculate the recovery of metals.

Firstly, leaching tests were carried out to evaluate gold dissolution directly by thiourea solution [9], with no preliminary treatment to remove copper that competes with gold for the formation of their respective complexes with thiourea. These tests were carried out on different particle-size fractions of the milled PCBs. After this, the two preliminary leaching treatments were studied to dissolve selectively copper from PCBs, before the successive thiourea leaching to recover gold left in the residue. These treatments can be represented by reactions (10) and (11).

$$Cu + H_2SO_4 + H_2O_2 \rightarrow CuSO_4 + 2\,H_2O \quad (10)$$

$$Cu + 4HNO_3 \rightarrow Cu(NO_3)_2 + 2\,NO_2 + 2\,H_2O \quad (11)$$

The H_2O_2/H_2SO_4 treatments [25] were carried out in two consecutive steps. In the first step, the −0.5-mm fraction was leached with a relatively low sulphuric acid concentration. The residue of leaching was vacuum filtered, washed and then leached with a higher concentration of sulfuric acid in the second step. Three treatments were carried out, the first with 1 mol/L and 2 mol/L, the second with 2 mol/L and 3 mol/L and the third with 3 mol/L and 4 mol/L sulfuric acid. In the HNO_3 treatment, the copper extraction from the −0.5 mm fraction, as a function of leach time and of the acid concentration in the range 0.1–6 mol/L, was investigated.

On the basis of the experimental data, the residues obtained by preliminary acid treatments were leached with thiourea and ferric sulphate to evaluate how the two preliminary treatments affected the gold dissolution.

The thiourea leaching tests on the residues of the two preliminary tests were planned according to a 2^3 full factorial design whose three investigated factors and two levels adopted (−1 and +1) are reported in Table 2. Three replicated tests were carried out to evaluate the purely experimental error at a middle level (0).

Table 2. Factors and levels investigated with the full factorial plan 2^3.

Factor	Level (−1)	Level (0)	Level (+1)
A, thiourea (g/L)	10	25	40
B, Fe^{3+} (% w/v)	0.4	0.6	0.8
C, temperature (°C)	25	52.5	80

These operating parameters were kept constant during the tests: liquid-solid ratio (L/S) 10, temperature (T) 25 °C, H_2SO_4 0.1 mol/L (pH 1), stirring 250 rpm, and leach time 1 h. The results were elaborated by analysis of variance [26]. Finally, on the basis of the results obtained by factorial experimentation, a further full factorial plan (2^2) was studied, with the aim to evaluate the effect of thiourea at higher concentration and of the L/S ratio (Table 3). Duplicated tests were carried out at middle level.

Table 3. Factors and levels investigated with the full factorial plan 2^2.

Factor	Level (−1)	Level (0)	Level (+1)
A, thiourea (g/L)	40	60	80
B, L/S ratio (v/w)	5	10	15

4. Results

4.1. Characterization of Waste PCBs Mobile Phones

The +0.5-mm fraction of the milled PCBs was 11.5% of the initial material. Visual inspection revealed the presence of metal lamellae not further grindable. XRF analysis showed that the main elements were Fe 51.4%, Cu 15.0% and Cr 13.7%, while no gold was detected.

Chemical composition of the −0.5 mm fraction is reported in Table 4.

Table 4. Chemical composition of PCBs mobile phones (−0.5 mm) determined by atomic absorption spectroscopy (AAS). Averages of six replicates.

Element	Concentration, %	Std Deviation, %
Cu	27.0	1.2
Si	9.66	0.48
Fe	3.22	0.37
Sn	2.85	0.23
Al	1.98	0.21
Ni	1.45	0.24
Zn	0.59	0.11
Cr	0.13	0.05
Precious metals	**Concentration, mg/kg**	**Std Deviation, mg/kg**
Au	439	18
Ag	336	15
Pd	18	2

The main component is Cu (27.0 ± 1.2%) followed by the major elements (Si, Fe, Sn Al). Regarding the precious metals, the following concentrations were detected: Au 439 mg/kg, Ag 336 mg/kg, and Pd 18 mg/kg. Similar values were reported by Ning et al. [27] and by Pietrelli et al. [7]. Pd concentration was low because the palladium-rich components were previously removed. Heavy metals were also present, like Ni, Zn and Cr in concentrations of 14,500 mg/kg, 5900 mg/kg and 1300 mg/kg, respectively, and it would be necessary to take account of this for the safe disposal of process residues.

The low standard deviations indicate the good homogeneity of the −0.5 mm fractions involved in characterization and in further leaching tests.

4.2. Thiourea Leaching with No Preliminary Treatment

Thiourea leaching tests were firstly carried out under the following operative conditions: thiourea 25 g/L, Fe^{3+} = 0.6% w/v, H_2SO_4 0.1 mol/L (pH 1), L/S ratio 10, T 25 °C, and stirring 250 rpm [9,18,28]. The effect of leach time (0.5, 1, 2 and 3 h) and of the size of PCBs (full board, half board, particle size of 1 cm, 1 mm and 0.5 mm) on gold dissolution was studied.

The results reported in Figure 2 show that gold extraction was negligible in any condition and did not reach 12%, contrary to what was reported by Li et al. [9].

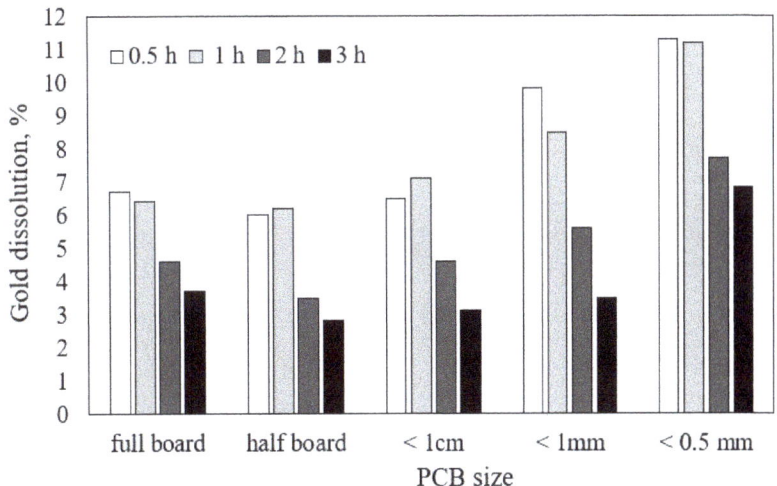

Figure 2. Leaching of PCB as a function of size and leach time.

The leach time and the size of PCB slightly affects the dissolution of gold. Between 0.5 h and 1 h of leach time, no consistent difference among dissolutions was observed and the effect of particle size seems to be irrelevant at size larger than 1 cm, while the higher gold dissolutions occurred for sizes smaller than 1 mm, due to the higher degree of liberation of the gold-containing particles. There is a decrease of gold dissolution with leaching times higher than 1 h and, after the same time, the solution turned from clear to turbid, probably owing to the degradation in acidic solution of thiourea into formamidine disulphide that produces elemental sulphur and cyanamide according to reaction (12) [29].

$$CS(NH_2)_2 + 2Fe^{3+} \rightarrow NH_2CN + 2Fe^{2+} + S + 2H^+ \qquad (12)$$

Therefore, it was assumed that leach times higher than 1 h are not suitable to maximize gold dissolution. On the basis of these tests, gold dissolution was evaluated on PCB particles of size −0.5 mm at higher concentrations of thiourea and Fe^{3+}. After 1 h, 10% of gold dissolution was achieved with 50 g/L thiourea and 1.2% w/v Fe^{3+}, while 15% was achieved with 130 g/L thiourea (close to the solubility) and 3.1% w/v Fe^{3+}.

Low gold dissolutions are due to the high content of copper (around 30%) in PCBs, which has a strong negative effect on the kinetics of the gold-leaching reaction, because copper competes with gold for thiourea with the formation of $Cu[CS(NH_2)_2]_2^{2+}$ that prevails over the formation of $Au[CS(NH_2)_2]_2^+$. This occurrence is less pronounced when gold is recovered from primary raw materials due to the negligible presence of copper in the siliceous or sulphide matrix where gold is generally present. For this reason, preliminary treatments with sulfuric or nitric acid were studied to firstly extract copper and ensure that the thiourea used for the subsequent leaching was mainly used to dissolve gold.

4.3. Sulfuric Acid and Hydrogen Peroxide Preliminary Treatment

In Figure 3, copper dissolution yields as a function of selected sulfuric acid concentration for each leaching stage are shown.

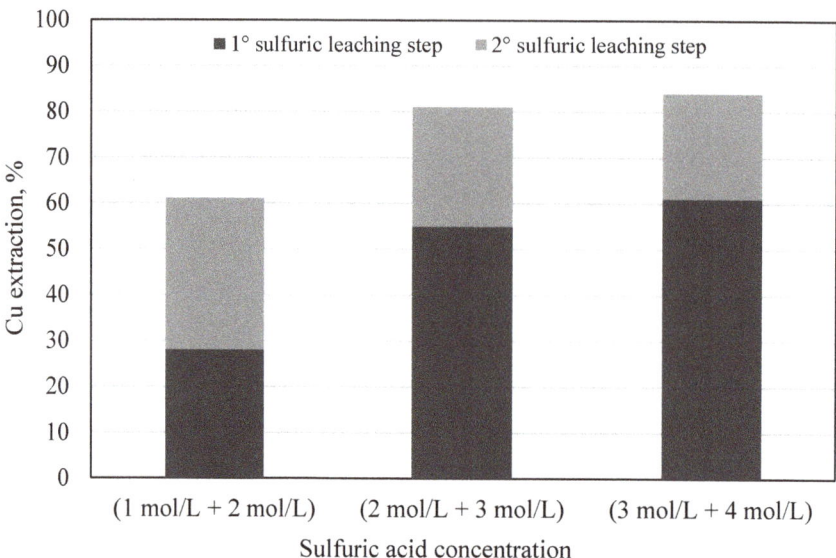

Figure 3. Copper dissolution yields as a function of sulfuric acid concentration in the two-step leaching. H_2O_2 20% v/v, L/S ratio 10, T 25 °C, leach time 1 h, particle size −0.5 mm, and stirring 250 rpm.

At the lowest sulfuric acid concentrations (1 and 2 mol/L), 28% of copper was dissolved in the first step and 33% in the second step for a total of 61%. At higher sulfuric acid concentrations (2 and 3 mol/L), most of the copper was dissolved in the first step (55%) and 26% in the second step for a total of 81%. Slightly better results were achieved with the highest concentration investigated (3 and 4 mol/L). Each copper extraction value was the average of three values, as each treatment was carried out three times. The other operative parameters were kept constant, namely H_2O_2 20% v/v, L/S ratio 10, T 25 °C, leach time 1 h, particle size −0.5 mm, and stirring 250 rpm. The recovery of copper as a metal can then be performed by electrodeposition directly from the sulphuric solution [30]. The residue of previous leaching was leached with thiourea with the same operative conditions as those used for the leaching carried on PCBs of size −0.5 mm without preliminary treatment: thiourea concentration 25 g/L, Fe^{3+} = 0.6% w/v, H_2SO_4 0.1 mol/L (pH 1), L/S ratio 10, T 25 °C, and leach time 2 h, under stirring. The results of thiourea dissolution with the preliminary treatment are reported in Table 5, where the results without preliminary treatment are also shown.

Table 5. Gold dissolution as a function of time with direct thiourea leaching (no preliminary treatment) and after preliminary treatment with H_2SO_4/H_2O_2.

Time, h	Gold Dissolution, %	
	Direct Thiourea Leaching	Thiourea Leaching after Sulfuric Acid—Hydrogen Peroxide Treatment
0.5	9.1	79
1.0	8.5	79
2.0	7.7	78

After 0.5 h leach time, the 79% of gold was dissolved and no further increase was achieved after 1 h leaching, while a slight decrease in dissolution was obtained after 2 h. This behaviour was probably tied to thiourea degradation as already discussed. These values were determined as an average of three replicated tests with low standard deviations (<1.4%).

4.4. Nitric Acid Preliminary Treatment

In Figure 4, the results of nitric acid tests are shown. For each test, the following operative conditions were kept constant, L/S ratio 10, T 25 °C. Tests were carried out with three replications, experimental error was evaluated to be in the range of 1–2%. All the copper was extracted with 3 mol/L acid concentration, after 1 h leach time.

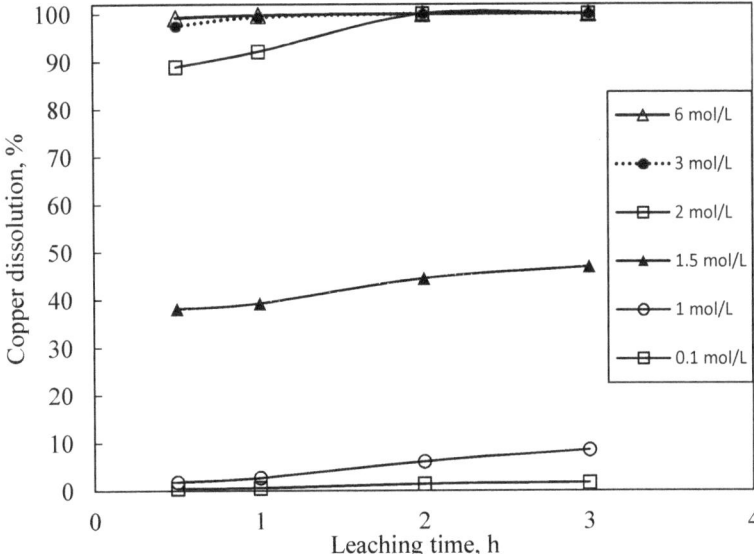

Figure 4. Copper dissolution yields as a function of nitric acid concentration and leach time. H_2O_2 20% v/v, L/S ratio 10, T 25 °C, particle size −0.5 mm, and stirring 250 rpm.

The leach time needed to achieve total dissolution of copper increased to 2 h when the acid concentration was lowered to 2 mol/L. Below 2 mol/L acid concentration, regardless of the leach time, no satisfactory extraction of copper was achieved. Therefore, on the basis of these results, only the residue of previous 3 mol/L HNO_3 treatment was leached with thiourea at the same operative conditions that were applied to the solid residues obtained from H_2O_2/H_2SO_4 treatment. Gold dissolution of 81% and 80% after 1 h and 2 h leach time, respectively, were achieved. Even though the HNO_3 treatment allowed to remove 100% of the copper present in PCBs, the increase of the gold dissolution was not significant with respect to the preliminary treatment with H_2O_2/H_2SO_4 that was able to dissolve as much as 81% of copper.

4.5. Influence of the Main Operative Variables on Dissolution of Gold

The influence of thiourea concentration (factor A), Fe^{3+} concentration (factor B) and temperature (factor C), was studied according to a 2^3 full factorial plan that is shown in Table 6 with the levels assigned to each factor and the gold dissolution yields.

Table 6. 2^3 full factorial design with replicated central point (A: thiourea, g/L; B: Fe^{3+}, % w/v; C: temperature, °C).

Test	Treatment	A	B	C	Gold Dissolution, %
1	(1)	10	0.4	25	73.7
2	a	40	0.4	25	79.6
3	b	10	0.8	25	68.6
4	ab	40	0.8	25	78.4
5	c	10	0.4	80	71.9
6	ac	40	0.4	80	80.2
7	bc	10	0.8	80	68.2
8	abc	40	0.8	80	81.8
9	I	25	0.6	52.5	73.5
10	II	25	0.6	52.5	78.8
11	III	25	0.6	52.5	77.4

Three replicated tests were carried out in the middle of the plan to determine the purely experimental error. The highest dissolution range 78.4–81.8% was achieved with the treatments "a", "ac" and "abc", at the highest level of thiourea concentration (factor A, 40 g/L). The lowest dissolution range, 68.2–71.9%, was achieved at the lowest level of thiourea concentration (factor A, 10 g/L). Factors B and C seem not to affect gold dissolution at 95% confidence level. This is confirmed by the results of analysis of variance reported in Table 7.

Table 7. ANOVA full factorial design (A: thiourea, g/L; B: Fe^{3+}, % w/v; C: temperature, °C).

Effects	Coefficient Value	Sum Square	F Value	Significance, %
A	9.40	176.62	23.43	96
B	−2.10	8.82	1.17	61
AB	2.30	10.58	1.40	64
C	0.45	0.41	0.05	16
AC	1.55	4.80	0.64	49
BC	1.05	2.20	0.29	36
ABC	0.35	0.24	0.03	13

There is a significant positive effect of thiourea concentration (factor A) on dissolution of gold while factors B, C, and all the interactions among the factors are not significant. Based on the results obtained by the plan, the following relation was obtained:

$$\text{Au dissolution (\%)} = 75.65 + 4.70 \times \text{thiourea concentration (g/L)}$$

Finally, a further full factorial plan (2^2) was studied in order to investigate the effect of higher thiourea concentrations on gold dissolution while the other already studied factors (Fe^{3+} concentration and temperature) were kept at a low level since they were not significant. The results are reported in Table 8.

Table 8. Results of 2^2 full factorial design with two replicated central points.

Treatment	Factor A, Thiourea (g/L)	Factor B, Liquid/Solid	Au Dissolution (%)
(1)	40	5	72.8
a	80	5	76.5
b	40	15	80.8
ab	80	15	84.5
I	60	10	85.8
II	60	10	84.6

The maximum gold extraction (85.2 ± 0.8%) was achieved at central point tests with the following conditions: thiourea 60 g/L (factor A), L/S ratio 10 (factor B). The analysis of the data, according to Yate's algorithm, showed that by the F-test method at 95% confidence level, factor A and factor B are not significant within the investigated value range.

5. Discussion

5.1. Gold Dissolution Treatments Comparison

A summary of the results obtained by the investigated treatments and a qualitative technical and economic feasibility are reported in Table 9.

Table 9. Technical and economic feasibility of the investigated treatments as a function of the recovery of metals (X: low impact; XX: medium impact; XXX: high impact).

	Direct Leaching Thiourea	Thiourea Leaching after Copper and Base Metals Leaching (H_2SO_4/H_2O_2)	Thiourea Leaching after Copper and Base Metals Leaching (HNO_3)
Au dissolution, %	11	79	81
Cu dissolution, %	8	81	100
	Technical and Economic Feasibility		
Emissions/wastewater	X	XX	XXX
Plant complexity	X	XXX	XXX
Number of chemicals	X	XXX	XX
Costs	X	XXX	XX

Recoveries of gold and copper are referred to thiourea leaching tests carried out with the same operative conditions. Direct thiourea leaching is not suitable to dissolve gold because of the high amount of copper in phone PCBs. Both preliminary treatments allow similar gold recovery in the successive thiourea leaching, but the treatment with nitric acid permits the total recovery of copper. Despite different copper recoveries being obtained by the two routes, gold dissolutions after thiourea leaching were similar, probably because after H_2SO_4/H_2O_2 preliminary treatment, non-leached copper is present as a refractory material and did not compete with gold for thiourea consumption. Moreover, the higher recovery of copper affects only 1% on the revenue of valuable metals, with respect to the sulfuric acid–hydrogen peroxide treatment.

The comparison between the two preliminary treatments must be made in terms of plant complexity, process management and disposal of waste. The use of hydrogen peroxide must be considered in terms of chemicals increase, complexity (storage) and, therefore, costs. The pre-treatment with HNO_3 will produce a large amount of NOx and exhausted NaOH and $Ca(OH)_2$ resulting from the treatment of off-gas. Despite nitric acid being less expensive than the solution sulfuric acid–hydrogen peroxide, the recovery and the grade of copper by electrodeposition from sulfuric acid solutions are higher than those obtained from nitric acid solutions [29].

5.2. Mass Balances

On the basis of experimental results, two flowsheets for the processing of 1 kg of PCBs from end-of-life mobile phones were described. The first scheme is reported in Figure 5.

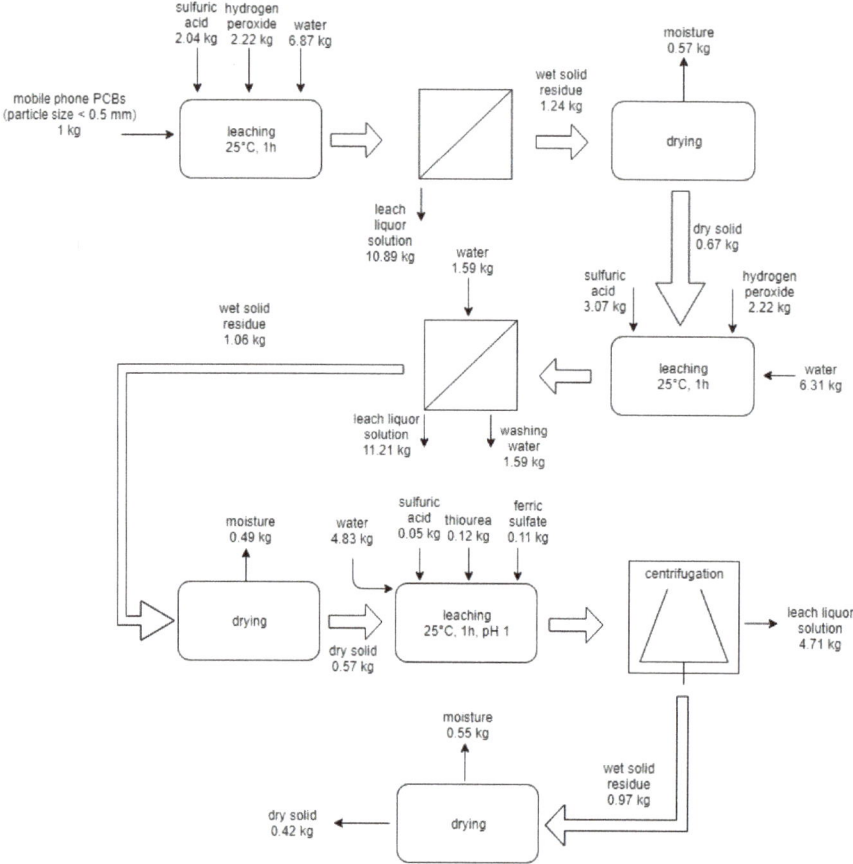

Figure 5. Flowsheet and mass balance of the first route process for the treatment of 1 kg of end-of-life mobile PCBs.

The flowsheet includes: First, leaching with sulfuric acid and hydrogen peroxide, filtration, drying of solid residue; and second, leaching with sulfuric acid and hydrogen peroxide, filtration and washing, drying of solid residue, leaching with thiourea, and centrifugation and drying of the solid residue. According to this process, the following products are obtained: two leach solutions enriched in copper (148 g of copper in the first and 70 g of copper in the second are dissolved) and a leach solution in which 347 mg of gold are dissolved. Copper and gold can be recovered by electrodeposition [31].

The second flowsheet is reported in Figure 6.

The flowsheet includes: the use of nitric acid leaching, filtration and washing, drying of solid residue, thiourea leaching, and centrifugation and drying of solid residue. The solid residue of both processes is mainly constituted by plastics and ceramics to dispose of or to be used as filler to reinforce the new polymer plastic items such as coating and carpets. The residual solutions can be reused, with the necessary make-up of chemicals, for a few cycles [32]; after that, the exhausted solutions can be sent to chemical-physical operations for abatement of pollutants and successive recycling or disposal.

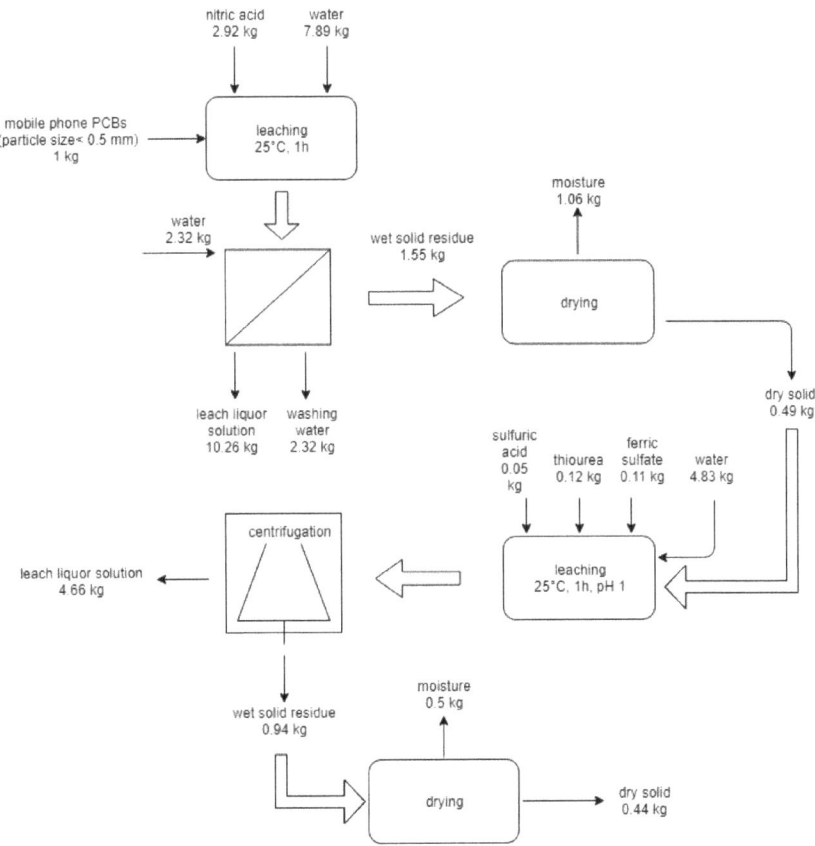

Figure 6. Flowsheet and mass balance of the second route process for the treatment of 1 kg of end-of-life mobile PCBs.

6. Conclusions

Printed circuit boards of spent mobiles contain 27% of Cu as the main component and precious metals like Au 439 mg/kg, Ag 336 mg/kg and Pd 18 mg/kg. In spite of the low content of Au with respect to copper, 90% of the value of board is given by gold. The recovery of gold from the board "as is" with thiourea leaching was not feasible due to the low liberation grade of the metal, and a recovery of only 15% was obtained. Direct leaching with thiourea on the board comminuted to 0.5 mm, and failed owing to the competition between copper and gold for thiourea to form the respective complexes. Therefore, two pre-treatments to solubilize copper and to facilitate the subsequent gold dissolution were tested. The first treatment with sulfuric acid and hydrogen peroxide led to the recovery of 81% of copper in the sulfuric solution after 0.5 h leaching and of 79% of gold in the successive thiourea leaching. The second pre-treatment with 3 mol/L nitric acid achieved dissolution of the totality of copper after 1 h leaching, and 81% of gold was then recovered in the successive thiourea leaching. Analysis of variance applied on the results of the experimental tests highlighted that only the concentration of thiourea is significative; ferric sulphate concentration, leach time, liquid-solid ratio, and temperature seem not to affect the recovery of gold, at least in the range of values investigated. In fact, an increase of thiourea concentration allowed to achieve 85% gold recovery.

Finally, taking into account the mass balance of copper and gold, the technical and economic feasibility and the lower environmental impact, the preliminary treatment with sulfuric acid and hydrogen peroxide seems to be the most promising process.

Author Contributions: Conceptualization, N.M.I., F.M., L.P. (Luigi Piga) and L.P. (Loris Pietrelli); methodology, N.M.I., F.M., L.P. (Luigi Piga) and L.P. (Loris Pietrelli); investigation, N.M.I.; resources, F.M., L.P. (Luigi Piga) and L.P. (Loris Pietrelli); data curation, N.M.I. and L.P. (Luigi Piga); writing-original draft preparation, N.M.I., F.M., L.P. (Luigi Piga) and L.P. (Loris Pietrelli); supervision, F.M., L.P. (Luigi Piga) and L.P. (Loris Pietrelli). All authors have read and agreed to the published version of the manuscript.

Funding: This research received no external funding.

Institutional Review Board Statement: Not applicable.

Informed Consent Statement: Not applicable.

Data Availability Statement: Data presented are openly available.

Acknowledgments: The authors want to thank Nike Srl for having supplied the spent mobiles phones on which this work was carried out. We are, also, grateful to Francesco Maria Arena and Oriana Gallaccio for laboratory work.

Conflicts of Interest: The authors declare no conflict of interest.

References

1. Associated Chambers of Commerce and Industry. India's WEEE Growing at 30% per Annum: ASSOCHAM-Kinetics Study. 2016. Available online: http://www.assocham.org/newsdetail.php?id=5725 (accessed on 21 August 2019).
2. Pathak, P.; Srivastava, R.R. Assessment of legislation and practices for the sustainable management of waste electrical and electronic equipment in India. *Renew. Sustain. Energy Rev.* **2017**, *78*, 220–232. [CrossRef]
3. European Commission. Report from the Commission to the European Parliament, the Council, the European Economic and Social Committee and the Committee of the Regions. Bruxelles, 26/01/2017. 2017. Available online: https://ec.europa.eu/environment/circular-economy/implementation_report.pdf. (accessed on 14 November 2019).
4. Yamane, L.H.; Tavares de Moraes, V.; Romano Espinosa, D.C.; Soares Tenorio, J.A. Recycling of WEEE: Characterization of spent printed circuit boards from phones and computers. *Waste Manag.* **2011**, *31*, 2553–2558. [CrossRef] [PubMed]
5. Cui, H.; Anderson, C.G. Literature Review of Hydrometallurgical Recycling of Printed Circuit Boards (PCBs). *J. Adv. Chem. Eng.* **2016**, *6*, 142. [CrossRef]
6. Jeon, S.; Baltazar Tabelin, C.; Park, I.; Nagata, Y.; Ito, M.; Hiroyoshi, N. Ammonium thiosulfate extraction of gold from printed circuit boards (PCBs) of end-of-life mobile phones and its recovery from pregnant leach solution by cementation. *Hydrometallurgy* **2020**, *191*, 105–214. [CrossRef]
7. Pietrelli, L.; Ferro, S.; Vocciante, M. Eco-friendly and cost-effective strategies for metals recovery from printed circuit boards. *Renew. Sustain. Energy* **2019**, *112*, 317–323. [CrossRef]
8. Syed, S. Recovery of gold from secondary: Sources—A review. *Hydrometall* **2012**, *115–116*, 30–51. [CrossRef]
9. Li, J.; Xu, X.; Liu, W. Thiourea leaching gold and silver from the printed circuit boards of waste mobile phones. *Waste Manag.* **2012**, *32*, 1209–1212. [CrossRef]
10. Camelino, S.; Rao, J.; Lopez Padilla, R.; Lucci, R. Initial studies about gold leaching from printed circuit boards (PCB's) of waste cell phones. *Proc. Mater. Sci.* **2015**, *9*, 105–112. [CrossRef]
11. Petter, P.M.H.; Veit, H.M.; Bernardes, A.M. Evaluation of gold and silver leaching from printed circuit board of cellphones. *Waste Manag.* **2014**, *34*, 475–482. [CrossRef]
12. Chiang, H.L.; Lin, K.H.; Lai, M.H.; Chen, T.C.; Ma, S.Y. Pyrolysis characteristics of integrated circuit boards at various particle sizes and temperatures. *J. Hazard. Mater.* **2007**, *149*, 151–159. [CrossRef]
13. Xia, M.; Bao, P.; Liu, A.; Wang, M.; Shen, L.; Yu, R.; Liu, Y.; Chen, M.; Li, J.; Wu, X.; et al. Bioleaching of low-grade waste printed circuit boards by mixed fungal culture and its community structure analysis. *Resour. Conserv. Recycl.* **2018**, *136*, 267–275. [CrossRef]
14. Srivastava, R.R.; Ilyas, S.; Kim, H.; Choi, S.; Trinh, H.B.; Ghauri, M.A.; Ilyas, N. Biotechnological recycling of critical metals from waste printed circuit boards. *J. Chem. Technol. Biotechnol.* **2020**, *95*, 2796–2810. [CrossRef]
15. Ilyas, S.; Srivastava, R.R.; Kim, H.; Das, S.; Singh, V.K. Circular bioeconomy and environmental benigness through microbial recycling of e-waste: A case study on copper and gold restoration. *Waste Manag.* **2021**, *121*, 175–185. [CrossRef] [PubMed]
16. Ilyas, S.; Lee, J. *Gold Metallurgy and the Environment*; CRC Press: Boca Raton, FL, USA, 2018.
17. Zhang, Y.; Liu, S.; Xie, H.; Zeng, X.; Li, J. Current status on leaching precious metals from waste printed circuit boards. *Proc. Environ. Sci.* **2012**, *16*, 560–568. [CrossRef]

18. Gurung, M.; Adhikari, B.B.; Kawakita, H.; Ohto, K.; Inoue, K.; Alam, S. Recovery of gold and silver from spent mobile phones by means of acid thiourea leaching followed by adsorption using biosorbent prepared from persimmon tannin. *Hydrometallurgy* **2013**, *133*, 84–93. [CrossRef]
19. Mironov, I.V.; Tsvelodub, L.D. Complexation of copper (I) by thiourea in acid aqueous solution. *J. Solut. Chem.* **1996**, *25*, 315–325. [CrossRef]
20. Birloaga, I.; De Michelis, I.; Ferella, F.; Buzatu, M.; Vegliò, F. Study on the influence of various factors in the hydrometallurgical processing of waste printed circuit boards for copper and gold recovery. *Waste Manag.* **2013**, *33*, 935–941. [CrossRef]
21. Mironov, I.V.; Kal'nyi, D.B.; Kokovkin, V.V. On gold(I) complexes and anodic gold dissolution in sulphite-thiourea solutions. *J. Solut. Chem.* **2017**, *46*, 989–1003. [CrossRef]
22. Van Staden, P.J.; Laxen, P.A. In-stope leaching with thiourea. *J. S. Afr. Inst. Min. Metall.* **1989**, *89*, 221–229.
23. Zazycki, M.A.; Tanabe, E.H.; Bertuol, D.A.; Dotto, G.L. Adsorption of valuable metals from leachates of mobile phone wastes using biopolymers and activated carbon. *J. Environ. Manag.* **2017**, *188*, 18–25. [CrossRef]
24. Gaspar, V.; Mejerovich, A.S.; Meretukov, M.A.; Schmiedl, J. Practical application of potential-pH diagrams for Au-CS(NH$_2$)$_2$-H$_2$O and Ag-CS(NH$_2$)$_2$-H$_2$O systems for leaching gold and silver with acidic thiourea solution. *Hydrometallurgy* **1994**, *34*, 369–381. [CrossRef]
25. Behnamfard, A.; Salarirad, M.M.; Vegliò, F. Process development for recovery of copper and precious metals from waste printed circuit boards with emphasize on palladium and gold leaching and precipitation. *Waste Manag.* **2013**, *33*, 2354–2363. [CrossRef] [PubMed]
26. Montgomery, D.C. Response Surface Methods and Other Approaches to Process Optimization. In *Design and Analysis of Experiments*; Montgomery, D.C., Ed.; John Wiley & Sons: New York, NY, USA, 1997; pp. 427–510.
27. Ning, C.; Lin, C.S.K.; Hui, D.C.W.; McKay, G. Waste Printed Circuit Board (PCB) Recycling Techniques. *Top. Curr. Chem.* **2017**, *375*, 21–56. [CrossRef]
28. Birloaga, I.; Vegliò, F. Study of multi-step hydrometallurgical method to extract the valuable content of gold, silver and copper from waste printed circuit boards. *J. Environ. Chem. Eng.* **2016**, *4*, 20–29. [CrossRef]
29. Kai, T.; Hagiwara, T.; Haseba, H.; Takahashi, T. Reduction of thiourea consumption in gold extraction by acid thiourea solutions. *Ind. Eng. Chem. Res.* **1997**, *36*, 2757–2759. [CrossRef]
30. Schlesinger, M.E.; King, M.J.; Sole, K.C.; Davenport, W.G. *Extractive Metallurgy of Copper*, 5th ed.; Elsevier: Oxford, UK, 2011; pp. 349–371.
31. Da Silva, M.S.B.; De Melo, R.A.C.; Lopes-Moriyama, A.L.; Souza, C.P. Electrochemical extraction of tin and copper from acid leachate of printed circuit boards using copper electrodes. *J. Environ. Manag.* **2019**, *246*, 410–417. [CrossRef]
32. Guo, X.; Zhang, L.; Tian, Q.; Qin, H. Stepwise extraction of gold and silver from refractory gold concentrate calcine by thiourea. *Hydrometallurgy* **2020**, *194*, 105–330. [CrossRef]

Article

Biochars from Post-Production Biomass and Waste from Wood Management: Analysis of Carbonization Products

Wojciech Kosakowski [1], Malgorzata Anita Bryszewska [2,*] and Piotr Dziugan [3]

1 Polmos Żyrardów Sp. z o.o., ul. Mickiewicza 1-3, 96-300 Żyrardów, Poland; wkosakowski@belvederevodka.pl
2 Institute of Natural Products and Cosmetics, Faculty of Biotechnology and Food Sciences, Lodz University of Technology, ul. Stefanowskiego 4/10, 90-924 Lodz, Poland
3 Department of Environmental Biotechnology, Faculty of Biotechnology and Food Sciences, Lodz University of Technology, Wolczanska 171/173, 90-924 Lodz, Poland; piotr.dziugan@p.lodz.pl
* Correspondence: malgorzata.bryszewska@p.lodz.pl; Tel.: +48-426-313-425

Received: 2 October 2020; Accepted: 31 October 2020; Published: 4 November 2020

Abstract: Waste biomass can be used as an alternative source of energy. However, such use requires prior treatment of the material. This paper describes the physicochemical characteristics of biochar obtained by the thermochemical decomposition of six types of agricultural waste biomass: residues from the production of flavored spirits (a pulp of lime, grapefruit and lemon), beetroot pulp, apple pomace, brewer's spent grain, bark and municipal solid waste (bark, sawdust, off-cuts and wood chips). The biomass conversion process was studied under conditions of limited oxygen access in a reactor. The temperature was raised from 450 to 850 °C over 30 min, followed by a residence time of 60 min. The solid products were characterized in terms of their elemental compositions, mass, energy yield and ash content. The gaseous products from pyrolysis of the biomass were also analyzed and their compositions were characterized by GCMS (Gas Chromatography–Mass Spectrometry). The carbonization process increased the carbon content by, on average, 1.7 times, from an average percentage of 46.09% ± 3.65% for biomass to an average percentage of 74.72% ± 5.36% for biochars. After carbonization, the biochars were found to have a net calorific value of between 27 and 32 MJ/kg, which is comparable or even higher than good-quality coal (eco pea coal 24–26 MJ/kg). The net calorific values show that the volatile products can also be considered as a valuable source of energy.

Keywords: pyrolysis; biochars; agricultural waste; biomass

1. Introduction

Recycling biomass and various types of organic waste is a way of increasing the share of renewable sources in energy production [1]. The Sustainable Development Goals (SDGs) set out by the United Nations highlight renewable energy as key to the success of Agenda 2030 [1]. Possible sources of bioenergy include energy crops, biomass residues and organic wastes. However, direct use of biomass as a heat source may be inefficient and difficult. Complications may arise even at the storage stage, when high humidity is associated with microbiological biomass degradation. The co-combustion of biomass and coal can raise technical and economic issues. Wet biomass may cause instability in the combustion process itself. Incomplete combustion reduces the efficiency of the whole process and leads to energy losses. Incomplete combustion may also make it impossible to maintain the required emission parameters. Given the limited possibilities for using unprocessed biomass, pre-treatment is necessary to improve its energy properties. Various methods of initial biomass preparation are described in the literature, which enable co-combustion with coal in power boilers [2,3]. The process of thermal conversion of biomass to biofuel may include combustion, gasification, biocarbonization, torrefaction, dry distillation and pyrolysis [4–8].

Pyrolysis is a thermochemical treatment, involving the extensive thermal decomposition of organic material under oxygen-limited conditions or in an atmosphere of inert gases. The gas phase contains two kinds of compounds. The first are volatiles that condense and form a dark brown, viscous liquid phase. The second are volatile compounds with low molecular weight (e.g., CO, CO_2, H_2, CH_4 and light hydrocarbons). These non-condensable gases remain in the gas phase. The physical process and chemical reactions that occur in pyrolysis are highly complex, and both the conditions of pyrolysis and the organic matrix (which may originate from different sources) affect the quality of the biochar and its eventual properties. These parameters may be helpful when ranking waste materials as potential sources of biocarbon, and for assessing their suitability for co-firing. These parameters may also be used to evaluate the possibility of using condensing and non-condensing gas products for energy generation.

The aim of this research was to convert biomass into a biofuel with properties similar to those of coal. We used waste from the agricultural industry and municipal management as feedstock. The properties of biochars obtained by biomass carbonization were determined in single-factor experiments. We also characterized the main products of the gas products and condensates. The results could contribute to the development of strategies for biomass treatment and the reduction of emissions, improving the sustainability of biomass conversion processes at an industrial scale.

2. Materials and Methods

2.1. Materials

Six agricultural waste biomass materials were used in the study: flavored spirits production waste (FSW) (lime, grapefruit and lemon), apple pomace (A.pomace), beetroot pulp (B.pulp), brewer's spent grain (BSG), bark (pine bark) and municipal solid waste (bark, off-cuts, wood chips, sawdust (MSW)). The analyzed biomasses were pre-prepared by drying and grinding.

2.2. Volatile Component Analysis

Volatile component analysis was carried out with use of a GCMS (Gas Chromatography–Mass Spectrometry) (Termo Science Trace GC Ultra) and an RTX—1.60 m × 0.25 mm × 0.25 µm capillary column (Restek, Saunderton, UK), combined with a DSQ-II (Dual-Stage Quadrupole) detector (Thermo Scientific, Austin, TX, USA). All the samples were analyzed in duplicate at a pyrolysis temperature of 550 °C with a heating rate of 20 °C ms^{-1}. The samples were collected in Tedlar bags (Merck, Darmstadt, Germany Tedlar® PLV- Push Lock Valve Gas Sampling Bag).

2.3. Moisture and Ash Content, Chemical Composition

The total moisture content in the tested biomass was determined using the thermogravimetric approach. The materials were dried at 110 °C until the complete removal of moisture. The ash content was determined using the slow ashing method, in which combustion and annealing of the analytical sample occurs in two stages, differing in temperature and duration. Dry ashing was performed in open inert crucibles in a muffle furnace. The samples (1 ± 0.1 g of biomass powdered by a broom mill) were placed in the cold furnace and combustion was performed with a constant increase in temperature up to 500 °C for 30 min. The temperature was gradually increased to 815 °C over 30 min. Complete ashing was achieved after thermal decomposition for 90 min.

The chemical composition (content of carbon, sulfur, nitrogen) was determined using an elementary analyzer (CE Instruments, Milan, Italy.) and Eager 200 software (amlyzer type 2500), using methionine or BBOT (2,5-(bis(5-tert-butyl-2-benzo-oxazol-2-yl) thiophene) as the reference material.

2.4. Combustion Heat and Net Calorific Values, Energy Yield

The calorific value was determined using a 6100 compensated jacket calorimeter (Parr Instruments GmbH, Moline, IL, USA). Net calorific values were calculated on the basis of the amount of heat generated during the complete combustion of a mass unit (1 g) of biomass in an oxygen atmosphere using the formula

$$\text{LHV} = 2.326 \times (\text{HHV} - 91.23 \times C_H) \tag{1}$$

where LHV is the net calorific value (Lower Heating Value) (J/g), HHV is the combustion heat and C_H is the hydrogen content of the sample (%) [9]. The net calorific value (LHV) of biomass is calculated as the heat of combustion reduced by the heat of water evaporation, obtained from the fuel in the process of combustion and from hygroscopic moisture. The energy densification ratio describes the ratio of the HHV of the dry product to the dry raw material.

$$\text{Energy densification} = \text{HHV}_{\text{biochar}}/\text{HHV}_{\text{biomass}} \tag{2}$$

The energy yields were calculated using the equation

$$\text{Energy yield} = (\text{mass}_{\text{biochar}}/\text{mass}_{\text{biomass}}) \times (\text{HHV}_{\text{biochar}}/\text{HHV}_{\text{biomass}}) \times 100\% \tag{3}$$

Mass yields were calculated using the equation

$$\text{Mass Yield} = \text{mass}_{\text{biochar}}/\text{mass}_{\text{biomass}} \tag{4}$$

2.5. Carbonization Process

Thermal carbonization (pyrolysis) was performed in a CZYLOK reactor (Jastrzebie-Zdroj, Poland; model FCF 2R) modified in the laboratory to enable the collection of pyrolytic gases. An accurately weighed sample was placed in the oven at room temperature. Thermal decomposition was then performed under conditions of limited oxygen access. The temperature in the reactor was gradually increased to 850 °C over 30 min. Pyrolysis was continued for 60 min at a constant temperature. Subsequently, the sample was left in the oven until the temperature fell. The solid residue after the process was weighed and stored for further analysis. The experiments were carried out in triplicate with seven gas collection points (450, 515, 585, 650, 715, 785 and 850 °C).

3. Results

3.1. Physico-Chemical Properties and Chemical Composition of the Biomass

The results from proximate and elemental analysis of all the waste samples are given in Table 1. The moisture contents of the tested biomasses in the working state were from 9.65% to 16.54%. The lowest water content in the biomass was observed in waste from vodka production, which can be explained by the fact that these wastes had been dried in the factory before being delivered for analysis. The difference in the moisture content (6.02%) between the bark samples and municipal waste was due to the weather conditions under which the biomasses had been processed and then stored.

The content of mineral substances was similar in most of the analyzed biomasses, ranging from 2.57% to 4.37%. An exception was apple pomace, which contained only 1.05% of inorganic components. This value was in line with the data presented in the literature [10,11]. The ash content of MSW was in line with the average values described in the literature for wood biomass, in the range of 0.3–7.4% [12,13]. These results also indicate that the samples were not contaminated with soil. Our values are much lower than those presented in the literature for coal, i.e., 5–45% (8.5–10.5% on average) [13]. The chlorine content in the analyzed biomass was in the range of 0.02–0.20% and such values are consistent with the literature data for woody biomass, the chlorine content of which ranges from <0.005% to 0.057%. In cultivated plants, the chlorine content may even be >1.00% [2,14]. High chlorine content in crops is often associated with the use of potassium fertilizer.

The total sulfur content was highest for agricultural crops, B.pulp 0.75%, which probably results from the application of fertilizers for agrotechnical treatments. The nitrogen content was similar in all the analyzed biomasses (1.35% ± 0.26%) and did not exceed 1.50%, with the exception of BSG for which it was almost four-fold higher. However, this result is in accordance with the literature data, according to which the nitrogen content in crops can be up to 6.45%. The high content of nitrogen in crops is correlated to their high content of proteins, which can represent up to 40% of the dry mass [2].

The bulk density of the analyzed biomasses varied widely, from 130 to 307 kg/m^3. This was lower than that of coal, which is on average in the range of 800–1000 kg/m^3. Low bulk density is uneconomical from the point of view of storage and transport. Therefore, biomass pre-treatment such as grinding, pressing or palletization should be considered.

3.2. Characterization of Biochars

3.2.1. Morphology

Applying pyrolysis to the waste biomass resulted in carbonization. Figure 1 presents graphical images of the biomass and the carbonized material. Carbonization occurred in the whole mass, and was as high on the surface as at the core. The transformations throughout the whole volume were probably a consequence of biomass fractionation, generated either by the production process (A.pomace, B.pulp, BSG) or by the grinding applied in this study (MSW, bark, FSW). Carbonization was not followed by major changes in the structure and mass density. The values for most of the carbonizates were similar to those determined for the biomass (Tables 1 and 2). An increase in mass density occurred only in two samples, apple pomace and MSW. These results are a consequence of pre-treatment, including grinding.

3.2.2. Material and Energy Balance

The material balances after pyrolysis were investigated using the gravimetric method. The results are presented in Figure 2. The solid residue (i.e., biochars) represented from 26.65% to 40.85% of the initial weight of the organic substance. However, this wide variation in yield was caused mainly by one sample, the bark. When the highest value is excluded, the values were very close, with a mean value of 27.98% ± 2.08%. The decrease in mass can be attributed to two causes. The first is moisture loss and the second is organic matter decomposition, with the formation of volatile products such as CO, CO_2, CH_4 and many others. During the process of carbonization, variable amounts of liquid and gas products were formed, in proportions that depended on the biomass. Volatile matter, including water and both gas and oil fractions, composed up to 73% of the initial mass of the samples, with the exception of bark, in which these fractions comprised 59.15%. These differences were probably due to the high content of hemicellulose, cellulose and lignin in bark.

The liquid fraction collected by condensation was composed of oil and water. Based on the mass of the liquid fraction, the oil fraction was calculated by diminishing the weight of the condensate by the moisture content. In general, the content of water was lower in the biomass from fruit waste (9.65% FSW, 12.56% A.pomace). The highest value was obtained for biomass from bark (16.54%). The greatest changes in the amounts of volatile compounds were recorded during heating up to 450 °C.

Table 2 presents the results of proximate and elemental analysis of the biochars obtained from pyrolysis. The carbonization process increased the carbon content on average 1.7-fold, from an average percentage of 46.09% ± 3.65% for biomass to an average percentage of 74.72% ± 5.36% for carbonizates. As a consequence, the C/H ratio also increased, reaching a value seven times higher.

Figure 1. Photographs of the raw and carbonized materials. Thermal decomposition performed under conditions of limited oxygen access with gradually increased temperature up to 850 °C over 30 min and continued for 60 min at a constant temperature.

Table 1. Compositional and elemental analysis of biomass, expressed as average values (±standard deviation). FSW—flavored spirits production waste (lime, grapefruit and lemon), B.pulp—beetroot pulp, A.pomace—apple pomace, BSG—brewer's spent grain, bark, MSW—municipal solid waste.

Biomass	Moisture Content	Ash	Values of Combustion Heat	Net Calorific Values	Mass Density	Elemental Analysis					
	(%)	(%)	(kJ/kg)	(kJ/kg)	(kg/m³)	C	H	N	S	O	Cl
						(%)					
FSW	9.65	3.85 ± 0.34	16,142 ± 600	14,904.3 ± 553.8	243.7 ± 10.0	41.94	5.83	1.32	0.42	37.24	0.09
A.pomace	12.56	1.05 ± 0.01	21,229 ± 135	19,775.2 ± 125.8	155.2 ± 9.3	51.39	6.85	1.44	0.30	36.64	0.07
B.pulp	12.33	2.93 ± 0.02	16,372 ± 28	15,169.2 ± 25.8	155.3 ± 3.0	41.83	5.67	1.32	0.75	35.50	0.20
BSG	11.74	3.59 ± 0.06	20,288 ± 59	18,912.8 ± 55.3	290.1 ± 11.6	46.92	7.12	5.45	0.10	25.83	0.08
Bark	16.54	4.37 ± 0.10	19,523 ± 551	18,477.1 ± 521.5	126.4 ± 7.2	47.23	6.30	1.18	0.18	39.94	0.03
MSW	10.52	2.57 ± 0.19	18,744 ± 104	17,407.3 ± 96.8	130.2 ± 10.1	47.23	6.30	1.18	0.15	40.49	0.02

Table 2. Compositional and elemental analysis of biochar, expressed as average values (±standard deviation). FSW—flavored spirits production waste (lime, grapefruit and lemon), B.pulp—beetroot pulp, A.pomace—apple pomace, BSG—brewer's spent grain, bark, MSW—municipal solid waste, d.m.—dry mass.

Biochars	Mass Yield	Ash	Values of Combustion Heat	Net Calorific Values	Mass Density	Energy Densification Ratio	Energy Yield	Elemental Analysis		
	(%)	(%)	(kJ/kg)	(kJ/kg)	(kg d.m./m³)		(%)	C	H	N
								(%)		
FSW	26.65	13.44	27,836 ± 78	26,598 ± 75	238.9 ± 4.9	1.724	46.0	71.47	1.50	1.61
A.pomace	25.89	5.38	32,402 ± 176	30,948 ± 168	218.0 ± 6.9	1.526	39.5	80.18	1.82	2.08
B.pulp	30.18	16.95	26,775 ± 146	25,572 ± 139	120.9 ± 10.9	1.635	49.4	68.54	1.19	1.87
BSG	26.92	14.43	27,840 ± 26	26,465 ± 25	240.3 ± 4.8	1.372	36.9	69.75	1.63	6.42
Bark	40.85	9.35	29,188 ± 26	28,142 ± 26	138.2 ± 3.4	1.495	61.1	79.34	1.58	0.52
MSW	30.26	8.51	30,311 ± 215	28,974 ± 206	236.0 ± 3.4	1.617	48.9	79.06	1.62	1.55

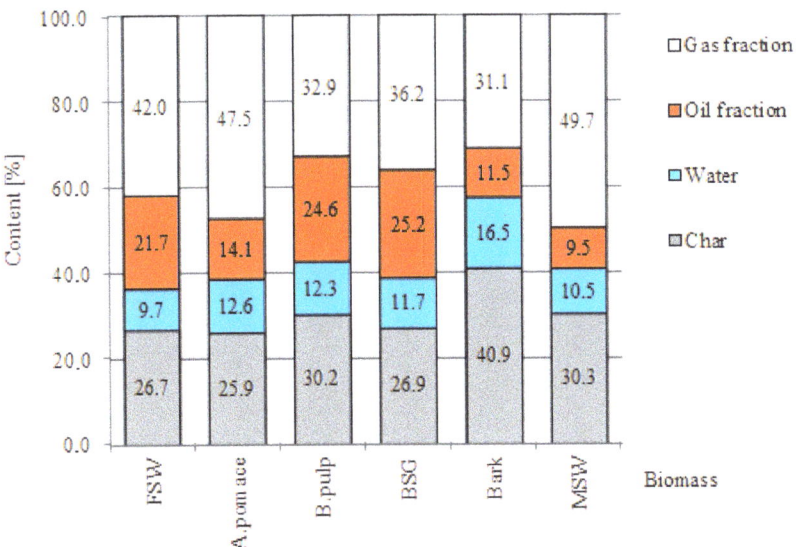

Figure 2. Solid, liquid and gaseous product distribution (wt.%), expressed as average values, obtained with carbonization (850 °C).

Ash content reached values from 5.38% for A.pomace to 16.95% for B.pulp. Relative to the ash content in the raw materials, these values represent increments of 4.33% and 14.02%, respectively. Similar results from the same thermochemical processes have been described by other authors, using municipal solid waste and biomass [15,16]. High ash biochars are of limited use as fuels in the combustion process, since they can cause excessive ash deposition or slag and contamination phenomena, leading to operational difficulties.

Differences in the elemental compositions of the biomasses and biochars resulting from carbonization clearly illustrate the effect of fuel refining, which involves the removal of water and the elimination of oxygen in the form of oxidized volatile compounds through decarboxylation, decarbonization and dehydration reactions.

To determine the energetic yield of the analyzed process, the energy densification ratio and energy yield were calculated (Equations (2) and (3)). The energy densification ratio, indicating the elevation in HHV during carbonization, differed in a narrow range from 1.372 to 1.724, with a mean value of 1.5615 ± 0.124. However, energy densification does not indicate true changes in the energy value of the product, because it does not take into account the reduction in mass that occurs as a result of the process. When mass loss is considered, greater differences between the biomasses are revealed. The energy yield varied from 36.9% to 61.1% for BSG and bark, respectively. The mean value of the energy yield was 46.97% ± 8.572%. The biochars resulting from carbonization had a net calorific value of between 27 and 32 MJ/kg. These values are comparable or even higher than good-quality milled coal (eco pea coal 24–26 MJ/kg, grain diameter 5–25 mm, produced from specially selected hard coal species to obtain fuel with low contents of sulfur and ash).

3.2.3. Composition Analysis of Non-Condensable Gases

Gaseous products are released during pyrolysis by the evaporation or thermal decomposition of the raw material. The amounts of emissions produced during thermal decomposition depend on the composition of the raw material, the heating rate, the temperature and the residence time. Table 3 shows the variations in the content and composition of volatile fractions. Figure 3 shows the variations in the content of hydrocarbons. The major gases produced from biomass carbonization were carbon dioxide and carbon monoxide. The content of CO_2 decreased gradually, whereas the content of carbon oxide increased with the pyrolysis temperature (i.e., the time of the process), which is assumed to be the result of thermal decomposition in an oxygen-poor atmosphere. The high content of carbohydrates, cellulose, hemicellulose and lignin in the tested materials implies the formation of carbon dioxide, carbon monoxide and water, as a result of decarboxylation, decarbonization and dehydration reactions during thermal decomposition. More CO is produced at elevated temperatures and with longer residence time. The formation of CO is strongly affected by secondary reactions of low molecular weight products (especially aldehyde-type compounds) and CO_2 is presumably produced in the early stage of cellulose pyrolysis, primarily in decarboxylation reactions [17–19].

Table 3. Chemical composition of non-condensable gases produced during the pyrolysis of biomass, expressed as average values. Content of compounds expressed as relative peak area (%) of gases found in the fraction. C_xH_y—hydrocarbons, FSW—flavored spirits production waste (lime, grapefruit and lemon), B.pulp—beetroot pulp, A.pomace—apple pomace, BSG—brewer's spent grain, bark, MSW—municipal solid waste.

Biomass	Product	450	515	585	650	715	785	850
					(°C)			
FSW	H_2	0.37	0.81	0.89	1.11	0.42	0.23	0.10
	N_2	7.34	2.60	2.83	4.14	4.46	3.54	5.51
	CO	35.82	39.9	39.46	35.29	38.34	43.49	42.95
	CO_2	43.08	27.91	27.54	25.51	31.95	26.84	17.42
	CH_4	6.33	14.69	16.16	19.06	14.62	15.06	16.52
	C_xH_y	6.12	13.03	12.33	14.29	9.55	10.52	16.70
A.pomace	H_2	0.52	1.03	1.77	3.66	2.87	1.11	0.87
	N_2	9.34	5.44	4.11	4.34	3.53	4.62	4.51
	CO	28.22	22.49	23.20	23.88	32.27	35.63	39.85
	CO_2	44.78	42.23	40.44	35.37	32.00	26.66	20.34
	CH_4	8.87	14.66	15.11	15.22	15.33	16.66	17.01
	C_xH_y	7.16	13.23	14.56	16.98	13.44	14.89	17.21
B.pulp	H_2	1.01	1.46	2.45	4.66	2.11	2.01	1.13
	N_2	7.44	4.23	4.32	4.34	4.67	4.87	4.55
	CO	19.59	13.44	12.57	13.95	24.36	24.86	22.39
	CO_2	50.66	48.78	45.67	39.76	35.44	30.35	29.88
	CH_4	10.6	15.38	17.66	18.67	20.11	21.46	22.65
	C_xH_y	9.47	15.60	16.34	17.87	12.62	15.88	18.96
BSG	H_2	0.66	1.89	2.22	3.13	3.45	2.04	0.84
	N_2	7.74	4.21	4.41	4.35	4.76	3.45	3.02
	CO	28.09	29.33	30.55	31.95	34.57	37.59	43.33
	CO_2	49.78	42.05	36.76	30.64	29.94	25.98	23.55
	CH_4	6.25	10.07	11.01	12.15	12.77	13.53	13.88
	C_xH_y	6.88	10.76	13.84	15.77	12.89	15.58	14.37
Bark	H_2	0.56	0.72	1.75	2.14	2.35	1.82	0.84
	N_2	7.32	3.71	4.05	3.15	3.40	3.86	3.23
	CO	26.21	35.11	29.54	27.85	28.72	32.75	38.37
	CO_2	43.21	31.65	28.77	25.00	25.94	23.06	20.29
	CH_4	11.08	14.28	17.84	19.8	20.99	19.58	18.26
	C_xH_y	10.02	13.92	16.59	19.66	17.78	18.06	17.92
MSW	H_2	0.64	1.27	1.89	2.53	1.06	0.62	0.21
	N_2	8.97	3.51	3.41	3.27	3.76	4.02	4.5
	CO	29.25	33.34	31.75	30.89	33.84	37.59	40.59
	CO_2	46.32	34.23	34.12	31.26	31.95	25.98	21.30
	CH_4	7.25	13.25	14.78	15.85	15.23	16.29	16.87
	C_xH_y	7.14	14.08	13.58	15.04	12.34	14.28	15.65

The yields of N_2 and CH_4 were much lower than those for CO and CO_2. Production of N_2 was the highest in the initial part of the process of carbonization. At temperatures from 515 °C, only slight fluctuations were observed. There was also a small amount of methane, the concentration of which practically did not change at temperatures above 515 °C. Methane may be formed by methanation (the reaction of carbon with hydrogen oxide to obtain methane and water) at higher temperatures [20].

Based on the chemical compositions of the non-condensing gases, the values for combustion heat and net calorific value were calculated (Table 4). The LHV of the biomass increased with rising pyrolysis temperature. The most intense increase of LHV was observed in the first phase of the process, when the temperature reached 515 °C. Bark and BSG were exceptions, in that a steady increase in LHV was observed up to 650 °C. Further temperature changes up to 715 °C did not have a major effect on LHV, which remained stable with some fluctuations. This was due to the pyrolysis reaction that occurred at higher temperatures. Following this stage of relative stabilization, LHV increased slightly again. These changes were the least pronounced in the case of bark and BGS. In the process of biomass thermal treatment, the major energy loss due to the release of volatile products took place in the torrefaction phase (up to 450 °C). The results for LHV show that non-condensing volatile products can be a valuable source of energy.

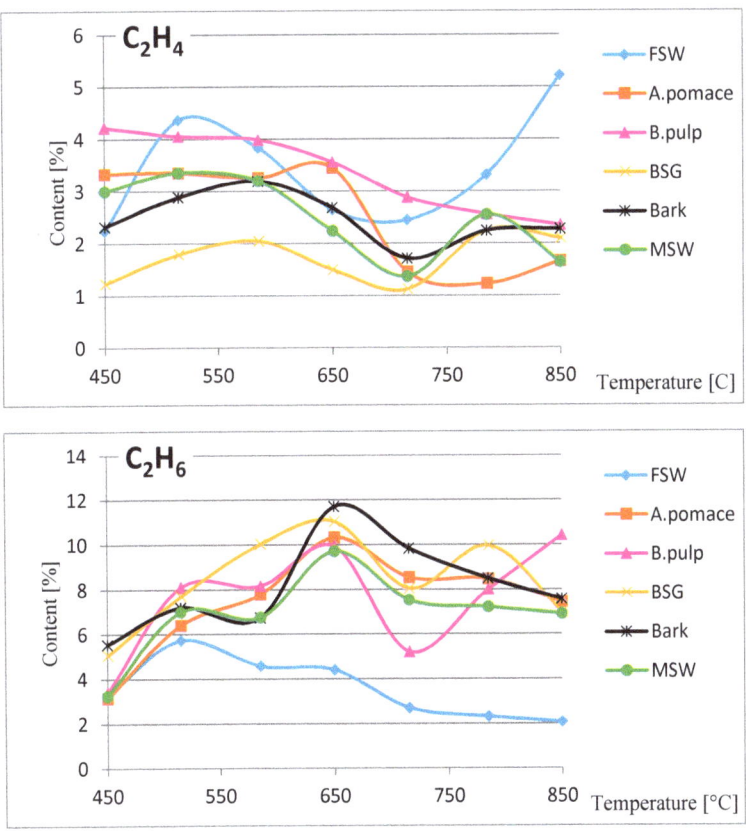

Figure 3. *Cont.*

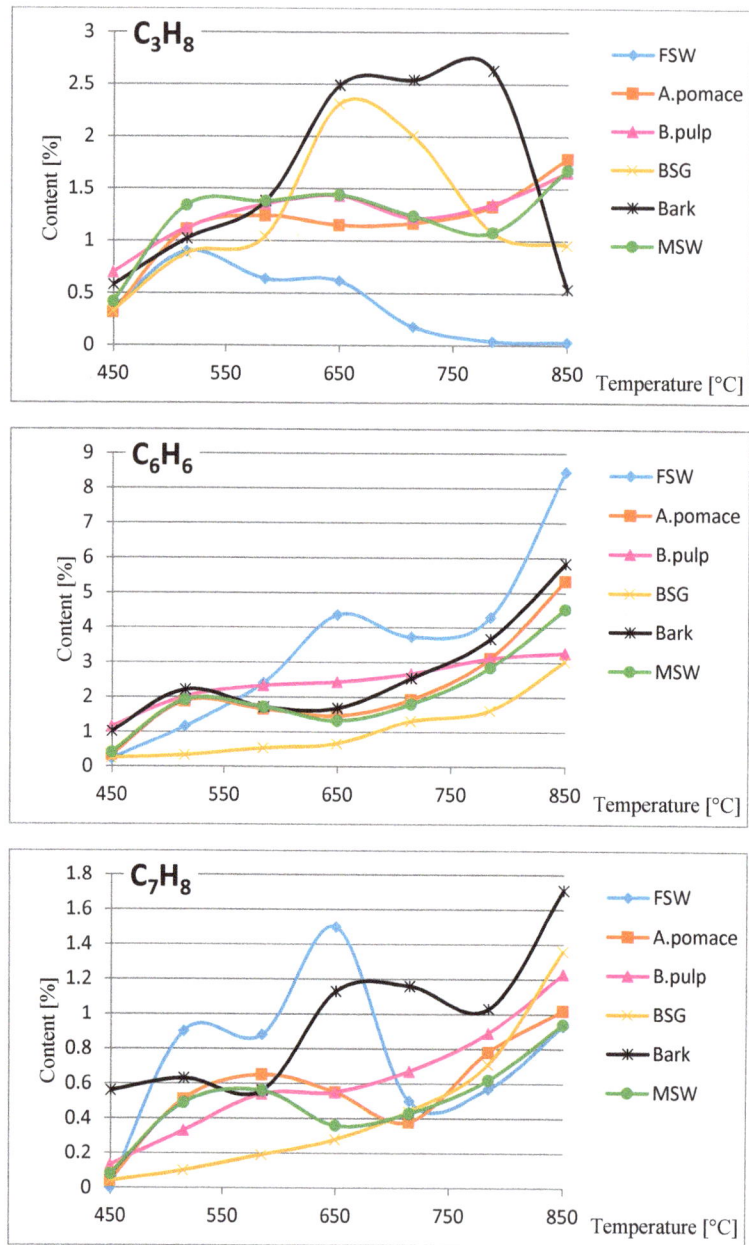

Figure 3. Volatile fractions produced during biomass pyrolysis as a function of temperature. FSW—flavored spirits production waste (lime, grapefruit and lemon), B.pulp—beetroot pulp, A.pomace—apple pomace, BSG—brewer's spent grain, bark, MSW—municipal solid waste.

Table 4. Net calorific value (Lower Heating Value) (LHV) of non-condensable gases from pyrolysis of the analyzed biomasses, expressed as average values. C_xH_y—hydrocarbons, FSW—flavored spirits production waste (lime, grapefruit and lemon), B.pulp—beetroot pulp, A.pomace—apple pomace, BSG—brewer's spent grain, bark, MSW—municipal solid waste.

Temperature	Biomass					
	FSW	A.pomace	B.pulp	BSG	Bark	MSW
°C	MJ/Nm^3	MJ/Nm^3	MJ/Nm^3	MJ/Nm^3	MJ/Nm^3	MJ/Nm^3
450	10.905	11.589	13.388	10.512	15.135	11.284
515	20.552	18.746	19.302	14.892	20.983	20.253
585	21.471	19.942	21.114	17.628	21.088	20.286
650	24.939	21.475	22.993	20.115	26.394	21.112
715	19.507	20.565	21.481	19.486	26.486	19.806
785	21.333	23.565	24.708	21.909	27.338	22.745
850	29.170	27.729	27.259	23.568	29.158	25.970

3.2.4. Composition Analysis of Condensable Gases—Liquid Products

The color and consistency of the condensates varied depending on the biomass. Images of the fractions are shown in the Supplementary Data (Figure S1). The condensates collected after A.pomace carbonization were distinctive. Their color was lighter and they did not look oily. Liquid phases collected from FSW and MSW were clearly darker than the others. They were oily and with a thicker consistency. As shown in Figure 2, the yield of condensates varied greatly, from 9.7% to 25.2% for MSW and BSG, respectively. The composition of the condensate fractions from the thermal decomposition of biomass appeared to be very complex. Gas chromatography (GC) analysis revealed the presence of over 250 organic compounds (Figure S2, Supplementary Data). The main compounds were cresols, phenols, aromatic hydrocarbons (toluene, benzene, xylene), nitrous aromatic hydrocarbons, aliphatic ketones, furan derivatives and aromatic polycyclic hydrocarbons. The cresol and phenol contents were similar, ranging from 5.2% to 9.2% and from 5.3% to 8.7%, respectively. The total content of benzene, toluene and xylenes ranged from 2.4% to 7.1%. None of the organic compounds from the remaining groups occurred in quantities greater than 1%. Polycyclic aromatic hydrocarbons were probably the product of the condensation of aromatic compounds, forming polycyclic structures [21]. The chemical pathways for cellulose pyrolysis and decomposition have been studied extensively [19,22,23]. According to the systematic review presented by Shen and Gu [19], furan and its furan derivatives are formed as a result of direct ring-opening and rearrangement reactions in cellulose molecules. As our results show, the condensates were composed of different organic compounds that can be recirculated and used as additives for other fuels, producing a positive impact on the environment. The elemental chemical composition values for bio-oils have been extensively reported in the literature [24]. The typical ranges for the products of fast pyrolysis are 50–60% for carbon, 6–9% for hydrogen, 30–40% for oxygen, <0.5% for nitrogen and <0.05% for sulfur [25].

4. Conclusions

Carbonization of agricultural waste biomass can be used to generate products with the composition and properties of alternative fuels. In this study, the thermochemical decomposition of waste biomaterials during rapid pyrolysis resulted in homogeneous biochar with yields of up to nearly 41% of the original mass. The biochars had much higher combustion heat and calorific values compared to the biomasses from which they were made. This was due to the fact they no longer contained water, which is physically and chemically bound to these biomasses and alternates in its chemical composition into products with higher carbon content. The biomass transformation process applied in this study enabled the moisture content to be lowered to such an extent that the addition of products no longer had a negative impact on the stability of the combustion process or the total combustion. As a consequence, the efficiency of the entire process was not affected. Both the non-condensable and condensate gaseous products were composed of various organic compounds with sufficiently high combustion heat and net calorific values that they could be recycled and used as additives in other fuels.

Supplementary Materials: The following are available online at http://www.mdpi.com/1996-1944/13/21/4971/s1, Figure S1: Photograph of the condensable gases—liquid products collected after pyrolysis of agricultural waste biomass. FSW—flavored spirits production waste (lime, grapefruit and lemon); B.pulp—beetroot pulp; A.pomace—Applepomace; BSG—brewer's spent grain; bark; MSW—municipal solid waste., Figure S2: GC–MS separation of organic compounds extracted to chloroform from the condensates obtained by carbonization of biomass. FSW—flavored spirits production waste (lime, grapefruit and lemon); B.pulp—beetroot pulp; A.pomace—Apple pomace; BSG—brewer's spent grain; bark; MSW—municipal solid waste.

Author Contributions: Conceptualization W.K. and P.D.; methodology, P.D. and M.A.B.; formal analysis M.A.B. and P.D.; investigation, M.A.B. and P.D.; resources, M.A.B.; data curation M.A.B.; writing—original draft preparation, M.A.B.; writing—review and editing, M.A.B. and P.D.; project administration W.K. and P.D.; funding acquisition, W.K. and P.D. All authors have read and agreed to the published version of the manuscript.

Funding: This study was financially supported by a grant from NCBiR (Polish National Centre for Research and Development) implemented within the project POIR.01.01.01-00-0374/17; Smart Growth Operational Programme -POIR in 2017; Priority axis I: Increased R&D activity of enterprises (Badania przemysłowe i prace rozwojowe realizowane przez przedsiębiorstwa; Konkurs nr 2/1.1.1/2017). Titled: Innowacyjny system zwiększenia wykorzystania potencjału energetycznego paliwa oparty o układ skojarzonej gospodarki cieplnej i elektrycznej. zagospodarowujący energię odpadową niezbędną do zasilenia instalacji odbiorczych o różnych stanach energetycznych.

Conflicts of Interest: The authors declare no conflict of interest.

References

1. Goal 7: Sustainable Development Knowledge Platform. Available online: https://sustainabledevelopment.un.org/sdg7 (accessed on 9 December 2019).
2. Tripathi, M.; Sahu, J.N.; Ganesan, P. Effect of process parameters on production of biochar from biomass waste through pyrolysis: A review. *Renew. Sustain. Energy Rev.* **2016**, *55*, 467–481. [CrossRef]
3. Elkhalifa, S.; Al-Ansari, T.; Mackey, H.R.; McKay, G. Food waste to biochars through pyrolysis: A review. *Resour. Conserv. Recycl.* **2019**, *144*, 310–320. [CrossRef]
4. Lewandowski, W.M.; Ryms, M.; Kosakowski, W. Thermal Biomass Conversion: A Review. *Processes* **2020**, *8*, 516. [CrossRef]
5. Wilk, M.; Magdziarz, A. Hydrothermal carbonization, torrefaction and slow pyrolysis of Miscanthus giganteus. *Energy* **2017**, *140*, 1292–1304. [CrossRef]
6. Matali, S.; Rahman, N.A.; Idris, S.S.; Yaacob, N.; Alias, A.B. Lignocellulosic Biomass Solid Fuel Properties Enhancement via Torrefaction. *Procedia Eng.* **2016**, *148*, 671–678. [CrossRef]
7. Proskurina, S.; Heinimö, J.; Schipfer, F.; Vakkilainen, E. Biomass for industrial applications: The role of torrefaction. *Renew. Energy* **2017**, *111*, 265–274. [CrossRef]
8. Ciolkosz, D.; Wallace, R. A review of torrefaction for bioenergy feedstock production. *Biofuels Bioprod. Biorefin.* **2011**, *5*, 317–329. [CrossRef]
9. Available online: https://www.parrinst.com/wp-content/uploads/downloads/2013/07/483M_Parr_Intro-to-Bomb-Calorimetry.pdf (accessed on 29 October 2020).
10. Gowman, A.C.; Picard, M.C.; Rodriguez-Uribe, A.; Misra, M.; Khalil, H.; Thimmanagari, M.; Mohanty, A.K. Physicochemical analysis of Apple and Grape Pomaces. *BioResources* **2019**, *14*, 3210–3230.
11. Cantero-Tubilla, B.; Cantero, D.A.; Martinez, C.M.; Tester, J.W.; Walker, L.P.; Posmanik, R. Characterization of the solid products from hydrothermal liquefaction of waste feedstocks from food and agricultural industries. *J. Supercrit. Fluids* **2018**, *133*, 665–673. [CrossRef]
12. Wilk, B. Określenie zależności wartości opałowej od wybranych właściwości fizykochemicznych biomasy (Determination of dependence of the calorific value on selected physicochemical properties of biomass). In *Mat. Seminar. Techniki Analityczne i Procedury Badawcze w Zastosowaniu do Nowych Uwarunkowań Prawnych w Energetyce (Analytical Techniques and Research Procedures for Application to New Legislative Conditions in the Energy Sector)*; IChPW: Zabrze, Poland, 2006.
13. Tumuluru, J.S.; Hess, J.R.; Boardman, R.D.; Wright, C.T.; Westover, T.L. Formulation, pretreatment, and densification options to improve biomass specifications for Co-firing high percentages with coal. *Ind. Biotechnol.* **2012**, *8*, 113–132. [CrossRef]

14. Jagustyn, B.; Bątorek-Giesa, N.; Wilk, B. Ocena właściwości biomasy wykorzystywanej do celów energetycznych. *Chemik* **2011**, *65*, 557–563.
15. Nobre, C.; Alves, O.; Longo, A.; Vilarinho, C.; Gonçalves, M. Torrefaction and carbonization of refuse derived fuel: Char characterization and evaluation of gaseous and liquid emissions. *Bioresour. Technol.* **2019**, *285*, 121325. [CrossRef]
16. Białowiec, A.; Pulka, J.; Stępień, P.; Manczarski, P.; Gołaszewski, J. The RDF/SRF torrefaction: An effect of temperature on characterization of the product—Carbonized Refuse Derived Fuel. *Waste Manag.* **2017**, *70*, 91–100. [CrossRef]
17. Shafizadeh, F.; Lai, Y.Z. Thermal Degradation of l,6-Anhydro-β-D-glucopyranose. *J. Org. Chem.* **1972**, *37*, 278–284. [CrossRef]
18. Li, S.; Lyons-Hart, J.; Banyasz, J.; Shafer, K. Real-time evolved gas analysis by FTIR method: An experimental study of cellulose pyrolysis. *Fuel* **2001**, *80*, 1809–1817. [CrossRef]
19. Shen, D.K.; Gu, S. The mechanism for thermal decomposition of cellulose and its main products. *Bioresour. Technol.* **2009**, *100*, 6496–6504. [CrossRef]
20. Wahid, F.R.A.A.; Harun, N.H.H.M.; Rashid, S.R.M.; Samad, N.A.F.A.; Saleh, S. Physicochemical property changes and volatile analysis for torrefactionn of oil palm frond. *Chem. Eng. Trans.* **2017**, *56*, 199–204.
21. Richter, H.; Howard, J.B. Formation of polycyclic aromatic hydrocarbons and their growth to soot-a review of chemical reaction pathways. *Prog. Energy Combust. Sci.* **2000**, *26*, 565–608. [CrossRef]
22. Piskorz, J.; Radlein, D.; Scott, D.S. On the mechanism of the rapid pyrolysis of cellulose. *J. Anal. Appl. Pyrolysis* **1986**, *9*, 121–137. [CrossRef]
23. Kawamoto, H.; Morisaki, H.; Saka, S. Secondary decomposition of levoglucosan in pyrolytic production from cellulosic biomass. *J. Anal. Appl. Pyrolysis* **2009**, *85*, 247–251. [CrossRef]
24. Oyebanji, J.A.; Okekunle, P.O.; Lasode, O.A.; Oyedepo, S.O. Chemical composition of bio-oils produced by fast pyrolysis of two energy biomass. *Biofuels* **2018**, *9*, 479–487. [CrossRef]
25. Broumand, M.; Albert-Green, S.; Yun, S.; Hong, Z.; Thomson, M.J. Spray combustion of fast pyrolysis bio-oils: Applications, challenges, and potential solutions. *Prog. Energy Combust. Sci.* **2020**, *79*, 100834. [CrossRef]

Publisher's Note: MDPI stays neutral with regard to jurisdictional claims in published maps and institutional affiliations.

© 2020 by the authors. Licensee MDPI, Basel, Switzerland. This article is an open access article distributed under the terms and conditions of the Creative Commons Attribution (CC BY) license (http://creativecommons.org/licenses/by/4.0/).

Review

Extraction of Value-Added Minerals from Various Agricultural, Industrial and Domestic Wastes

Virendra Kumar Yadav [1], Krishna Kumar Yadav [2], Vineet Tirth [3,4], Govindhan Gnanamoorthy [5], Nitin Gupta [6], Ali Algahtani [3,], Saiful Islam [7], Nisha Choudhary [6], Shreya Modi [8] and Byong-Hun Jeon [9,*]

[1] Department of Microbiology, School of Sciences, P P Savani University, Kosamba, Surat 394125, Gujarat, India; yadava94@gmail.com
[2] Faculty of Science and Technology, Madhyanchal Professional University, Ratibad, Bhopal 462044, India; envirokrishna@gmail.com
[3] Mechanical Engineering Department, College of Engineering, King Khalid University, Abha 61411, Asir, Saudi Arabia; vtirth@kku.edu.sa (V.T.); alialgahtani@kku.edu.sa (A.A.)
[4] Research Center for Advanced Materials Science (RCAMS), King Khalid University, Guraiger, Abha 61413, Asir, Saudi Arabia
[5] Department of Inorganic Chemistry, University of Madras, Chennai 660025, Tamil Nadu, India; gnanadrdo@gmail.com
[6] School of Nanosciences, Central University of Gujarat, Gandhinagar 382030, Gujarat, India; nitinkgupta1988@gmail.com (N.G.); nishanaseer03@gmail.com (N.C.)
[7] Civil Engineering Department, College of Engineering, King Khalid University, Abha 61413, Asir, Saudi Arabia; sfakrul@kku.edu.sa
[8] Department of microbiology, Shri Sarvajanik Science College, Mehsana 384001, Gujarat, India; shreyamodi20@gmail.com
[9] Department of Earth Resources and Environmental Engineering, Hanyang University, Seoul 04763, Korea
* Correspondence: bhjeon@hanyang.ac.kr

Abstract: Environmental pollution is one of the major concerns throughout the world. The rise of industrialization has increased the generation of waste materials, causing environmental degradation and threat to the health of living beings. To overcome this problem and effectively handle waste materials, proper management skills are required. Waste as a whole is not only waste, but it also holds various valuable materials that can be used again. Such useful materials or elements need to be segregated and recovered using sustainable recovery methods. Agricultural waste, industrial waste, and household waste have the potential to generate different value-added products. More specifically, the industrial waste like fly ash, gypsum waste, and red mud can be used for the recovery of alumina, silica, and zeolites. While agricultural waste like rice husks, sugarcane bagasse, and coconut shells can be used for recovery of silica, calcium, and carbon materials. In addition, domestic waste like incense stick ash and eggshell waste that is rich in calcium can be used for the recovery of calcium-related products. In agricultural, industrial, and domestic sectors, several raw materials are used; therefore, it is of high economic interest to recover valuable minerals and to process them and convert them into merchandisable products. This will not only decrease environmental pollution, it will also provide an environmentally friendly and cost-effective approach for materials synthesis. These value-added materials can be used for medicine, cosmetics, electronics, catalysis, and environmental cleanup.

Keywords: waste; agricultural waste; value-added materials; calcium oxide; eggshell; incense sticks

1. Introduction

Every day we come across various types of waste in our life either in houses, working areas, industries, farms, etc. These waste materials can be kitchen waste, agricultural waste, industrial waste, or poultry waste. Although there are also several other types of waste, they are outside the scope of the present study. The waste produced in our houses, such as kitchen waste and incense stick ash can be categorized as household waste [1], whereas the waste produced from the agricultural practices, such as wheat straw [2,3], wheat husks [4], rice straw [5], rice husks [6], coconut shells [7], palms and dates [8], lemon peel [9], almond shells [10], etc., can be considered to be agricultural waste [11]. The waste generated by industry, such as coal fly ash (CFA) [12,13], red mud [14,15], gypsum waste [16,17], sewage sludge [18], iron tailing [19], etc., fall under the category of industrial waste [20]. Eggshells [21–23] and poultry litter [24,25] can be categorized as poultry waste; however, most investigators have considered eggshells to also belong to industrial waste. Among the above-mentioned types of waste, agricultural and industrial wastes raise major concerns for the environment, as millions of tons are produced every year around the world. Most of these waste materials are still disposed of via landfill [26,27], dumping, or disposal in water, creating pollution of the environment. The dumping of such waste may further deteriorate fertile agricultural soil [28], produce a foul odor [29] that may attract pests and mosquitos, and lead to health issues for human beings and animals [30]. Some types of industrial waste, like CFA [31] and red mud [32], are considered hazardous because of their high concentration of toxic heavy metals [33]. Heavy metals from dumping areas can leach into the surrounding soil, and when it rains, they may percolate into water bodies, thus leading to water pollution [34].

Nowadays, with the advent of new technologies, such waste can be processed into value-added minerals, especially in metallurgy. They can also be applied in the fields of agriculture [35], adsorbents [36,37], geo-polymers [38,39], ceramics [40], and environmental cleanup [41]. This article reviews and discusses a variety of waste materials in detail, including silica, alumina, calcium oxides, carbonates, etc., along with their properties, applications, and methods for their recovery.

2. Industrial Waste

Waste can be classified into several classes depending on its origin, for instance, industrial waste (CFA, paper and pulp waste), agricultural waste (coconut coir, sugarcane bagasse, and lemon peel) [42–49]. Similarly, waste materials produced in the home are referred to as domestic waste, including kitchen waste and incense stick ash, while eggshells and waste from poultry are regarded as poultry waste. The major categories of waste are shown below in Figure 1.

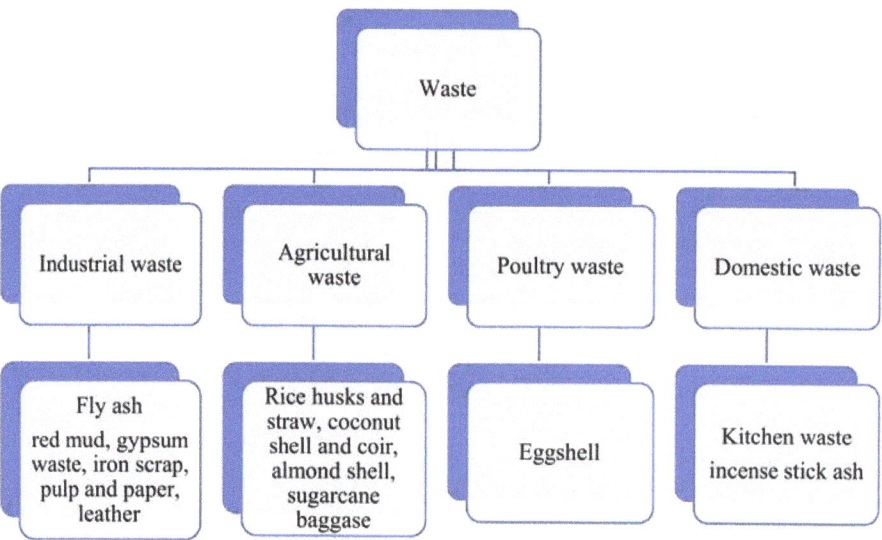

Figure 1. Classifications of the different valuable waste materials.

Industrial waste refers to byproducts, e.g., CFA, red mud, gypsum waste, and other municipal waste, generated in the industrial sector. Among the various types of waste, this is one of the major sources that requires special attention due to its hazardous nature.

2.1. Fly Ash

Coal fly ash consists of fine powders produced from the pulverized coal used in the thermal power plants during the generation of electricity. It is heterogeneous in nature, and is made up of a glassy amorphous phase and a crystalline quartz phase, such as mullite and magnetite. Fly ash is produced in large amounts in TPPs, and is mainly considered waste, although its mineral value makes it a useable material.

2.1.1. Recovery of Minerals from CFA

CFA can be applied in several ways in the fields of agriculture, remediation (as an economical adsorbent), ceramics, etc. [50]. Formerly, CFA was considered a hazardous waste [51], but today, it is considered to be a useable material. CFA has high amounts of micro- and macro-nutrients that can be used as a source of nutrients for plants [52,53]. It is used in agriculture (as a fertilizer) [54], forestation [55], reclamation of wasteland [56], and for maintaining the pH of acidic soil [56]. CFA is also used as an adsorbent for the removal of pollutants—mainly dyes [57,58], pesticides [57,58], and heavy metals [59,60]—from wastewater. It is used for making nanocomposites [61] that are applicable in the defense industry [62,63] and the production of lightweight materials [64,65]. CFA can be used for making blended cement [66] tiles, bricks [67], blocks [56], RCC, kitchen panels, and geo-polymers. CFA can also be used as fillers in rubbers and tires [68], and can also be used in the mining industry [56,69] for the recovery of ferrous metals [70], cenospheres [55], mullite [71], silica [72], zeolites [73], and alumina. It can also be applied for the recovery of unburnt carbon, soot, and carbon nanomaterials like carbon nanotubes, fullerenes, and graphene. These carbonaceous materials are formed due to the burning of the organic content of the coal, while soot and unburnt carbon are due to the incomplete burning of coal. Higher grades of coal like anthracite and bituminous coal have higher carbon contents, so after the burning of such coals in thermal power plants, the resulting ash will also have a high content of carbon in the form of unburnt carbon, soot, and carbon nanomaterials like carbon nanotubes, fullerene, etc.

Silica

Silica accounts for up to 40–60% of CFA [74], depending on the source of the CFA, the type of coal, the operating parameters of the thermal power plant, the furnace temperature, etc. Silica is present in CFA in either crystalline or amorphous form [75]. The crystalline form of silica is present mainly as quartz, sillimanite, and mullite [76], whereas the glassy amorphous form is the only form of amorphous silica. Crystalline silica is mainly inert, and does not easily react with acids or bases. Therefore, silica can only be extracted from the amorphous form, as it reacts easily with strong bases. In addition, silica can easily be extracted from CFA through treatment with strong bases like NaOH, KOH, and sodium bicarbonate [77]. Silica is mainly extracted using either the alkali treatment [78] method or the alkali fusion method [79]. The complete extraction of silica using the alkali treatment method is depicted in Figure 2, while in alkali fusion method, the silica source is calcinated with NaOH at high temperature in order to form a new mineral (such as nephaline), from which it is easier and more efficient to extract the silica. Yadav et al. reported the extraction of nanosilica from fly ash tiles using the alkali fusion method [80]. Silica can be used in the production of glass and ceramics [80], in drug delivery [81], and used in medicine, foundries, adsorbents, catalytic processes [82], and molecular sieves [83]. Silica is nontoxic and mesoporous, meaning it can be readily used in various industries. The silica extracted from the CFA is nanosized (20–60 nm), but aggregates together to form lumps.

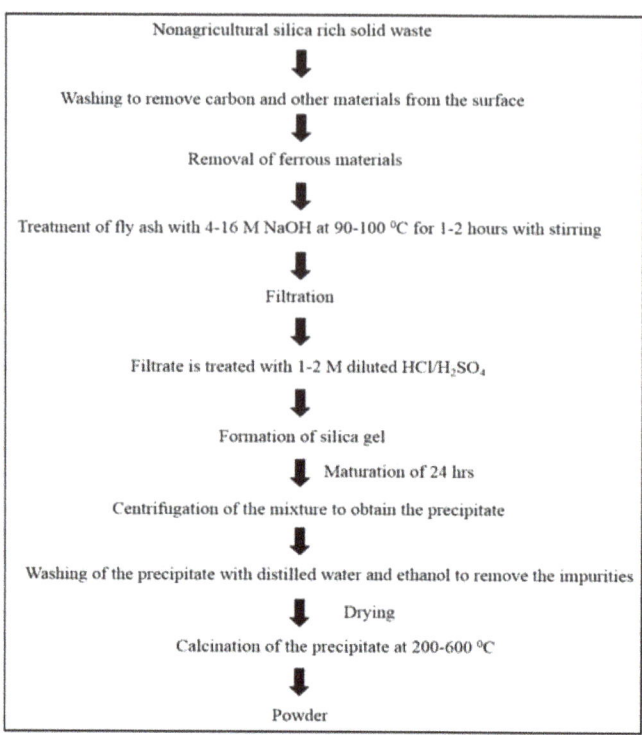

Figure 2. Schematic diagram of the synthesis of silica from silica-rich waste materials.

Alumina

Alumina, which is amphoteric in its nature, can be obtained from CFA or other alumina-rich materials using both acids and bases. Depending on the CFA type and the source of the coal, the alumina content in CFA can range between 20 and 40% [84]. In CFA, alumina can mostly be found in the form of crystalline aluminosilicates [85] such as mullite. These aluminates are highly inert, and rarely react with acids; thus, only low yields can be obtained. They can be treated with strong mineral acids that can be directly converted to powder by thermal decomposition. They can also be extracted using alkali like NaOH at high temperatures, i.e., 125–1100 °C. Alumina forms in numerous meta-stable phases, including gamma [γ]-, epsilon [η]-, delta [d]-, theta [θ]-, kappa [κ]- and χ-alumina. While the most stable phase of alumina is [a] alpha. Of all of the phases, the gamma-phase of alumina is the most important and widely used nanosized material. Gamma alumina is widely used in the petroleum and automobile industries as a catalyst and catalyst substrate, to produce ceramics [86] and glass, adsorbents, spacecraft materials, microelectronics, thermal-resistant materials, and as a coating material for thermal wear and abrasives [87], optoelectronics, and metallurgy. Alumina nanoparticles have high compression strength, high chemical resistance, a high degree of refractoriness, high thermal shock resistance, high abrasion strength, and high dielectric strength. Alumina is also applied in ceramics, adsorbents, fire-retardant materials, acid, and alkali-resistant materials [88].

Cenospheres

Cenospheres are aluminosilicate spheres [89,90], the size of which varies by microns. They are lightweight nanostructured materials that are created when coal is combusted within a furnace. These are used as lightweight materials for aircraft, in the field of defense, in fire-proof materials [55], and for acid and alkali wear-resistant materials [91]. Today, cenospheres that are tiny hollow spheres with diameter of approximately 10–1000 μm are among the most desirable byproducts that can be obtained from coal combustion processes. They normally comprise 1–2% of CFA obtained during the process of coal combustion. Cenospheres possess properties such as very high mechanical strength and low density; thus, they are regarded as a highly significant issue in coal-fired power plants. A number of parameters affect cenosphere properties, including the grinding operations, the nature of the coal used, the combustion parameters, and withdrawal when generating electricity. These materials are mainly possess a glassy surface and a crystalline matrix such as mullite with a nano-film covering with a thickness of 30–50 nm. Their form is similar to that of a shell, with a thickness varying between 2 and 30 microns. They are extracted using the density-based centrifugation method, during which finer lightweight cenospheres float at the top of the slurry, while the particles of heavier ferrous metals settle at the bottom [55]. The cenospheres are collected from the top and dried before use. The formation mechanism of cenospheres during pulverized coal combustion is complex and is highly dependent on fuel properties and combustion parameters. The formation mechanism of cenospheres is very similar to the procedure of glass blowing [92]. Therefore, it would be beneficial to take a closer look into glass formation principles.

The literature presents two approaches for the extraction of cenospheres from fly ash, namely dry separation and wet separation. The conventional cenosphere extraction method is primarily performed using wet processes, namely simple sedimentation and flotation [93]. There are two methods for estimating the degree of separation when recovering cenospheres from coal CFA, namely the float method and the sink method. To separate cenospheres from CFA, various liquids, viz., water (1 g/cc) and acetone (0.789 g/cc), can be used. Fly ash is kept in a vessel and water is added to it. The complete mass is stirred for four hours; afterwards, it is allowed to settle for ten hours. Then, all cenospheres with densities lower than 1 g/cc will float up and can be separated [94].

Mullite

Mullite, consisting of micron-sized particles (1–1.5 microns in length and 0.3–0.5 microns in width), is a rarely observed crystalline mineral that contains aluminum silicate ($3Al_2O_3 \cdot 2SiO_2$) [95] and is mainly made up of Al, Si, and O; however, its composition can be quite variable. In the process of combusting aluminosilicate raw materials, mullite is created. This material is a key component in porcelains, ceramic whiteware, high temperature insulation, refractory materials and traditional ceramics [96]. Mullite is a compositional orthorhombic aluminosilicate, and generally possesses the composition $Al_2(Al_{2+2x}Si_{2-2x})O_{10-x}$. Mullites are non-stoichiometric compounds whose structure is similar to magnetite containing impurities. It is rarely formed in nature, because of its formation conditions, which require high temperature in combination with low pressure. Synthesis of mullites is possible from high-silico-aluminous CFA only, as they are rich in silica and alumina, with lower contents of iron oxides. Iron oxides have a negative effect on mullite. Generally, for the purpose of synthesizing mullites from CFA, cenospheres are the most effective materials, since they have lower contents of iron oxides.

Mullites are formed in CFA from the organic and inorganic materials present in coal as a result of different melting and firing processes. After the extraction of mullites using a hydrofluoric process, it can be characterized by XRD and SEM-EDS in CFA. A clearer picture of mullite can be obtained using XRD and NMR. Both of these techniques can be used to efficiently investigate the value of x and the oxygen hole rate in the general formula, thus obtaining the mullite composition. Mullites can be used as refractory materials because of their high melting point (1840 °C).

Mullites are aluminosilicate minerals with the general formula $Al_{4+2x}Si_{2+2x}O_{10-x}$ (with the value of x varying between 0.17 and 0.59). Mullites are capable of forming two stoichiometric forms, namely $3Al_2O_3 \cdot 2SiO_2$ and $2Al_2O_3 \cdot SiO_2$. Mullite is known to be the only stable binary phase of the Al_2O_3-SiO_2 system that exists under ambient conditions. From an empirical perspective, its chemical compositions include 71.8 wt%

Al$_2$O$_3$ and 28.2 wt% SiO$_2$, designated as 3/2-mullite (3Al$_2$O$_y$2SiO$_2$). Mullite has two common morphologies: a platelet shape and a needle shape. In the platelet shape, it has a low aspect ratio, whereas in the needle shape, it possesses a high aspect ratio. In addition, mullite has low thermal conductivity, low thermal expansion, high thermal stability, high corrosion stability, high strength, and high fracture toughness. It has excellent creep resistance, acceptable thermal shock and stress resistance, and acceptable strength wear resistance, and it can be used at high temperatures. Mullites can be applied as an effective replacement for platinum in diesel engines, furnace liners, electrical insulators, protection tubes, kiln furniture, rollers, heat exchanger components, heat insulation parts, pressed parts, and isostatically pressed parts [97–100].

Yadav et al., 2021 reported the recovery of needle-shaped mullite, 90–300 nm in size, extracted from CFA using 16 M HF acid. An optimum ratio of CFA and HF was mixed and kept for interaction in an incubator shaker. The CFA was collected from the Gandhinagar and Gujarat thermal power plants. The source of the coal was anthracite/bituminous coal, i.e., higher grades of coal. The detailed mechanism for recovery of mullites from CFA is given below in Figure 3 [94].

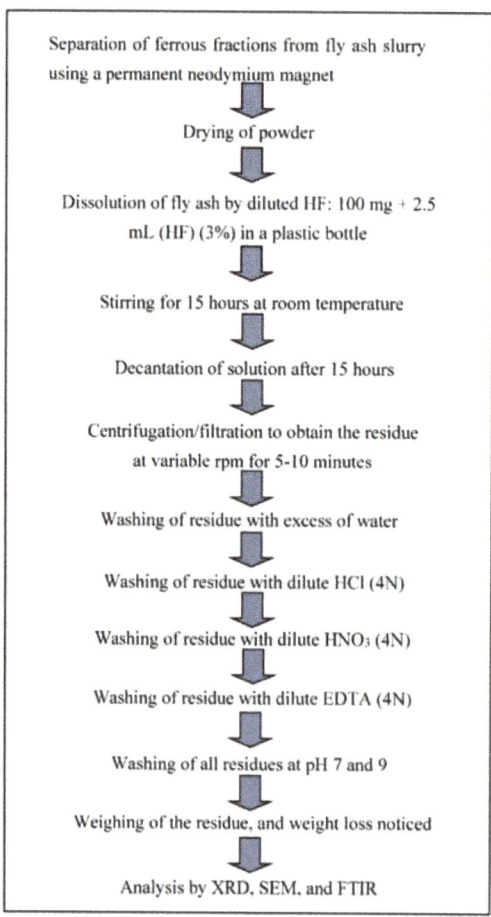

Figure 3. Schematic diagram for the recovery of mullite from CFA adopted from Yadav et al., open access journal Crystals, 2021 [94].

Zeolites Synthesis from CFA

Zeolites are another class of materials that are widely used in industry for petroleum cracking, cation exchange resins, water softening processes, etc. These zeolites are also hydrates of alumina and silica, mainly containing Al, Si, cations from gp II (Ca, Mg, Na) and water molecules. Due to the presence of 40–60% silica and 20–40% alumina, the CFA serves as an economical resource for zeolite material synthesis. To date, zeolites have been synthesized from CFA by using 4–8 M, NaOH treatment along with continuous heating at 90–105 °C, for 1 h to 24 h with rigorous stirring. The pH, temperature, heating and stirring time decides the morphology and class of zeolites to be synthesized. There are several reports in the literature in which CFA from different parts of the globe has been used for the synthesis of zeolites. Yadav et al., 2019 and 2021, reported the synthesis of zeolites from CFA collected from the Gandhinagar and Gujarat TPPs in India. The size of the synthesized zeolites varied between 80 nm and 180 nm in width, while their length varied from 120 nm to 300 nm [99].

2.2. Red Mud

Red mud is a hazardous byproduct of the bauxite industry [101], and is produced at the time of extraction of alumina from bauxite using Bayer's method [102]. It is considered hazardous because of its high alkalinity and the presence of various toxic heavy metals; however, it is also a rich source of titanium, iron, and aluminum (composition is shown in Table 1). In addition, red mud is recognized as a bauxite residue [103], as it is what is left after the extraction of all the extractable alumina from bauxite. The resulting mud is a mixture of the insoluble fraction of solid and metallic oxides, and ore, which remains after the extraction of the aluminum-containing components [104]. It is typically disposed of as a slurry with a solid concentration ranging between 10 and 30%, pH in the range of 13, and high ionic strength [14]. This disposal problem is compounded by the fact that typical bauxite processing produces up to three times as much toxic red mud as aluminum. Figure 4 shows a typical SEM micrograph of red mud. The particles of red mud aggregate together to form lumps.

Table 1. Elemental composition of red mud.

Composition	Percentage
Fe_2O_3	30–60%
Al_2O_3	10–20%
SiO_2	3–50%
Na_2O	2–10%
CaO	2–8%
TiO_2	2–5%

Approximately 44 million tons [MTs] of primary aluminum are produced annually around the world [105]; by that count, roughly 132 MTs of red mud enter retention ponds and some dry stack tailing areas annually [106]. The alumina plants in the Indian context have an annual capacity of 1.692 MTs; they produce 0.6 MTs of metal, and approximately 2 MTs of red mud per year [107]. The chemical analysis of RM reveals the presence of silica, aluminum, iron, calcium, and titanium, as well as an array of minor constituents, namely Na, K, Cr, V, Ni, Ba, Cu, Mn, Pb, Zn, etc. [15,108]. There is a high variance in chemical composition among various red muds generated around the world [14].

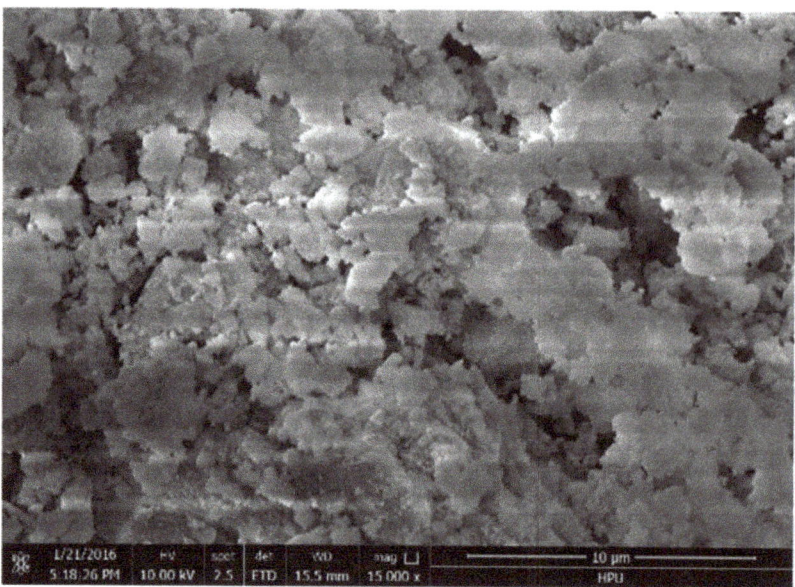

Figure 4. SEM micrograph of red mud, adopted from Zhang et al., open access journal Applied Sciences, 2018 [109].

Currently, in the alumina industry, alumina is obtained by carrying out Bayer's process on bauxite. The byproduct of such processes is red mud, which is generally regarded as dangerous waste, because it contains high concentrations of some heavy metals, iron oxide, and other metal oxides. Industry does not recycle it, and there is no available disposal method for it. Currently, companies collect this waste in their vicinity after the extraction of alumina. This continuous piling up of red mud constitutes a major and global threat to the environment [110,111]. This red mud contains high amounts of iron oxides and heavy metals, which can leach out into water bodies and cause damage to aquatic fauna. In India, two major alumina-producing companies are NALCO and HINDALCO, who generate a huge amount of waste every year in the form of red mud. This causes water pollution through its entry into water streams, leaching toxic heavy metals into water bodies. This also affects the aquatic flora and fauna. In addition, it has adverse effects on the biological oxygen demand (BOD) and chemical oxygen demand (COD) of the water bodies, resulting in a lowering of these parameters. This waste changes the community structure of water bodies. The alkalinity of water bodies is increased as a result of releasing NaOH into the water. In any alumina refinery, large amounts of land are required for the handling of this waste.

2.2.1. Recovery of Alumina from Red Mud

The red mud-based recovery of alumina involves several steps in series—recovery of ferrous particles, leaching of dried red mud with strong mineral acids, filtration to obtain the filtrate, crystallization of the leachate, the recovery of acids, and finally calcination—in order to obtain the alumina powder. A complete flow chart depicting the extraction of alumina from red mud is given below in Figure 5.

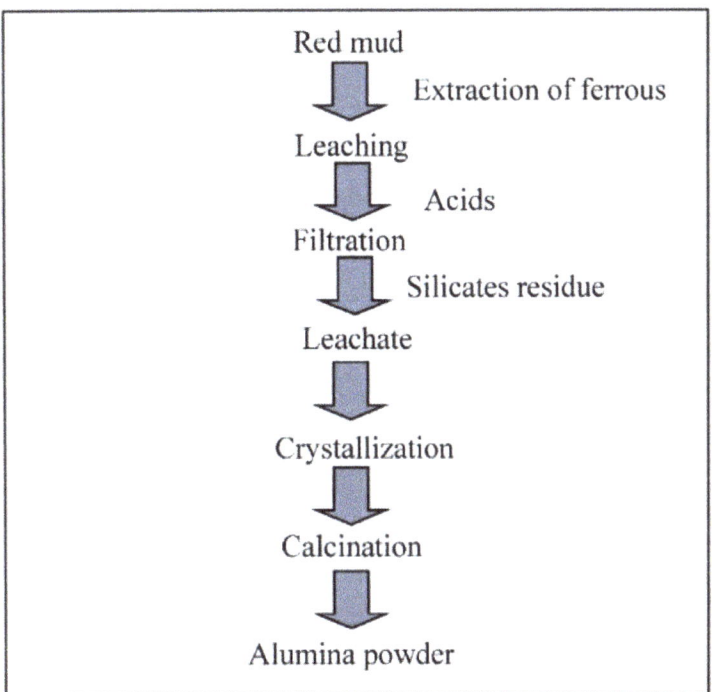

Figure 5. Flow chart for the recovery of alumina from red mud/CFA.

The recovery of alumina and ferric oxides from iron-rich red mud has been reported using the reduction sintering technique [112]. Different experiments have shown that up to 89.71% of alumina can be extracted, with a Fe recovery rate of 60.67%, under optimal conditions [113]. Zhang et al. [114] investigated andradite-grossular hydrogarnet formation in the hydrothermal process with the aim of examining its effect on alumina and alkali recovery from Bayer red mud. For the evaluation of the parameters with the highest impact on the recovery process, they took into consideration the batch experiments and parameters such as caustic ratio (molar ratio of Na_2O to Al_2O_3 in sodium solution), reaction temperature, residence time, and sodium concentration. Zhu et al. [115] reported the recovery of alumina and alkali from red mud using a novel calcification–carbonation method under mild reaction conditions. Batch experiments were performed, and the effects of temperature, pressure, and additive addition on the extraction efficiency of alumina were examined, and the extraction efficiency of alumina was 95.2%. In another study, Meher [116] reported the extraction of alumina from red mud using a calcium carbonate and sodium carbonate sintering process. They studied the impacts of Na_2CO_3 and $CaCO_3$ additives, sintering time and temperature, and leaching time on the effectiveness of alumina extracted from red mud. The alumina extraction was up to 97.64% at a sintering temperature of 1100 °C for 4 h with red mud.

2.2.2. Applications of Red Mud

Many studies have confirmed the benefits of red mud in the process of treating water and removing inorganic anions (e.g., fluoride, phosphate, and nitrate), toxic heavy metals and metalloid ions, as well as organic substances (e.g., phenolic compounds, dyes, and bacteria) [117,118]. Moreover, red mud can be employed as an effective catalyst in processes such as hydrocarbon oxidation, hydrodechlorination, and hydrogenation [119]. The broader areas of application of red mud are given in Figure 6.

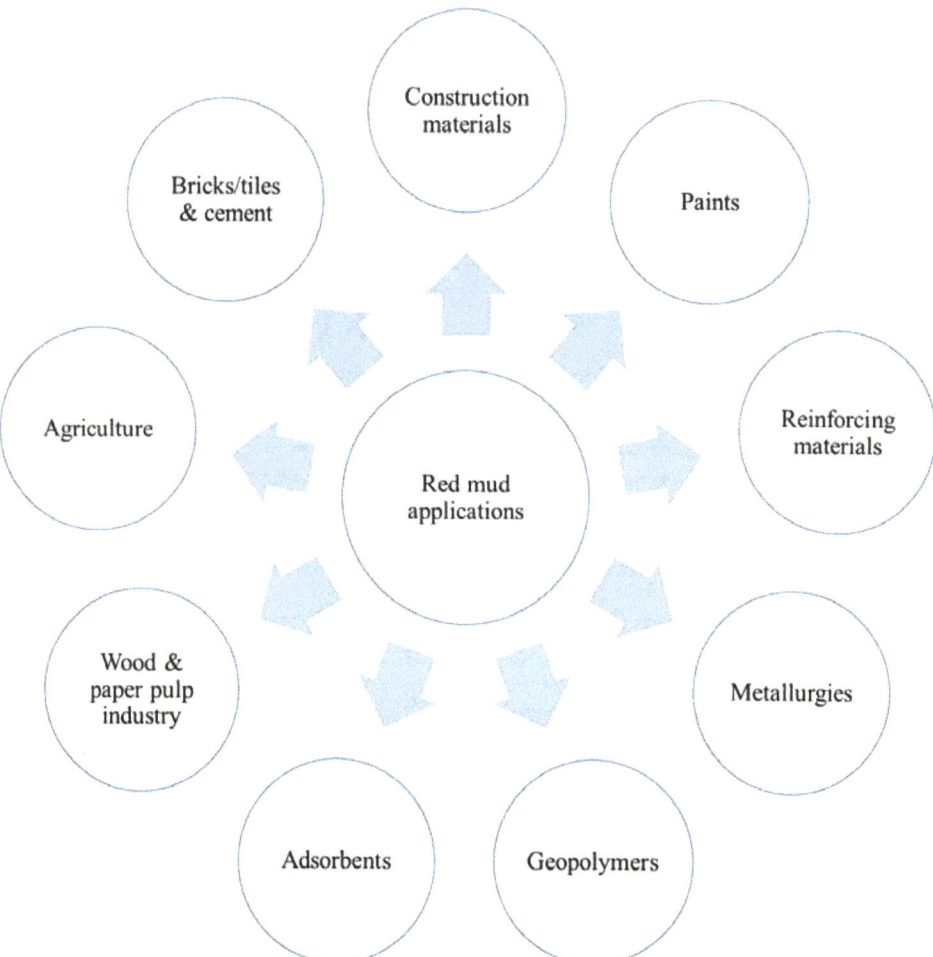

Figure 6. Diagrammatic representation of red mud in their broader applications.

2.3. Iron Slags/Scraps

There are several iron-based industries, like the steel industry, where iron particles are generated as waste materials [120]. These iron materials can be processed for the recovery of highly pure iron, zero-valent iron, or iron oxides. These can be used directly as filings or in coke industries [121]; thus, they have drawn much attention. These iron particles can be recovered using strong magnets in either wet or slurry form. These iron particles can further be treated with acids to obtain iron-rich leachates that can be used as precursor materials for the synthesis of different types of iron oxide particles. The synthesized iron oxide particles can be recovered by precipitation, chemical precipitation, and calcination. These iron oxides are so highly pure that they can act as an adsorbent in processes like the treatment of wastewater, environmental cleanup, in ceramics, and in the steel industry [122,123]. The major advantages of such iron oxide particles include their ability to be easily recovered from the reaction site, their recyclability, and their easy external manipulation using strong magnets [124]. The synthesis of iron oxide particles from waste is cost-effective and environmentally friendly.

Tang et al. [125] attempted a coal-based smelting reduction method for the recovery of Fe, Ni, and Cr from pickling sludge waste. Their findings showed that the Fe recovery was 98.1% under optimized conditions. Tang et al. [126] reported the recovery of iron from iron ore tailings. They used it to develop a concrete composite mixture. They investigated the impacts of different parameters upon the extraction of ferrous particles. Up to 83.86 wt% of iron was recovered from a feed iron grade of 12.61 wt%. Zhang et al. [127] attempted to recover iron from the waste slag of pyrite processing using a reduction roasting magnetic separation method. The iron content of the concentrate was initially 57%, of which 87% was extracted using the aforementioned method. It should also be noted that, through further treatment using chlorinated segregation–magnetic separation, the iron content in the slag was increased to 83%.

In the study carried out by Wang Yu et al. [128], the co-precipitation and magnetic separation methods were adopted with the aim of recovering iron from waste ferrous sulphate. They investigated the impacts of various reaction parameters on the iron recovery, and also examined the impacts of milling time and magnetic induction intensity on the separation of magnetic particles. The mixed magnetic particles were wet-milled for 20 min before magnetic separation. The grade and recovery rate of iron in the magnetic concentrate drastically increased from 51.41% to 62.05%, and from 84.15% to 85.35%, respectively.

2.4. *Gypsum Waste ($CaSO_4 \cdot 2H_2O$)*

Gypsum waste is a byproduct of the gypsum industry. It is widely used in dental applications, with its disposal representing a potential threat to the environment [129]. There is a possibility that when disposing of gypsum waste in landfill, there might be a reaction between the gypsum and biodegradable waste, which may produce poisonous and odorous hydrogen sulfide gas [130,131]. Gypsum waste can be processed for the recovery of CaS, which ultimately changes to calcium carbonate. A schematic diagram of the process of recovering calcium carbonate from gypsum waste is shown below in Figure 7, as reported by Yadav et al., 2021 [129].

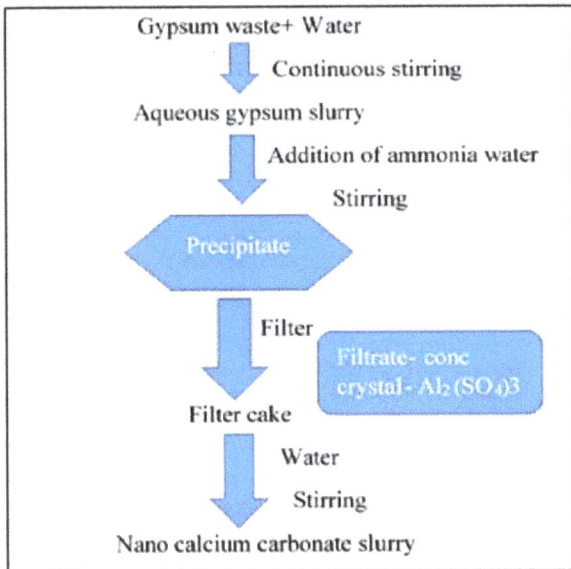

Figure 7. Steps involved in the synthesis of calcium carbonate nanoparticles from gypsum waste, adapted from Yadav et al., open access journal Applied Sciences, 2021 [129].

Beer et al. [132] attempted to synthesize calcium carbonate nanoparticles from gypsum waste, producing elemental sulfur as a byproduct. In this process, the first step was the thermal reduction of the gypsum waste into calcium sulfide (CaS), followed by its direct aqueous carbonation, yielding low-grade carbonate products (i.e., 99 mass% as $CaCO_3$) or precipitated calcium carbonate (PCC). The carbonate product was found to be predominantly composed of calcite (99.5%) with only 0.5% quartz. Calcite was the only $CaCO_3$ polymorph obtained in the experimental process, while no vaterite or aragonite was found [133]. In addition, Mulopo and Radebe [129] investigated the batch recovery of calcium carbonate from gypsum waste slurry using sodium carbonate under ambient conditions. The results were applied to the pre-treatment of acid mine drainage (AMD) from coal mines. US patent no 2013/0288887 A1 reports a simple, cost-effective, and novel method for the recovery of nano-calcium carbonate from gypsum waste slurry [134]. Okumura et al. [135] extracted calcium oxide particles from the gypsum waste using reductive decomposition in a $CO-CO_2-N_2$ atmosphere. They also investigated the reductive decomposition of spent $CaSO_4$ using a packed-bed reactor. The $CaSO_4$ was used for the production of calcium oxide. Ramachandran and Maniam [136] reported a two-step method comprising chemical and thermal reactions for the regeneration of calcium oxide from gypsum waste. The chemical changes and confirmation of the formation of calcium oxide were determined using XRD. Mbhele et al. [137] attempted to recover sulphur from gypsum waste using the following sequence of steps: (1) reduction of gypsum to CaS_2 (2) stripping of the sulphide with carbon dioxide gas, and finally (3) the production of S.

2.5. Agricultural Waste

A great deal of solid agricultural waste is generated each year, presenting a major challenge for disposal processes, as it creates a foul odor and attracts pests, resulting in health issues for living beings. Some of the most commonly generated types of agricultural waste include rice husks [138], rice straw [139], wheat husks and straw, coconut shells [140], sugarcane bagasse [141], corn cobs, and almond shells. These types of agricultural waste are rich sources of carbon, silica, calcium, and other trace elements like Fe, Al, etc. Even today, only a small fraction of such agricultural waste is applied as a fuel, additive, and filler in the construction industry. A major fraction of agricultural waste is disposed of using three main techniques, thermal treatment, landfill, and decomposition, and these have been reported to negatively affect the environment [26]. For example, the thermal treatment of waste results in the release of numerous noxious gases like CO_2, CO, Cn Hm, SOx, NOx, ash, etc., many of which are classified as greenhouse gases. As a result, many scholars have raised objections over combustion as a method of agricultural waste disposal.

Agricultural waste such as sugarcane bagasse and rice husks constitute a rich source of silica. Both rice husks and sugarcane bagasse can act as potential candidates for silica extraction. The silica synthesized from such wastes can provide an alternative, renewable source, minimizing the pollution resulting from such materials. Meanwhile, coconut shell husks are a rich source of calcium [26]. The possibility of recovering of such valuable minerals from waste not only provides alternative precursor materials, it also provides an environmentally friendly, cost-effective approach to the problem. The use of such agricultural waste also minimizes environmental pollution. All of these waste products are biological materials; thus, they are rich sources of carbons that can be used for the synthesis of activated carbons or biochars.

There are certain plants that possess a high accumulation of silica in their leaves, stems, fruits, etc. The silica is taken up from the soil by the roots of the plants and distributed to the other plant parts. In plants, silica is mainly present as silicon. Silica can potentially be found in the solid waste used extensively in the industrial sector. Two factors, namely, the availability and quantity of silicon in the soil, affect silica deposition in agricultural residues [142]. Several plants consist of silica, including wheat, rice, sunflowers, corn, and bamboo [143]. In general, silica can be extracted from leaves, stems, and other parts of a plant with a yield of between 0.1 and 10 wt%. On the other hand, the quantity of silica in agricultural residues is dependent on the season, species, maturity, and geographical characteristics of the given farm.

2.5.1. Rice Husks (RH) and the Recovery of Silica

Rice husks are agricultural waste products that are produced during processing. During the hulling process, rice husks are obtained, which are mainly used as fodder for cattle. Rice husks and straw are major agricultural products, with an annual global production as high as one million tons. These waste products are common in rice-producing countries like India, Vietnam, and Japan. RH are rich is silica, but at the same time they also contain organic compounds that may interfere with the final purity of the silica. Therefore, RH have to first be calcinated at high temperature (400–1000 °C) in a muffle furnace. Furthermore, they have to be washed with phosphoric acid, which will remove the organic content. Once the RH ash is free from organic content, it is dried and treated with 4–16 M NaOH, along with 90–95 °C, for 60–90 min along with continuous stirring. NaOH will react with NaOH and will form sodium silicate, which will be further treated with 1–2 N HCl, resulting in silica gel. This silica gel is further washed and dried to obtain pure silica or nanosilica. Some authors have also reported calcination at 400–600 °C for 2–6 h in order to obtain nanosilica with the desired shaped [144].

2.5.2. Sugarcane Bagasse and the Recovery of Silica

Similarly, sugarcane bagasse is also one of the major byproducts of the sugarcane industry, and is produced after the extraction of sugarcane juice. Sugarcane bagasse is a rich source of carbohydrates and other minerals besides silica. It is also produced in huge quantities in sugar-producing countries. Sugarcane bagasse (SB) is considered to be a non-biodegradable solid material, and primarily consists of crystalline silica [145]. A major problem in the sugarcane industry is how to dispose of SB. At present, it is used in the production of ceramic tiles, soil fertilizers, and fodder in some parts of the world [146]. In comparison with other agricultural residues, sugarcane bagasse ash (SBA) possesses a very high quantity of silica. Table 2 presents the complete elemental composition of SBA. Several parameters affect the quantity of silica extractable from bagasse, including the nature of the soil, the surrounding environment, the harvesting process, and the time of harvesting.

Table 2. Elemental composition of sugarcane bagasse ash (SBA).

Elements	Raw Sample	Sample after Acid Treatment
SiO_2	53.10	88.13
MgO	20.72	3.04
CaO	3.77	0.57
SO_3	11.20	4.69
P_2O_5	7.36	1.15
K_2O	1.26	0.50

Drummond et al. [147] reported silica extraction from different preparations of SBA by performing natural burning (SBA-NB) and laboratory burning (SBA-LP) at 700 °C for two hours using muffle furnace, which was followed by alkaline extraction. The experiments revealed that silica extraction of about 94.47% was achieved using natural burning, and 96.93% of the silica was obtained using laboratory burning. In another project, Harish et al. [148] attempted to recover silica from SBA and silica fumes as low-cost precursors for the synthesis of silica gel using a sodium hydroxide-based alkali treatment method. Norsuraya et al. [149] reported the synthesis of Santa Barbara Amorphous-15 from SBA. XRF revealed that raw sugarcane ash contained 53.10% silica, while acid-treated ash contained 88.13% silica.

Channoy et al. [141] conducted a study aiming to synthesize silica gel from SBA by treating the ash with 1.5, 2, and 2.5 N NaOH. The particle size of the obtained silica gel varied, with values of 120, 100, and 80 nm, respectively, as confirmed by SEM. SEM revealed that as the concentration of NaOH increased, the particle size decreased. Rovani et al. [150] made use of SBA waste to synthesize highly pure silica with a high adsorption capacity. The synthesized silica nanoparticles were characterized by FTIR, SEM, TEM, XRD, ICP, etc., and it was found that the particle size was 20 nm and the purity was about 88%.

To obtain silica from SBA, several steps are needed: washing of the collected SB, shredding it into small pieces, pyrolysis at high temperature, alkali treatment of ash, acidic treatment of the sodium silicate leachate, formation of the silica gel, washing, precipitation, and calcination.

2.5.3. Coconut Shells

Coconut (Cocos nucifera) is the major plantation crop of coastal countries such as African countries [151]. A large number of coconuts are produced around the world, and these are mainly used for food and cosmetics. The soft part of the fruit is consumed, while the coir and shell are left behind as waste. Coir is used to make mattresses, while the shell is mainly disposed of into the environment. However, in African countries, coconut waste is conventionally used as a source of fuel and applied in burning processes [152]. Nowadays, with recent research advances, coconut shell waste can be used as a source of activated carbon and/or other value-added minerals. Coconut shell has high silica content and can act as a potential candidate as a source of silica and activated carbon. The use of coconut shell as a source of both silica and activated carbon presents cost-effective green synthesis strategy. Coconut shell can act as another substance for the synthesis of silica from renewable sources. Roughly 33–35% of a coconut is composed of husk, playing the role of the mesocarp of the fruit. Nowadays, coconut husk is used as a favorable source of fuel for coconut processing and domestic fuel, and as a fiber source for the production of mats, ropes, etc. [153]. The extraction of silica from agricultural wastes involves three sequential steps: acid leaching, mixing alkaline treatment, and precipitation with acid [154]. A schematic diagram is shown in Figure 8. The chemical composition of coconut shell is given below in Table 3. It mainly contains calcium oxide, followed by silica, along with traces of Al, K, Fe, and P, in the form of either oxides or chlorides.

Synthesis of Carbon Nanotubes from Coconut Shell Husk Ash

Carbon nanotubes have gained considerable attention from scientific fields of study such as medicine, drug delivery, photo dynamic therapy, and environmental cleanup. They have numerous advantages over other nanoparticles due to their high mechanical properties, high tensile strength, high electrical conductivity, high aspect ratio, high thermal conductivity, and ultra-light weight. Generally, these nanotubes are synthesized using the physical vapor deposition method or the chemical vapor deposition method; both of these methods are costly and energy intensive. As a result, the synthesis of carbon nanotubes from carbon-rich waste materials like coconut shell husk ash is an economical and environmentally friendly approach. There are several approaches in which coconut shell has been used for the synthesis of carbon nanotubes.

Anuar et al. [153] reported the synthesis of silica nanoparticles from coconut shell ash. In their study, the coconut husks were burned under different temperatures, and were analyzed using XRF to identify their elemental composition. The composition of silica varies from 8–11% in the husk ash. The husks and silica were characterized by SEM-EDS, XRD for confirmation and to determine the optoelectronic properties of the silica. In another project, Sivasubramanian and Sravanthi (2015) attempted to synthesize silica nanoparticles from coconut shell ash using the NaOH-based alkali treatment method. Ash was treated with 2.5 N sodium hydroxide to obtain sodium silicate. Finally, silica was formed by treatment with HCl, and was confirmed by SEM-EDS, FTIR, TEM, and XRD [155]. Melati and Hidayati [156] reported the synthesis of multi-walled carbon nanotubes from coconut shell in two steps. First, the coconut shell was activated by treating it at 500–600 °C, and it was subsequently converted into carbon nanotubes by applying pyrolysis and a wet CVD process. The characterization of the nanotubes revealed the properties of the MWCNTs. It was used for the detection of cancer in mammalian cell lines.

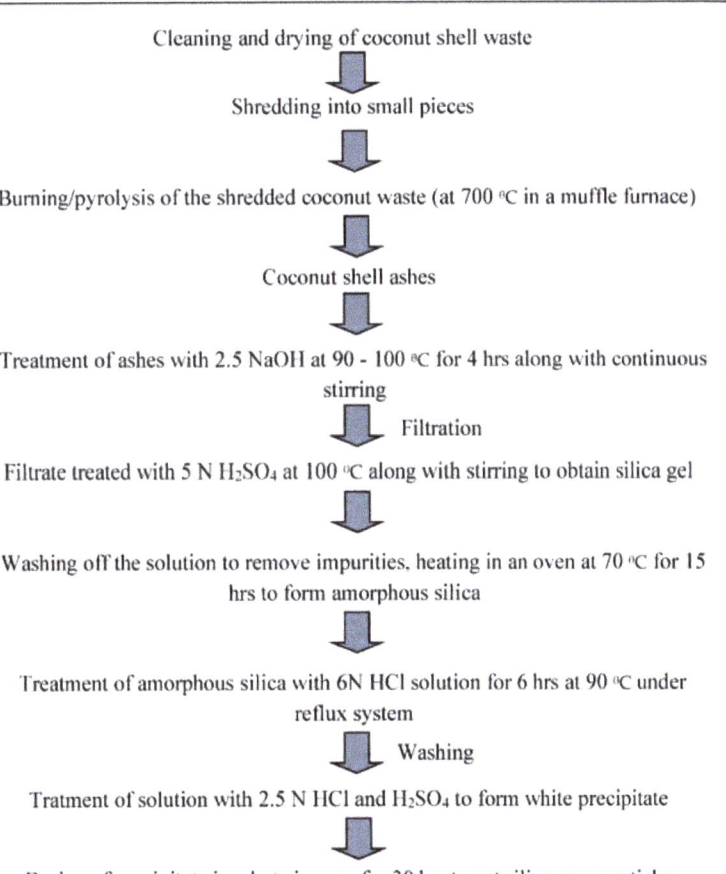

Figure 8. General method for the synthesis of silica from coconut shell waste.

Table 3. Elemental composition of coconut husk ash (CHA).

Elements	Composition (%)
SiO_2	8–12
CaO	27–31.5
K_2O	17–20
Al_2O_3	0.3–0.8
SO_3	2–3.5
Fe_2O_3	0.3–1.0
P_2O_5	0.05–0.3
Cl	35–38

In the study conducted by Adewumi et al. [157], carbon nanospheres were synthesized from low-cost coconut fibers in three sequential steps: pyrolyzation, physical activation, and ethanol vapor treatment. The analysis of the samples revealed that the spherical-shaped particles had a diameter of 30–150 nm. Hakim et al. [158] applied an easy, environmentally friendly approach called the one-step water-assisted (quenching) synthesis method to obtain carbon nanotubes using coconut shell husk ash. The chemical and physical properties of the carbon nanotubes were analyzed using sophisticated instruments, and it was found that the average diameter was 123 nm; the nanotubes were finally applied for the remediation of Pb^{2+} ions from wastewater. Araga and Sharma [158] synthesized PECV-assisted multiwalled carbon nanotubes (MWCNTs) over coconut shell-derived charcoal pyrolyzed at 900 °C in a process with only a single step. They used the mineral content in the source material as the catalyst for carbon nanotube (CNT) growth.

Corn Cobs as a Source of Activated Carbon

Maize, or corn (*Zea mays*), is a popular cereal crop cultivated in many parts of the world [159]. During the processing and production of corn, several waste products are generated, including corn cobs and corn husks [160]. Corn cob waste is a rich source of carbon, and can act as a potential candidate for reparing carbon with ultra-high specific surface area [161]. On the other hand, there are various waste materials that can be applied as activated carbon sources, for instance, date and palm waste, coconut shell waste, corn stalks, and corn cobs. Corn cob is most preferable source of activated carbon, as it is produced in huge amounts around the world. Corn cobs are agricultural waste materials produced in abundance around the world. After extracting the corn, the major fraction of the corn is disposed of as waste, although it is actually a rich source of various minerals. The cob is mainly made up of carbons; therefore, it can be used as a source of biochars or activated carbon.

Activated carbon (AC) refers to carbonaceous materials that possess an internal surface area (that is extremely developed) as well as porosity [162]. The large surface area (an area between 250 m^2/g and 2000 m^2/g) offers a significant ability to adsorb chemicals from liquids/gases, and permit application as a versatile adsorbent under various conditions. AC has been widely used to produce adsorbents, supporting materials, textiles, fabrics, animal foods, etc. [163]. AC is known to be an effective material as a result of its low density, well-developed porosity, accessibility, chemical stability, and low cost [164]. In the last few years, a major emphasis has been placed on the development of AC with ultra-high specific surface area from both renewable and non-renewable sources using chemical approaches or with the use of chemical vapor deposition (CVD). In the process of water treatment, a substantial amount of AC is used for the purpose of removing organic and other compounds that could change the water odor and taste [165].

Tsai et al. [166] reported the synthesis of AC from corn cob waste through treatment with different physical and chemical activators such as NaOH, KOH, K_2CO_3, and CO_2. They studied the effect of impregnation time, impregnation ratio, activation temperature, and soaking time on carbon dioxide. The surface area of the AC was analyzed using the BET analyzer. The total pore volume and BET surface area were roughly 1.0 cm^3/g and 2000 m^2/g, respectively. The findings of this research demonstrate that corn cob activation with KOH/K_2CO_3 and CO_2 is able to appropriately prepare large-surface-area ACs. Furthermore, Sai et al. [167] attempted to activate corn cob using potassium salts, and subsequently gasified it with carbon dioxide. The obtained AC had a large surface area, as measured by BET. In the study carried out by Kazmierczak et al. [168], AC was developed from corn cob by activating it using chemical and physical methods. They further studied the sorption properties of the activated carbon. The final product consisted of microporous activated carbon with a high surface area, varying from m 337 to 1213 m^2/g, and showing diverse acid–base characteristics on the surface [169,170]. It was also assessed for the adsorption of different materials from the aqueous solutions. Figure 9 shows the schematic diagram for the formation of activated carbon from corn cob.

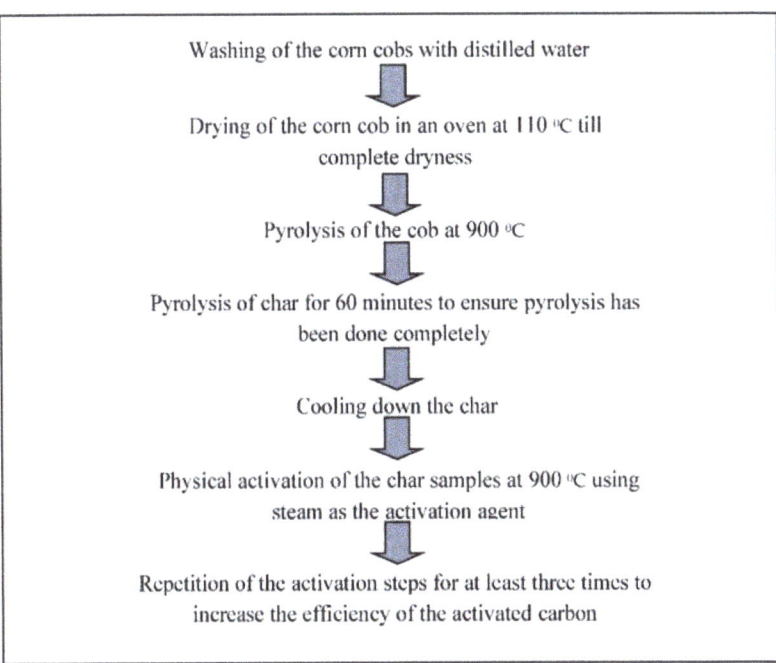

Figure 9. Schematic diagram for the formation of activated carbon from corn cob.

3. Domestic Waste

Domestic waste refers to the materials produced in houses, mainly kitchens. There are several wastes produced in houses that offer potential for the recovery of value-added minerals. Calcium carbonate is a highly important particle that is used in every aspect of our life, as well as in a number of different industries [171]. Calcium carbonates are alkaline earth materials [172] that are present in our environment and which are widely applied in industries like papers [173], paints [174], coating agents, cosmetics [175,176], pharmaceuticals and medicine [177], agriculture, automobiles and textiles [178], and reinforcements, fillers, bio-nanocomposites and bio-ceramics in dentistry [178,179].

Calcium carbonates have excellent properties, including biodegradability, biocompatibility, pH sensitivity, safety, and cost-effectiveness, and they exhibit polymorphicity [179]. They are very lightweight; thus, they can be used for making lightweight materials. Calcium carbonate exists in three polymorphs in nature, i.e., calcite, aragonite, and vaterite [180]. All these three polymorphs vary with respect to their thermodynamic stability and morphology. Among them, calcite is thermodynamically the most stable, while aragonite shows an intermediate stability, and vaterite (μ-$CaCO_3$) is the least stable polymorph, due to which it is less commonly present in nature. Vaterite can be rapidly transformed into the aragonite and calcite in aqueous solution. It has been experimentally proved that vaterite can transform to aragonite in 60 min at 60 °C and to calcite in 24 h at room temperature [181]. While calcite is rhombohedral in structure, aragonite is spindle- or needle-shaped, while vaterite is octahedral [182]. There are several waste materials that are rich sources of calcium oxide and calcium carbonates, including incense sticks, eggshells, cockle shell waste [183], and gypsum waste; however, the following sections only discuss domestic waste, i.e., incense stick ash and eggshell waste.

3.1. Incense Stick Ash (ISA)

Incense stick ash is one of the most unexplored products, and are generally produced at religious places, i.e., temples, churches, mosques, etc. The burning of incense sticks is a ritual practiced in every religion. However, in South-Asian countries and zones, e.g., China, Taiwan, India, and Japan, large amounts of incense sticks are consumed [184,185]. Incense sticks are cylindrical in shape, and are fragranced and intended to be burned in order to purify the air [186]. After the burning of incense sticks, the ash is left behind in the form of a residue, which is mainly disposed of in rivers and other water bodies, as can be frequently observed in India, where this practice is considered to be holy. Moreover, due to its sacred value in Hinduism, incense is even applied on the forehead or eaten as Prasad. These incense sticks mainly contain calcium oxide, silica, ferrous, alumina, rutile, Mg, and traces of oxides of K, Na, and Mn. The sticks contain 45–60% calcium oxides or carbonates, 10–20% silica, and 5–7% Mg, with less than 5% consisting of other materials. The disposal of incense stick ash into water pollutes it by increasing the concentration of Ca and Mg, ultimately increasing the water hardness. These two elements also play a role in increasing the pH of water by forming precipitates of hydroxides in it. Due to the high calcium content, incense stick ash can act as a potential source material for the recovery of calcium oxides and carbonates. The complete schematic diagram for the synthesis of calcium carbonate from incense stick ash is given below in Figure 10.

3.2. Eggshell Waste

Eggshells are another waste product produced by houses and industry, and are still classified as a byproduct of the poultry industry. Approximately one million tons of eggshell are generated per year globally. Currently, eggshell waste is dumped into landfills. Such practices may lead to the deterioration of agricultural land [187]. This waste is mainly made up of calcium oxides that are meant to provide safety to growing chicks. Due to their high calcium content (composition shown in Table 4), eggshells can be used as a potential alternative source of calcium oxide or carbonates [188,189].

The recovery of calcium oxides and carbonates from incense stick ash and poultry is easy, economical, and environmental friendly. Recovery from such wastes focuses on the use of renewable sources without affecting our natural resources. Consequently, it also minimizes pollution in the form of solid waste. Hassan et al. [190,191] reported the synthesis of $CaCO_3$ NPs from chicken eggshell waste, which involved the following steps: cleaning and size reduction of eggshells, followed by surface modification using the sonochemical method to achieve enhanced dispersion [190,191]. In another study, Hariharan et al. [192] attempted to perform the synthesis of calcite nanoparticles from eggshell waste using gelatin. Chicken eggshells were used to obtain the calcite polymorph of calcium carbonate using gelatin by means of the precipitation method. Nanocalcite was confirmed using FTIR, XRD, UV-Visible spectroscopy, and SEM. The particles were identified as calcite polymorphs with a particle size of 25 nm. The results obtained in the FTIR experiments confirmed the creation of calcite, with characteristic absorption bands being observed at 712, 876, and 1410 cm^{-1}, corresponding to the bending and stretching vibrations of CO_3^{2-} ions. A comparison was also made between the obtained results and calcium carbonate synthesized with no gelatin.

Calcium oxides are currently produced at commercial levels from calcium-based precursor materials like calcium nitrate [193], calcium hydroxide [194], calcium sulphate, and calcium chloride [193] using techniques such as thermal decomposition [194], microwave irradiation [195], sol–gel [196], co-precipitation [197,198], hydrogen plasma–metal reaction [199], and sonochemical synthesis [191,200]. However, the use of the above-mentioned calcium precursors for the synthesis of calcium oxide nanoparticles using the techniques described above makes the whole process costly and energy intensive, and also requires the use of hazardous chemicals. As a result, for the synthesis of calcium oxide nanoparticles, there is a need to switch to green methods. Some of the most common calcium-rich waste materials are cockle shells [201], eggshells [202], gypsum waste, and incense stick ash. Among these, incense stick ash is the most underestimated calcium-rich waste material, and is produced abundantly at religious places and in houses. Such materials can act as potential substitutes for various calcium precursors used at commercial level [203,204].

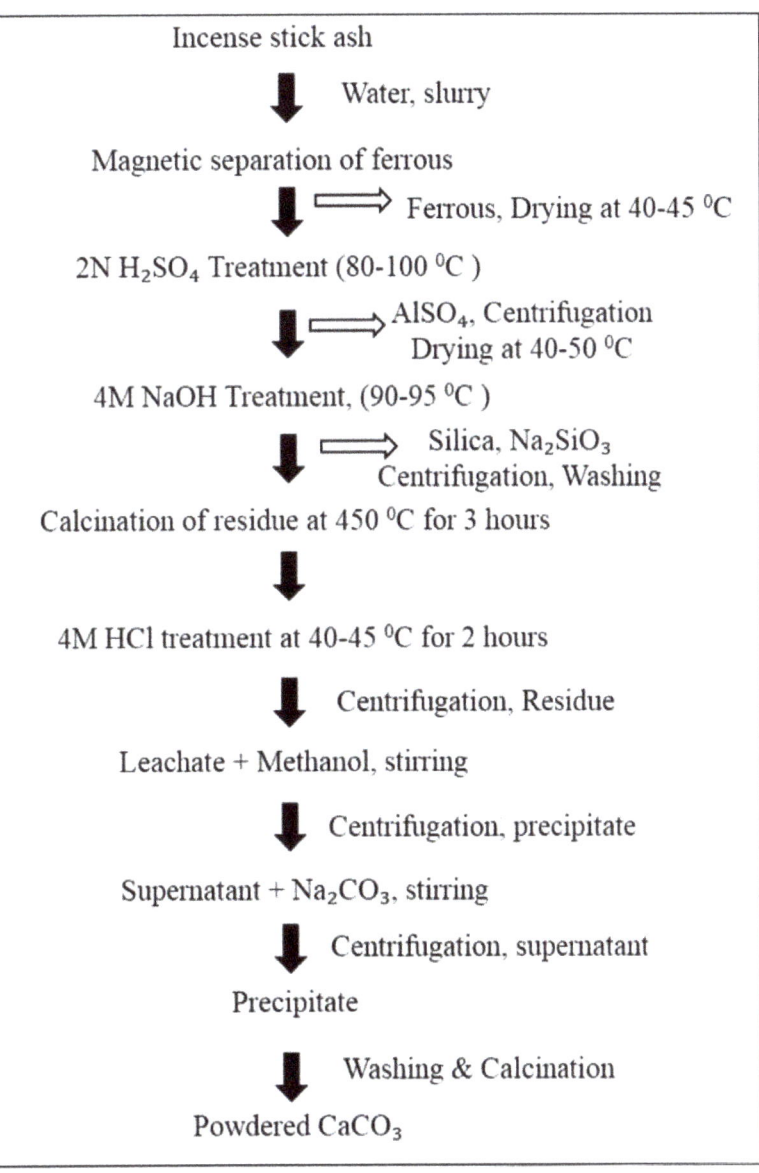

Figure 10. Flow chart for the synthesis of $CaCO_3$ from incense stick ash, adopted from Yadav et al., open access journal Applied Sciences, 2021 [129].

Table 4. The chemical composition of chicken eggshells.

Chemical Elements	Concentration (mg/L)
Ca	2296–2304
Mg	849–852
Na	33–35
K	16–19
Fe	1.01–1.43
Zn	0.95–1.03
Cu	0.062–0.064

Tangboriboon et al. [205] synthesized calcium oxide nanoparticles from duck eggshell waste using the calcination technique and analyzed their properties. The duck eggshells and the calcined eggshells were analyzed using FTIR, STA, XRD, XRF, TEM, BET, a particle size analyzer, and an impedance analyzer. The microscopy revealed the good dispersion quality of the calcium oxide nanoparticles, which had a spherical shape, with a ceramic yield of 53%. In another study, Jirimali et al. [206] reported the synthesis of calcium oxide and hydroxyapaptite using eggshell for the development of LLDPE Polymer Nanocomposite [206]. Mohadi et al. [202] synthesized the calcium oxide nanoparticles from the chicken eggshell waste by calcination of the shells at different temperatures ranging from 600 to 1000 °C. The synthesized calcium oxide nanoparticles were analyzed using FTIR, SEM-EDS, XRD, and BET. The calcium oxides were porous in nature, with pore sizes of 6.6 nm, meaning they could be classified as mesoporous, and with a surface area of 68 m^2/g [202].

A typical TEM image of calcium oxide nanoparticles obtained from eggshell waste is shown in Figure 11 [204].

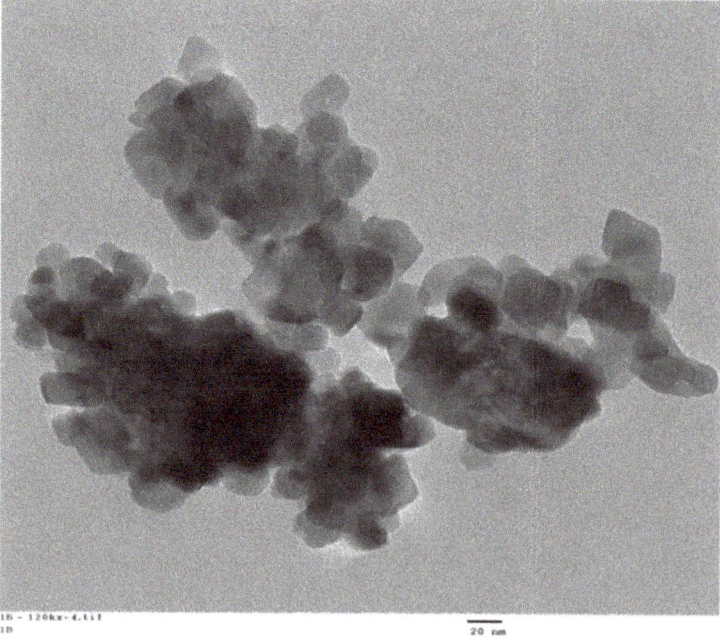

Figure 11. TEM image of calcium oxide nanoparticles obtained from eggshell waste adapted from (Render et al. [204]).

There are several reports of the synthesis of composite materials and fertilizers from plastic waste; however, research conducted in this field is very limited. A lot of work needs to be performed in this field in future, since plastics constitute a major global problem.

4. Conclusions

Waste management mainly addresses two subjects: resource recovery and final disposal. Individuals worldwide earn revenue from each stage, through the recovery of reusable materials and, to a lesser extent, the conversion of waste to energy. Turning waste into wealth not only makes sensible environmental sense, but also turns "trash" into "cash". Agricultural waste materials are mainly organic in nature; they are biodegradable, and can be used for the development of carbon-based materials and activated carbon. The burning of agricultural waste leads to the pollution of the air. In recent decades, waste management and technology awareness programs have successfully transformed hazardous materials (e.g., CFA and red mud) into useable value-added minerals. Both of these materials have been applied in the fields of ceramics, construction materials, and metallurgy, and have proved to be highly valuable. Poultry waste and domestic waste have been also been found to be important precursors of carbon and calcium oxides. The ash from incense sticks is among the emerging domestic byproducts produced at religious places, and have proved to be a valuable source of calcium oxide. All of these waste byproducts have gained importance with the advent and increasing significance of renewable energy sources. The recovery of minerals from such wastes is an economical and environmentally friendly method. Such materials act as alternative precursors, reducing the burden on industry. The use of such wastes reduces pollution at minimum cost, while developing materials and generating revenue.

Author Contributions: Conceptualization, G.G., S.M. and N.G.; Data curation, N.C., G.G., S.I. and S.M.; methodology, N.C., N.G. and V.T.; validation, K.K.Y., S.I., B.-H.J. and S.M.; formal analysis, V.K.Y., N.G. and A.A.; resources, K.K.Y., B.-H.J. and A.A.; writing—original draft preparation, V.K.Y., N.C. and K.K.Y.; writing—review and editing, V.K.Y., V.T., G.G., S.M. and A.A.; supervision, V.K.Y., S.I. and B.-H.J.; project administration V.K.Y., N.C., K.K.Y. and A.A.; Funding acquisition, B.-H.J., S.I., V.T. and A.A.; Investigation, V.K.Y. and V.T.; Software's, G.G., N.G. and B.-H.J.; Visualization, K.K.Y. and B.-H.J. All authors have read and agreed to the published version of the manuscript.

Funding: The authors gratefully acknowledge the Deanship of Scientific Research, King Khalid University (KKU), Abha-Asir, Kingdom of Saudi Arabia for funding this research work under the grant number RGP.2/58/42. This study was supported by the Korea Environment Industry & Technology Institute (KEITI) through Subsurface Environment Management (SEM) Projects, funded by Korea Ministry of Environment (MOE)(No.2020002480007).

Institutional Review Board Statement: Not applicable.

Informed Consent Statement: Not applicable.

Data Availability Statement: Not applicable.

Acknowledgments: The authors gratefully acknowledge the Deanship of Scientific Research, King Khalid University (KKU), Abha-Asir, Kingdom of Saudi Arabia for funding this research work under the grant number RGP.2/58/42. This study was supported by the Korea Environment Industry & Technology Institute (KEITI) through Subsurface Environment Management (SEM) Projects, funded by Korea Ministry of Environment (MOE)(No.2020002480007).

Conflicts of Interest: The authors declare no conflict of interest.

References

1. Lin, T.C.; Krishnaswamy, G.; Chi, D.S. Incense smoke: Clinical, structural and molecular effects on airway disease. *Clin. Mol. Allergy* **2008**, *6*, 3. [CrossRef]
2. Khan, T.; Mubeen, U. Wheat Straw: A pragmatic overview. *Curr. Res. J. Biol. Sci.* **2012**, *4*, 673–675.
3. Schnitzer, M.; Monreal, C.; Powell, E. Wheat straw biomass: A resource for high-value chemicals. *J. Environ. Sci. Health* **2014**, *49*, 51–67. [CrossRef]

4. Zhang, Z.; Li, C.; Davies, E.G.; Liu, Y. Agricultural Waste. *Water Environ. Res.* **2013**, *85*, 1377–1451. [CrossRef]
5. Domínguez-Escribá, L.; Porcar, M. Rice straw management: The big waste. *Biofuels Bioprod. Biorefining* **2010**, *4*, 154–159. [CrossRef]
6. Hasnain, M.H.; Javed, U.; Ali, A.; Zafar, M.S. Eco-friendly utilization of rice husk ash and bagasse ash blend as partial sand replacement in self-compacting concrete. *Constr. Build. Mater.* **2021**, *273*, 121753. [CrossRef]
7. Ganiron, T.U., Jr. Sustainable Management of Waste Coconut Shells as Aggregates in Concrete Mixture. *J. Eng. Sci. Technol. Rev.* **2013**, *6*, 7–14. [CrossRef]
8. Ahmad, T.; Danish, M.; Rafatullah, M.; Ghazali, A.; Sulaiman, O.; Hashim, R.; Ibrahim, M.N.M. The use of date palm as a potential adsorbent for wastewater treatment: A review. *Environ. Sci. Pollut. Res.* **2012**, *19*, 1464–1484. [CrossRef]
9. Gómez, B.; Gullón, B.; Yáñez, R.; Parajó, J.C.; Alonso, J.L. Pectic Oligosacharides from Lemon Peel Wastes: Production, Purification, and Chemical Characterization. *J. Agric. Food Chem.* **2013**, *61*, 10043–10053. [CrossRef]
10. Landge, A.; VShahade, I.; Topare, N. Application of Biosorbent Prunus Amygdalus (Almond) Nut Shell Carbon on Treatment of Cr (VI) Contaminated Waste Water. In Proceedings of the 66th Annual Session of Indian Institute of Chemical Engineers, Pune, Indian, 27–30 December 2013; pp. 1–5.
11. Zhang, Z.; Gonzalez, A.M.; Davies, E.; Liu, Y. Agricultural Wastes. *Water Environ. Res.* **2012**, *84*, 1386–1406. [CrossRef]
12. Mohomane, S.; Linganiso, L.Z.; Linganiso, E.C.; Motaung, T.E.; Songca, S.P. The application of fly ash as industrial waste material in building construction industries. In *"Waste-To-Profit"(W-T-P)*; Nova Science Publishers Inc.: New York, NY, USA, 2018; pp. 181–202.
13. Nordin, N.; Abdullah, M.M.A.B.; Tahir, M.F.M.; Sandu, A.V.; Hussin, K. Utilization of fly ash waste as construction material. *Int. J. Conserv. Sci.* **2016**, *7*, 161–166.
14. Rai, S.; Wasewar, K.L.; Lataye, D.H.; Mishra, R.S.; Puttewar, S.; Chaddha, M.J.; Mahindiran, P.; Mukhopadhyay, J. Neutralization of red mud with pickling waste liquor using Taguchi's design of experimental methodology. *Waste Manag. Res.* **2012**, *30*, 922–930. [CrossRef] [PubMed]
15. Wang, P.; Liu, D.Y. Physical and Chemical Properties of Sintering Red Mud and Bayer Red Mud and the Implications for Beneficial Utilization. *Materials* **2012**, *5*, 1800–1810. [CrossRef]
16. Geraldo, R.H.; Pinheiro, S.M.; Silva, J.S.; Andrade, H.M.; Dweck, J.; Gonçalves, J.; Camarini, G. Gypsum plaster waste recycling: A potential environmental and industrial solution. *J. Clean. Prod.* **2017**, *164*, 288–300. [CrossRef]
17. Romero-Hermida, I.; Morales-Flórez, V.; Santos, A.; Villena, A.; Esquivias, L. Technological Proposals for Recycling Industrial Wastes for Environmental Applications. *Minerals* **2014**, *4*, 746–757. [CrossRef]
18. Ahmadi, M.; Bayati, N.; Babaei, A.; Teymouri, P. Sludge characterization of an industrial wastewater treatmnt plant, Iran. *Iran. J. Health Sci.* **2013**, *1*, 10–18.
19. Shreekant, R.; Vardhan, A.M.H. Utilization of Iron Ore Waste in Brick Making For the Construction Industry. *Int. J. Earth Sci. Eng.* **2016**, *9*, 450–455.
20. Awuchi, C.G.; Igwe, S.V. Industrial waste management: Brief survey and advice to cottage, small, and medium scale industries in Uganda. *Int. J. Adv. Acad. Res.* **2017**, *3*, 26–43.
21. Fecheyr-Lippens, D.; Nallapaneni, A.; Shawkey, M. Exploring the Use of Unprocessed Waste Chicken Eggshells for UV-Protective Applications. *Sustainability* **2017**, *9*, 232. [CrossRef]
22. Meng, X.; Deng, D. Trash to Treasure: Waste Eggshells Used as Reactor and Template for Synthesis of Co_9S_8 Nanorod Arrays on Carbon Fibers for Energy Storage. *Chem. Mater.* **2016**, *28*, 3897–3904. [CrossRef]
23. Ummartyotin, S.; Manuspiya, H. A critical review of eggshell waste: An effective source of hydroxyapatite as photocatalyst. *J. Met. Mater. Miner.* **2018**, *28*, 124–135.
24. Adeoye, A.; Man, H.C.; Soom, M.; Thamer, A.M.; Oluwakunmi, A.C. Poultry waste generation, management and the environment: A case of Minna, North Central Nigeria. *J. Solid Waste Technol. Manag.* **2015**, *41*, 146–156. [CrossRef]
25. Bolan, N.S.; Szogi, A.A.; Chuasavathi, T.; Seshadri, B.; Rothrock, M.J.; Panneerselvam, P. Uses and management of poultry litter. *World's Poult. Sci. J.* **2010**, *66*, 673–698. [CrossRef]
26. Abdel-Shafy, H.I.; Mohamed-Mansour, M.S. Solid waste issue: Sources, composition, disposal, recycling and valorization. *Egypt. J. Pet.* **2018**, *27*, 1275–1290. [CrossRef]
27. Latake, T.; Kulkarni, A.; Bhosale, S. Disposal of Solid Waste For Landfilling In Karad City A Review. In *International Journal for Research in Applied Science and Engineering Technology*; Seventh Sense Research Group: Thanjavur, India, 2016; p. 4. ISSN 2395-0072.
28. Ali, S.M.; Pervaiz, A.; Afzal, B.; Hamid, N.; Yasmin, A. Open dumping of municipal solid waste and its hazardous impacts on soil and vegetation diversity at waste dumping sites of Islamabad city. *J. King Saudi Univ. Sci.* **2014**, *26*, 59–65. [CrossRef]
29. Sakawi, Z.; Mastura, S.; Jaafar, O.; Mahmud, M. Community perception of odour pollution from landfills. *Res. J. Environ. Earth Sci.* **2017**, *3*, 142–145.
30. De Moraes, C.M.; Stanczyk, N.M.; Betz, H.S.; Pulido, H.; Sim, D.G.; Read, A.F.; Mescher, M.C. Malaria-induced changes in host odors enhance mosquito attraction. *Proc. Natl. Acad. Sci. USA* **2014**, *111*, 11079–11084. [CrossRef]
31. Sijakova Ivanova, T.; Panov, Z.; Blazev, K.; Zajkova, V. Investigation of fly ash heavy metals content and physico chemical properties from thermal power plant, Republic of Macedonia. *Int. J. Eng. Sci. Technol. (IJEST)* **2011**, *3*, 8219–8225.
32. Ozden, B.; Brennan, C.; Landsberger, S. Investigation of bauxite residue (red mud) in terms of its environmental risk. *J. Radioanal. Nucl. Chem.* **2019**, *319*, 339–346. [CrossRef]

33. Nguyen, H.T.; Bui, T.H.; Pham, V.T.H.Q.; Do, M.Q.; Hoang, M.D.; Le, V.Q. Leaching Behavior and Immobilization of Heavy Metals in Geopolymer Synthesized from Red Mud and Fly Ash. In *Key Engineering Materials*; Trans Tech Publications Ltd.: Zurich, Switzerland, 2018; pp. 518–522.
34. Vongdala, N.; Tran, H.D.; Xuan, T.D.; Teschke, R.; Khanh, T.D. Heavy Metal Accumulation in Water, Soil, and Plants of Municipal Solid Waste Landfill in Vientiane, Laos. *Int. J. Environ. Res. Public Health* **2019**, *16*, 22. [CrossRef] [PubMed]
35. Lokeshwari, M.; Swamy, C.N. Waste to wealth-Agriculture solid waste management study. *Poll. Res.* **2010**, *29*, 513–517.
36. Jain, A.K.; Gupta, V.K.; Bhatnagar, A.; Bhatnagar, A. Utilisation of Industrial Waste Products as Adsorbents for the Removal of Dyes. *J. Hazard. Mater.* **2003**, *101*, 31–42. [CrossRef]
37. Sulyman, M.; Namiesnik, J.; Gierak, A. Low-cost Adsorbents Derived from Agricultural By-products/Wastes for Enhancing Contaminant Uptakes from Wastewater: A Review. *Pol. J. Environ. Stud.* **2017**, *26*, 479–510. [CrossRef]
38. Krishnappa, V. Properties and Application of Geopolymer Masonry Units. In *International Journal of Civil Engineering EFES*; Seventh Sense Research Group: Thanjavur, India, 2015; pp. 117–119. ISSN 2348-8352.
39. Mohajerani, A.; Suter, D.; Jeffrey-Bailey, T.; Song, T.; Arulrajah, A.; Horpibulsuk, S.; Law, D. Recycling waste materials in geopolymer concrete. *Clean Technol. Environ. Policy* **2019**, *21*, 493–515. [CrossRef]
40. Bernardo, E.; Dattoli, A.; Bonomo, E.; Esposito, L.; Rambaldi, E.; Tucci, A. Application of an Industrial Waste Glass in "Glass–Ceramic Stoneware". *Int. J. Appl. Ceram. Technol.* **2011**, *8*, 1153–1162. [CrossRef]
41. Skenderovic, I.; Kalac, B.; Becirovic, S. Environmental pollution and waste management. *Balk. J. Health Sci.* **2015**, *3*, 2–10.
42. Kowalski, Z.; Krupa-Żuczek, K. A model of the meat waste management. *Pol. J. Chem. Technol.* **2007**, *9*, 91–97. [CrossRef]
43. Mladenov, M.; Pelovski, Y. Utilization of wastes from pulp and paper industry. *J. Univ. Chem. Technol. Metall.* **2010**, *45*, 33–38.
44. Simão, L.; Hotza, D.; Raupp-Pereira, F.; Labrincha, J.A.; Montedo, O. Wastes from pulp and paper mills—A review of generation and recycling alternatives. *Cerâmica* **2018**, *64*, 443–453. [CrossRef]
45. Fela, K.; Wieczorek-Ciurowa, K.; Konopka, M.; Woźny, Z. Present and prospective leather industry waste disposal. *Pol. J. Chem. Technol.* **2011**, *13*, 53–55. [CrossRef]
46. Paramanandham, J.; Ronald Ross, P. Lignin and cellulose content in coir waste on subject to sequential washing. *J. Chem. Chem. Res. Vol.* **2011**, *1*, 10–13.
47. Noguera, P.; Abad, M.; Noguera, V.; Puchades, R.; Maquieira, A. Coconut coir waste, a new and viable ecologically-Friendly peat substitute. In *XXV International Horticultural Congress, Part 7: Quality of Horticultural Products 517*; International Society for Horticultural Science: Brussels, Belgium, 1998; pp. 279–286.
48. Baz, W.A.; Nawar, L.S.; Aly, M.M. Production of biofuel from sugarcane baggase wastes using Saccharomyces cerevisiae. *J. Exp. Biol.* **2017**, *5*, 6.
49. Mzimela, Z.; Mochane, M.; Tshwafo, M. Sugarcane bagasse waste management. In *"Waste-to-Profit"? (W-t-P): Value Added Products to Generate Wealth for a Sustainable Economy*; Nova Science Publishers: Hauppauge, NY, USA, 2018; Volume 1, pp. 293–302.
50. Xu, G.; Shi, X. Characteristics and applications of fly ash as a sustainable construction material: A state-of-the-art review. *Resour. Conserv. Recycl.* **2018**, *136*, 95–109. [CrossRef]
51. Lieberman, R.N.; Knop, Y.; Izquierdo, M.; Palmerola, N.M.; de la Rosa, J.; Cohen, H.; Querol, X. Potential of hazardous waste encapsulation in concrete with coal fly ash and bivalve shells. *J. Clean. Prod.* **2018**, *185*, 870–881. [CrossRef]
52. Sahu, G.; Bag, A.; Chatterjee, N.; Mukherjee, A. Potential use of flyash in agriculture: A way to improve soil health. *J. Pharmacogn. Phytochem.* **2017**, *6*, 873–880.
53. Sheoran, H.S.; Duhan, B.S.; Kumar, A. Effect of Fly Ash Application on Soil Properties: A Review. *J. Agroecol. Nat. Resour. Manag.* **2014**, *1*, 98–103.
54. Kalra, N.; Jain, M.C.; Joshi, H.C.; Choudhary, R.; Harit, R.C.; Vatsa, B.K.; Sharma, S.K.; Kumar, V. Flyash as a soil conditioner and fertilizer. *Bioresour. Technol.* **1998**, *64*, 163–167. [CrossRef]
55. Yadav, V.K.; Fulekar, M.H. Advances in Methods for Recovery of Ferrous, Alumina, and Silica Nanoparticles from Fly Ash Waste. *Ceramics* **2020**, *3*, 34. [CrossRef]
56. Yadav, V.K.; Pandita, R. Fly Ash Properties and Their Applications as a Soil Ameliorant. In *Amelioration Technology for Soil Sustainability*; IGI Global: Hershey, PA, USA, 2019; pp. 59–89.
57. Ahmaruzzaman, M. Role of fly ash in the removal of organic pollutants from wastewater. *Energy Fuels* **2019**, *23*, 1494–1511. [CrossRef]
58. Singh, V.K.; Singh, R.S.; Tiwari, N.; Singh, J.K.; Gode, F.; Sharma, Y.C. Removal of malathion from aqueous solutions and waste water using fly ash. *J. Water Resour. Prot.* **2010**, *2*, 322. [CrossRef]
59. Al-Harahsheh, M.S.; Alzboon, K.; Al-Makhadmeh, L.; Hararah, M.; Mahasneh, M. Fly ash based geopolymer for heavy metal removal: A case study on copper removal. *J. Environ. Chem. Eng.* **2011**, *3*, 1669–1677. [CrossRef]
60. Ge, C.J.; Yoon, K.S.; Choi, J.N. Application of Fly Ash as an Adsorbent for Removal of Air and Water Pollutants. *Appl. Sci.* **2018**, *8*, 1116. [CrossRef]
61. Yeole, K.; Kadam, P.; Mhaske, S. Synthesis and characterization of fly ash-zinc oxide nanocomposite. *J. Mater. Res. Technol.* **2014**, *3*, 186–190. [CrossRef]
62. Yao, Z.T.; Ji, X.S.; Sarker, K.; Tang, J.H.; Ge, L.Q.; Xia, M.S.; Xi, Y.Q. A comprehensive review on the applications of coal fly ash. *Earth-Sci. Rev.* **2015**, *141*, 105–121. [CrossRef]

63. Yu, J.; Li, X.; Fleming, D.; Meng, Z.; Wang, D.; Tahmasebi, A. Analysis on Characteristics of Fly Ash from Coal Fired Power Stations. *Energy Procedia* **2012**, *17*, 3–9. [CrossRef]
64. Abbas, S.; Saleem, M.A.; Kazmi, S.M.S.; Munir, M.J. Production of sustainable clay bricks using waste fly ash: Mechanical and durability properties. *J. Build. Eng.* **2017**, *14*, 7–14. [CrossRef]
65. Pawlik, T.; Michalik, D.; Sopicka-Lizer, M.; Godzierz, M. Manufacturing of Light Weight Aggregates from the Local Waste Materials for Application in the Building Concrete. In *Materials Science Forum*; Trans Tech Publications Ltd: Bäch, Switzerland, 2017; Volume 904, pp. 174–178. [CrossRef]
66. Bouzoubaâ, N.; Zhang, M.H.; Zakaria, M.; Malhotra, V.M.; Golden, D.M. Blended fly ash cements—A review. *ACI Mater. J.* **1999**, *96*, 641–650.
67. Nataatmadja, A.; Haigh, R.; Amaratunga, D. (Eds.) Development of low-cost fly ash bricks. In Proceedings of the CIB International Conference on Building Education and Research, Heritance Kandalama, Sri Lanka, 11–15 February 2008; pp. 831–843.
68. Bohara, N. Study of the Influence of Fly Ash and Its Content in Marshall Properties of Asphalt Concrete. *J. Sustain. Constr. Mater. Technol.* **2018**, *3*, 261–270. [CrossRef]
69. Mishra, M.K.; Karanam, U. Geotechnical characterization of fly ash composites for backfilling mine voids. *Geotech. Geol. Eng.* **2006**, *24*, 1749–1765. [CrossRef]
70. Yadav, V.K.; Fulekar, M.H. Isolation and charcterization of iron nanoparticles from coal fly ash from Gandhinagar Gujarat thermal power plant a mechanical-method-of-isolation. *Int. J. Eng. Res. Technol.* **2014**, *3*, 471–477.
71. Kumar, S.; Das, S.K.; Daspoddar, K. Synthesis of mullite aggregates from fly ash: Effect on thermomechanical behaviour of low cement castables. *Br. Ceram. Trans.* **2004**, *103*, 176–180. [CrossRef]
72. Yan, F.; Jiang, J.; Li, K.; Liu, N.; Chen, X.; Gao, Y.; Tian, S. Green Synthesis of Nanosilica from Coal Fly Ash and Its Stabilizing Effect on CaO Sorbents for CO_2 Capture. *Environ. Sci. Technol.* **2017**, *51*, 7606–7615. [CrossRef] [PubMed]
73. Ojha, K.; Pradhan, N.C.; Samanta, A.N. Zeolite from fly ash: Synthesis and characterization. *Bull. Mater. Sci.* **2004**, *27*, 555–564. [CrossRef]
74. Kumar, S.; Srivastava, V.; Agarwal, V.C. Utilization of fly ash and lime in PPC concrete. *Int. J. Eng. Tech. Res. (IJETR)* **2015**, *3*, 121–124.
75. Al-Shether, B.; Shamsa, M.; Al-Attar, T. Relationship between amorphous silica in source materials and compressive strength of geopolymer concrete. In *MATEC Web of Conferences*; EDP Sciences: Bhagdad, Iraq, 2018; Volume 162, p. 02019.
76. Havlíček, D.; Přibil, R.; Kratochvíl, B. Content of quartz and mullite in some selected power-plant fly ash in Czechoslovakia. *Atmos. Environ.* **1989**, *23*, 701–706. [CrossRef]
77. Ryu, G.; Lee, Y.B.; Koh, K.T.; Chung, Y.S. The mechanical properties of fly ash-based geopolymer concrete with alkaline activators. *Constr. Build. Mater.* **2013**, *47*, 409–418. [CrossRef]
78. Fuller, A.; Maier, J.; Karampinis, E.; Kalivodova, J.; Grammelis, P.; Kakaras, E.; Scheffknecht, G. Fly Ash Formation and Characteristics from (co-)Combustion of an Herbaceous Biomass and a Greek Lignite (Low-Rank Coal) in a Pulverized Fuel Pilot-Scale Test Facility. *Energies* **2018**, *11*, 1581. [CrossRef]
79. Inada, M.; Eguchi, Y.; Enomoto, N.; Hojo, J. Synthesis of zeolite from coal fly ashes with different silica–alumina composition. *Fuel* **2015**, *84*, 299–304. [CrossRef]
80. Yadav, V.K.; Suriyaprabha, R.; Khan, S.H.; Singh, B.; Gnanamoorthy, G.; Choudhary, N.; Kalasariya, H. A novel and efficient method for the synthesis of amorphous nanosilica from fly ash tiles. *Mater. Today Proc.* **2020**, *26*, 701–705. [CrossRef]
81. Colilla, M.; Baeza, A.; Vallet-Regí, M. Mesoporous Silica Nanoparticles for Drug Delivery and Controlled Release Applications. *Sol-Gel Handb.* **2015**, *1*, 1309–1344. [CrossRef]
82. Devi, L.R.; Singh, M.O. Application of silica based heterogeneous catalyst for the synthesis of bioactive heterocycles. In *Heterogeneous Catalysis*; Penoni, A., Penoni, A., Eds.; CRC Press: Boca Raton, FL, USA, 2014; Chapter 6; pp. 163–190.
83. Sharma, P.; Han, M.H.; Cho, C.H. Synthesis of Zeolite Nanomolecular Sieves of Different Si/Al Ratios. *J. Nanomater.* **2015**, *16*, 9. [CrossRef]
84. Singh, G.B.; Subramaniam, K. Characterization of Indian fly ashes using different Experimental Techniques. *Indian Concr. J.* **2018**, *92*, 10–23.
85. Zhao, L.; Xiao, H.; Wang, B.; Sun, Q. Characterization of Glasses in One Type of Alumina Rich Fly Ash by Chemical Digestion Methods: Implications for Alumina Extraction. *J. Chem.* **2016**, *2016*, 10. [CrossRef]
86. Ribeiro, M.J.; Abrantes, J.C.; Ferreira, J.M.; Labrincha, J.A. Recycling of Alrich industrial sludge in refractory ceramic pressed bodies. *Ceram. Int.* **2002**, *28*, 319–326. [CrossRef]
87. Paglia, G.; Buckley, C.E.; Rohl, A.L.; Hart, R.D.; Winter, K.; Studer, A.J.; Hunter, B.A.; Hanna, J.V. Boehmite derived γ-alumina system. 1. Structural evolution with temperature, with the identification and structural determination of a new transition phase, γ'-alumina. *Chem. Mater.* **2004**, *16*, 220–236.
88. Piconi, C. Alumina Ceramics for Biomedical Applications. In Proceedings of the 3rd International Conference on "High-Tech Aluminas and Unfolding their Business Prospects" (Aluminas-2013), Kolkata, West Bengal, India, 20 March 2013.
89. Fomenko, E.; Anshits, N.; Solovyov, A.L.; Mikhaylova, O.A.; Anshits, A.G. Composition and Morphology Of Fly Ash Cenospheres Produced from the Combustion of Kuznetsk Coal. *Energy Fuels* **2013**, *27*, 5440–5448. [CrossRef]

90. Fomenko, E.V.; Anshits, N.N.; Vasil'Eva, N.G.; Rogovenko, E.S.; Mikhaylova, O.A.; Mazurova, E.V.; Solovyev, L.A.; Anshits, A.G. Composition and structure of the shells of aluminosilicate microspheres in fly ash formed on the combustion of Ekibastuz coal. *Solid Fuel Chem.* **2016**, *50*, 238–247. [CrossRef]
91. Kruger, R.A. The Use of Cenospheres in Refractories. *Energeia* **1996**, *7*, 1–5.
92. Ghosal, S.; Ebert, J.L.; Self, S.A. Chemical composition and size distributions for fly ashes. *Fuel Process. Technol.* **1995**, *44*, 81–94. [CrossRef]
93. Alcala, J.C.; Davila, R.M.; Quintero, R.L. Recovery of cenospheres and magnetite from coal burning power plant fly ash. *Trans. Iron Steel Inst. Jpn.* **1987**, *27*, 531–538. [CrossRef]
94. Yadav, V.K.; Yadav, K.K.; Tirth, V.; Jangid, A.; Gnanamoorthy, G.; Choudhary, N.; Islam, S.; Gupta, N.; Son, C.T.; Jeon, B.-H. Recent Advances in Methods for Recovery of Cenospheres from Fly Ash and Their Emerging Applications in Ceramics, Composites, Polymers and Environmental Cleanup. *Crystals* **2021**, *11*, 1067. [CrossRef]
95. Rodrigo, D.D.; Boch, P. High purity mullite ceramics by reaction sintering. *Int. J. High Technol. Ceram.* **1985**, *1*, 3–30. [CrossRef]
96. Schneider, H.; Fischer, R.X.; Schreuer, J. Mullite: Crystal Structure and Related Properties. *J. Am. Ceram. Soc.* **2015**, *98*, 2948–2967. [CrossRef]
97. Schmucker, M.; Hildmann, B.; Schneider, H. Mechanism of 2/1-to 3/2-mullite transformation at 1650 C. *Am. Mineral.* **2002**, *87*, 1190–1193. [CrossRef]
98. Jiangfeng, C.H.E.N.; Longyi, S.H.A.O.; Jing, L.U. Synthesis of Mullite from High-alumina Fly Ash: A Case from the Jungar Power Plant in Inner Mongolia, Northern China. *Acta Geol. Sin. Engl. Ed.* **2008**, *82*, 99–104. [CrossRef]
99. Yadav, V.K.; Saxena, P.; Lal, C.; Gnanamoorthy, G.; Choudhary, N.; Singh, B.; Tavker, N.; Kalasariya, H.; Kumar, P. Synthesis and Characterization of Mullites From Silicoaluminous Fly Ash Waste. *Int. J. Appl. Nanotechnol. Res. (IJANR)* **2020**, *5*, 10–25. [CrossRef]
100. Virendra, K.Y.; Suriyaprabha, R.; Inwati, G.K.; Gupta, N.; Singh, B.; Lal, C.; Kumar, P.; Godha, M.; Kalasariya, H. A Noble and Economical Method for the Synthesis of Low Cost Zeolites From Coal Fly Ash Waste. *Adv. Mater. Process. Technol.* **2021**, 1–19. [CrossRef]
101. Wahyudi, A.; Kurniawan, W.; Husaini, A.A.M.; Hinode, H. Potential Application of Red Mud (Bauxite Residue) in Indonesia. In Proceedings of the Seminar-Workshop on Utilization of Waste Materials, Manila, Philippines, 11–15 February 2015; pp. 57–62.
102. Balomenos, E.; Giannopoulou, I.; Panias, D.; Paspaliaris, I.; Perry, K.; Boufounos, D. Efficient and complete exploitation of the bauxite residue (red mud) produced in Bayer Process. In Proceedings of the European Metallurgical Conference (EMC, 2011), Duesseldorf, Germany, 26–29 June 2011; Volume 3, pp. 745–757.
103. Evans, K.; Nordheim, E.; Tsesmelis, K. Bauxite Residue Management. In *Light Metals*; Springer: Cham, Switzerland, 2012; pp. 63–66.
104. Paramguru, R.K.; Rath, C.; Misra, V.N. Trends in red mud utilization—A review Mineral Processing and Extractive Metall. *Miner. Process. Extr. Metall. Rev.* **2005**, *26*, 1–29.
105. Tan, R.B.; Khoo, H.H. An LCA study of a primary aluminum supply chain. *J. Clean. Prod.* **2005**, *13*, 607–618. [CrossRef]
106. Pogue, A.I.; Lukiw, W.J. The mobilization of aluminum into the biosphere. *Front. Neurol.* **2014**, *5*, 262. [CrossRef]
107. Abhilash, S.S.; Meshram, P.; Pandey, B.D.; Behera, K.; Satapathy, B.K. Redmud: A secondary resource for rare earth elements. In International bauxite, alumina and aluminium symposium. *IBAAS Bind.* **2014**, *3*, 148–162.
108. Moise, G.; Capota, P.; Enache, L.; Neagu, E.; Dragut, D.; Mihaescu, D.; Sarbu, A. Material composition and properties of red mud coming from domestic alumina processing plant. In Proceedings of the 20th International Symposium "Environment and Industry" SIMI 2017, Section Pollution Control and Monitoring, Bucharest, Romania, 28–29 September 2017; pp. 279–289.
109. Zhang, Y.; Wang, Y.; Meng, X.; Zheng, L.; Gao, J. The Suppression Characteristics of $NH_4H_2PO_4$/Red Mud Composite Powders on Methane Explosion. *Appl. Sci.* **2018**, *8*, 1433. [CrossRef]
110. Dentoni, V.; Grosso, B.; Massacci, G. Environmental Sustainability of the Alumina Industry in Western Europe. *Sustainability* **2014**, *6*, 9477–9493. [CrossRef]
111. Milacic, R.; Zuliani, T.; Ščančar, J. Environmental impact of toxic elements in red mud studied by fractionation and speciation procedures. *Sci. Total. Environ.* **2012**, *426*, 359–365. [CrossRef]
112. Li, X.B.; Xiao, W.; Liu, W.; Liu, G.H.; Peng, Z.H.; Zhou, Q.S.; Qi, T.G. Recovery of alumina and ferric oxide from Bayer red mud rich in iron by reduction sintering. *Trans. Nonferrous Met. Soc. China* **2009**, *19*, 1342–1347. [CrossRef]
113. Salman, A.D.; Juzsakova, T.; Rédey, Á.; Le, P.C.; Nguyen, X.C.; Domokos, E.; Abdullah, T.A.; Vagvolgyi, V.; Chang, S.W.; Nguyen, D.D. Enhancing the Recovery of Rare Earth Elements from Red Mud. *Chem. Eng. Technol.* **2021**, *44*, 1768–1774. [CrossRef]
114. Zhang, R.; Zheng, S.; Ma, S.; Zhang, Y. Recovery of alumina and alkali in Bayer red mud by the formation of andradite-grossular hydrogarnet in hydrothermal process. *J. Hazard. Mater.* **2011**, *189*, 827–835. [CrossRef]
115. Zhu, X.F.; Zhang, T.A.; Wang, Y.X.; Lü, G.Z.; Zhang, W.G. Recovery of alkali and alumina from Bayer red mud by the calcification–carbonation method. *Int. J. Miner. Metall. Mater.* **2016**, *23*, 257–268. [CrossRef]
116. Meher, S.N. Alumina extraction from red mud by $CaCO_3$ and Na_2CO_3 sinter process. *Int. J. Chem. Stud.* **2016**, *4*, 122–127.
117. Wang, S.; Ang, H.M.; Tadé, M.O. Novel applications of red mud as coagulant, adsorbent and catalyst for environmentally benign processes. *Chemosphere* **2008**, *72*, 1621–1635. [CrossRef] [PubMed]

118. Zouboulis, A.I.; Kydros, K.A. Use of red mud for toxic metals removal: The case of nickel. *J. Chem. Technol. Biotechnol.* **1993**, *58*, 95–101. [CrossRef]
119. Eamsiri, A.; Jackson, W.R.; Pratt, K.C.; Christov, V.; Marshall, M. Activated red mud as a catalyst for the hydrogenation of coals and of aromatic compounds. *Fuel* **1992**, *71*, 449–453. [CrossRef]
120. Hodge, W.W. Wastes Problems of the Iron and Steel Industries. *Ind. Eng. Chem.* **1939**, *31*, 1364–1380. [CrossRef]
121. Sarkar, S. Solid waste management in steel industry-challanges and opportunity. *Int. J. Nucl. Energy Sci. Technol.* **2015**, *9*, 884–887.
122. Anjum, M.; Miandad, R.; Waqas, M.; Gehany, F.; Barakat, M.A. Remediation of wastewater using various nanomaterials. *Arab. J. Chem.* **2019**, *12*, 4897–4919. [CrossRef]
123. Weidner, E.; Ciesielczyk, F. Removal of Hazardous Oxyanions from the Environment Using Metal-Oxide-Based Materials. *Materials* **2019**, *12*, 927. [CrossRef] [PubMed]
124. Akbarzadeh, A.; Samiei, M.; Davaran, S. Magnetic nanoparticles: Preparattion, physical properties and applications in biomedicine. *Nanoscale Res. Lett.* **2012**, *7*, 144. [CrossRef]
125. Tang, Z.; Ding, X.; Yan, X.; Dong, Y.; Liu, C. Recovery of Iron, Chromium, and Nickel from Pickling Sludge Using Smelting Reduction. *Metals* **2018**, *8*, 936. [CrossRef]
126. Tang, C.; Li, K.; Ni, W.; Fan, D. Recovering Iron from Iron Ore Tailings and Preparing Concrete Composite Admixtures. *Minerals* **2019**, *9*, 232. [CrossRef]
127. Zhang, G.F.; Yang, Q.R.; Yang, Y.D.; Wu, P.; McLean, A. Recovery of iron from waste slag of pyrite processing using reduction roasting magnetic separation method. *Can. Metall. Q.* **2013**, *52*, 153–159. [CrossRef]
128. Wang, Y.U.; Peng, Y.L.; Zheng, Y.J. Recovery of iron from waste ferrous sulphate by co-precipitation and magnetic separation. *Trans. Nonferrous Met. Soc. China* **2017**, *27*, 211–219.
129. Yadav, V.K.; Yadav, K.K.; Cabral-Pinto, M.M.S.; Choudhary, N.; Gnanamoorthy, G.; Tirth, V.; Prasad, S.; Khan, A.H.; Islam, S.; Khan, N.A. The Processing of Calcium Rich Agricultural and Industrial Waste for Recovery of Calcium Carbonate and Calcium Oxide and Their Application for Environmental Cleanup: A Review. *Appl. Sci.* **2021**, *11*, 4212. [CrossRef]
130. Ko, J.H.; Xu, Q.; Jang, Y.C. Emissions and Control of Hydrogen Sulfide at Landfills: A Review. *Crit. Rev. Environ. Sci. Technol.* **2015**, *45*, 2043–2083. [CrossRef]
131. Xu, Q.; Townsend, T.; Bitton, G. Inhibition of hydrogen sulfide generation from disposed gypsum drywall using chemical inhibitors. *J. Hazard. Mater.* **2011**, *191*, 204–211. [CrossRef]
132. De Beer, M.; Doucet, F.J.; Maree, J.; Liebenberg, L. Synthesis of high-purity precipitated calcium carbonate during the process of recovery of elemental sulphur from gypsum waste. *Waste Manag.* **2015**, *46*, 619–627. [CrossRef]
133. Cuesta Mayorga, I.; Astilleros, J.M.; Fernández-Díaz, L. Precipitation of CaCO3 Polymorphs from Aqueous Solutions: The Role of pH and Sulphate Groups. *Minerals* **2019**, *9*, 178. [CrossRef]
134. Sufang, W.U.; Peiqiang, L.A.N. Method for Preparing a Nano-Calcium Carbonate Slurry from Waste Gypsum as Calciumsource, the Product and Use Thereof. United States Patent Application Publication. U.S. Patent 8,846,562, 30 September 2014.
135. Okumura, S.; Mihara, N.; Kamiya, K.; Ozawa, S.; Onyango, M.S.; Kojima, Y.; Iwashita, T. Recovery of CaO by Reductive Decomposition of Spent Gypsum in a $CO-CO_2-N_2$ Atmosphere. *Ind. Eng. Chem. Res.* **2003**, *42*, 6046–6052. [CrossRef]
136. Ramachandran, N.; Maniam, G. Regeneration of calcium oxide (CaO) from waste gypsum via two-step reaction. In Proceedings of the 8th Malaysian Technical Universities Conference on Engineering & Technology (MUCET 2014), Malacca, Malaysia, 10–11 November 2014.
137. Mbhele, N.R.; Van Der Merwe, W.; Maree, J.; Theron, D. Recovery of sulphur from waste gypsum. In Proceedings of the Abstracts of the International Mine Water Conference, Mpumalanga, South Africa, 19–23 October 2009; pp. 622–630.
138. Nagrale, S.D.; Hajare, H.; Modak, R. Utilization of rice husk ash. *Int. J. Eng. Res. Appl.* **2012**, *2*, 42.
139. Yadav, V.K.; Choudhary, N.; Tirth, V.; Kalasariya, H.; Gnanamoorthy, G.; Algahtani, A.; Yadav, K.K.; Soni, S.; Islam, S.; Yadav, S.; et al. A Short Review on the Utilization of Incense Sticks Ash as an Emerging and Overlooked Material for the Synthesis of Zeolites. *Crystals* **2021**, *11*, 1255. [CrossRef]
140. Tomar, R.; Kishore, K.; Singh Parihar, H.; Gupta, N. A comprehensive study of waste coconut shell aggregate as raw material in concrete. *Mater. Today Proc.* **2021**, *44*, 437–443. [CrossRef]
141. Channoy, C.; Maneewan, S.; Punlek, C.; Chirarattananon, S. Preparation and Characterization of Silica Gel from Bagasse Ash. In *Advanced Materials Research*; Trans Tech Publications Ltd.: Zurich, Switzerland, 2018; Volume 1145, pp. 44–48. [CrossRef]
142. Balasubramaniam, P.; Peera, S.P.G.; Mahendran, P.P.; Tajuddin, A. Release of silicon from soil applied with graded levels of fly ash with silicate solubilizing bacteria and farm yard manure. In Proceedings of the 5th International Conference on Silicon in Agriculture, Beijing, China, 13–18 September 2011.
143. Yadav, V.K.; Malik, P.; Khan, A.H.; Pandit, R.; Hasan, M.A.; Cabral-Pinto, M.M.S.; Islam, S.; Suriyaprabha, R.; Yadav, K.K.; Dinis, A.; et al. Recent Advances on Properties and Utility of Nanomaterials Generated from Industrial and Biological Activities. *Crystals* **2021**, *11*, 634. [CrossRef]
144. Nayak, P.P.; Datta, A. Synthesis of SiO_2-Nanoparticles from Rice Husk Ash and its Comparison with Commercial Amorphous Silica through Material Characterization. *Silicon* **2021**, *13*, 1209–1214. [CrossRef]
145. Mokhena, T.C.; Mochane, M.J.; Motaung, T.E.; Linganiso, L.Z.; Thekisoe, O.M.; Songca, S. Sugarcane Bagasse and Cellulose Polymer Composites. In *Sugarcane-Technology and Research*; IntechOpen: London, UK, 2018.

146. Schettino, M.A.S.; Holanda, J.N.F. Characterization of sugarcane bagasse ash waste for its use in ceramic floor tile. *Procedia Mater. Sci.* **2015**, *8*, 190–196. [CrossRef]
147. Drummond, A.R.F.; Drummond, W.I. Pyrolysis of Sugar Cane Bagasse in a Wire-Mesh Reactor. *Ind. Eng. Chem. Res.* **1996**, *35*, 1263–1268. [CrossRef]
148. Harish, R.; Aru, A.; Ponnusami, V. Recovery of silica from various low cost precursors for the synthesis of silica gel. *Pharm. Lett.* **2015**, *7*, 208–213.
149. Norsuraya, S.; Hamzah, F.; Rahmat, N. Sugarcane Bagasse as a Renewable Source of Silica to Synthesize Santa Barbara Amorphous-15 (SBA-15). *Procedia Eng.* **2016**, *148*, 839–846. [CrossRef]
150. Rovani, S.; Santos, J.J.; Corio, P.; Fungaro, D.A. Highly Pure Silica Nanoparticles with High Adsorption Capacity Obtained from Sugarcane Waste Ash. *ACS Omega* **2018**, *3*, 2618–2627. [CrossRef] [PubMed]
151. Manjula, C.; Kukkamgai, S.; Rahman, S.; Rajesh, M.K. Characterization of Kuttiyadi ecotype of coconut (*Cocos nucifera* L.) using morphological and microsatellite markers. *J. Plant. Crops* **2014**, *42*, 301–315.
152. Omer, A.M. Agricultural residues for future energy option in Sudan: An analysis. *Ann. Adv. Chem.* **2016**, *2*, 17–31. [CrossRef]
153. Anuar, M.F.; Fen, Y.W.; Zaid, M.H.M.; Matori, K.A.; Khaidir, R.E.M. Synthesis and structural properties of coconut husk as potential silica source. *Results Phys.* **2018**, *11*, 1–4. [CrossRef]
154. Rodgers, K.; Hursthouse, A.; Cuthbert, S. The Potential of Sequential Extraction in the Characterisation and Management of Wastes from Steel Processing: A Prospective Review. *Int. J. Environ. Res. Public Health* **2015**, *12*, 11724–11755. [CrossRef] [PubMed]
155. Sivasubramanian, S.; Sravanthi, K. Synthesis and characterisation of silica nano particles from coconut shell. *Int. J. Pharma Bio Sci.* **2015**, *6*, 530–536.
156. Melati, A.; Hidayati, E. Synthesis and characterization of carbon nanotube from coconut shells activated carbon. *J. Phys. Conf. Ser.* **2016**, *694*, 012073. [CrossRef]
157. Adewumi, G.A.; Revaprasadu, N.; Eloka-Eboka, A.C.; Inambo, F.L.; Gervas, C. A facile low-cost synthesis of carbon nanosphere from coconut fibre. In Proceedings of the World Congress on Engineering and Computer Science, San Francisco, CA, USA, 25–27 October 2017; pp. 577–582.
158. Hakim, Y.Z.; Yulizar, Y.; Nurcahyo, A.; Surya, M. Green Synthesis of Carbon Nanotubes from Coconut Shell Waste for the Adsorption of Pb(II) Ions. *Acta Chim. Asiana* **2018**, *1*, 6–10. [CrossRef]
159. Kumar, S.; Bhatt, B. Status and production technology of maize. In *Status of Agricultural Development in Eastern India*; Bhatt, B.P., Sikka, A., Mukharjee, J., Islam, A., Dey, A., Eds.; ICAR RCER: Patna, India, 2012; pp. 151–167.
160. Berber-Villamar, N.K.; Netzahuatl-Muñoz, A.R.; Morales-Barrera, L.; Chávez-Camarillo, G.M.; Flores-Ortiz, C.M.; Cristiani-Urbina, E. Corncob as an effective, eco-friendly, and economic biosorbent for removing the azo dye Direct Yellow 27 from aqueous solutions. *PLoS ONE* **2018**, *13*, e0196428. [CrossRef]
161. Sun, Y.; Webley, P.A. Preparation of activated carbons from corncob with large specific surface area by a variety of chemical activators and their application in gas storage. *Chem. Eng. J.* **2010**, *162*, 883–892. [CrossRef]
162. Sabir, M.; Zia-ur-Rehman, M.; Hakeem, K.R.; Saifullah, U. Phytoremediation of Metal-Contaminated Soils Using Organic Amendments: Prospects and Challenges. In *Soil Remediation and Plants*; Hakeem, K.R., Sabir, M., Öztürk, M., Mermut, A.R., Eds.; Academic Press: London, UK, 2015; pp. 503–523.
163. Dias, J.M.; Alvim-Ferraz, M.C.; Almeida, M.F.; Rivera-Utrilla, J.; Sánchez-Polo, M. Waste Materials for Activated Carbon Preparation and Its Use in Aqueous-phase Treatment: A Review. *J. Environ. Manag.* **2007**, *85*, 833–846. [CrossRef]
164. Wang, J.; Kaskel, S. KOH activation of carbon-based materials for energy storage. *J. Mater. Chem.* **2012**, *22*, 23710–23725. [CrossRef]
165. Rashed, M.N. Adsorption Technique for the Removal of Organic Pollutants from Water and Wastewater. *Org. Pollut. Monit. Risk Treat.* **2013**, *7*, 167–194.
166. Tsai, W.T.; Chang, C.Y.; Wang, S.Y.; Chang, C.F.; Chien, S.F.; Sun, H.F. Utilization of agricultural waste corn cob for the preparation of carbon adsorbent. *J. Environ. Sci. Health* **2011**, *36*, 677–686. [CrossRef]
167. Tsai, W.T.; Chang, C.Y.; Wang, S.Y.; Chang, C.F.; Chien, S.F.; Sun, H.F. Preparation of activated carbons from corn cob catalyzed by potassium salts and subsequent gasification with CO_2. *Bioresour. Technol.* **2001**, *78*, 203–208. [CrossRef]
168. Kaźmierczak-Raźna, J.; Nowicki, P.; Pietrzak, R. Sorption properties of activated carbons obtained from corn cobs by chemical and physical activation. *Adsorption* **2013**, *19*, 273–281. [CrossRef]
169. Hashemian, S.; Salari, K.; Yazdi, Z.A. Preparation of activated carbon from agricultural wastes (almond shell and orange peel) for adsorption of 2-pic from aqueous solution. *J. Ind. Eng. Chem.* **2014**, *20*, 1892–1900. [CrossRef]
170. Ratan, J.K.; Kaur, M.; Adiraju, B. Synthesis of activated carbon from agricultural waste using a simple method: Characterization, parametric and isotherms study. *Mater. Today Proc.* **2018**, *5*, 3334–3345. [CrossRef]
171. Zhu, T.; Dittrich, M. Carbonate Precipitation through Microbial Activities in Natural Environment, and Their Potential in Biotechnology: A Review. *Front. Bioeng. Biotechnol.* **2016**, *4*, 4. [CrossRef]
172. Al Omari, M.M.H.; Rashid, I.S.; Qinna, N.A.; Jaber, A.M.; Badwan, A.A. Calcium Carbonate. In *Profiles of Drug Substances, Excipients and Related Methodology*; Brittain, H.G., Ed.; Academic Press: London, UK, 2016; Volume 41, pp. 31–132.
173. Cho, S.H.; Park, J.K.; Lee, S.K.; Joo, S.M.; Kim, I.H.; Ahn, J.-W.; Kim, H. Synthesis of Precipitated Calcium Carbonate Using a Limestone and its Application in Paper Filler and Coating Color. In *Materials Science Forum*; Trans Tech Publications Ltd.: Zurich, Switzerland, 2007; pp. 881–884.

174. Huwald, E. Calcium carbonate-pigment and filler. In *Calcium Carbonate: From the Cretaceous Period into the 21st Century*; Tegethoff, F.W., Ed.; Birkhäuser Basel: Basel, Switzerland, 2001; pp. 160–170.
175. Boyjoo, Y.; Pareek, V.K.; Liu, J. Synthesis of micro and nano-sized calcium carbonate particles and their applications. *J. Mater. Chem. A* **2014**, *2*, 14270–14288. [CrossRef]
176. Thomas, S.; Sharma, H.; Mishra, P.; Talegaonkar, S. Ceramic Nanoparticles: Fabrication Methods and Applications in Drug Delivery. *Curr. Pharm. Des.* **2015**, *21*, 6165–6188. [CrossRef] [PubMed]
177. Dizaj, S.M.; Barzegar-Jalali, M.; Zarrintan, M.H.; Adibkia, K.; Lotfipour, F. Calcium carbonate nanoparticles as cancer drug delivery system. *Expert Opin. Drug Deliv.* **2015**, *12*, 1649–1660. [CrossRef]
178. Hua, K.H.; Wang, H.C.; Chung, R.S.; Hsu, J.C. Calcium carbonate nanoparticles can enhance plant nutrition and insect pest tolerance. *J. Pestic. Sci.* **2015**, *40*, 208–213. [CrossRef]
179. Aframehr, W.M.; Molki, B.; Heidarian, P.; Behzad, T.; Sadeghi, M.; Bagheri, R. Effect of calcium carbonate nanoparticles on barrier properties and biodegradability of polylactic acid. *Fibers Polym.* **2017**, *18*, 2041–2048. [CrossRef]
180. Trushina, D.B.; Bukreeva, T.V.; Kovalchuk, M.V.; Antipina, M.N. CaCO3 vaterite microparticles for biomedical and personal care applications. *Mater. Sci. Eng. C* **2014**, *45*, 644–658. [CrossRef]
181. Grasby, S.E. Naturally precipitating vaterite (μ-CaCO$_3$) spheres: Unusual carbonates formed in an extreme environment. *Geochim. Et Cosmochim. Acta* **2013**, *67*, 1659–1666. [CrossRef]
182. De Leeuw, N.H.; Parker, S.C. Surface Structure and Morphology of Calcium Carbonate Polymorphs Calcite, Aragonite, and Vaterite: An Atomistic Approach. *J. Phys. Chem. B* **1998**, *102*, 2914–2922. [CrossRef]
183. Buasri, A.; Chaiyut, N.; Loryuenyong, V.; Worawanitchaphong, P.; Trongyong, S. Calcium Oxide Derived from Waste Shells of Mussel, Cockle, and Scallop as the Heterogeneous Catalyst for Biodiesel Production. *Sci. World J.* **2013**, *2013*, 7. [CrossRef] [PubMed]
184. Dewangan, S.; Chakrabarty, R.; Zielinska, B.; Pervez, S. Emission of volatile organic compounds from religious and ritual activities in India. *Environ. Monit. Assess.* **2013**, *185*, 9279–9286. [CrossRef] [PubMed]
185. Hazarika, P.; Dutta, N.B.; Biswas, S.C.; Dutta, R.C.; Jayaraj, R.S.C. Status of agarbatti industry in India with special reference to Northeast. *Int. J. Adv. Res. Biol. Sci.* **2018**, *5*, 173–186.
186. Settimo, G.; Tirler, W. Incense, sparklers and cigarettes are significant contributors to indoor benzene and particle levels. *Ann. Ist. Super. Sanita.* **2015**, *51*, 28–33.
187. Abdulrahman, I.; Tijani, H.I.; Mohammed, B.A.; Saidu, H.; Yusuf, H.; Jibrin, M.N.; Mohammed, S. From garbage to biomaterials:an overview on eggshell based hydroxyapatite. *J. Mater.* **2014**, *2014*, 802467.
188. Asri, N.P.; Podjojono, B.; Fujiani, R. Utilization of eggshell waste as low-cost solid base catalyst for biodiesel production from used cooking oil. In *IOP Conference Series: Earth and Environmental Science*; IOP Publishing: Bristol, UK, 2017; Volume 67, p. 012021.
189. Ummartyotin, S.; Tangnorawich, B. Utilization of eggshell waste as raw material for synthesis of hydroxyapatite. *Colloid Polym. Sci.* **2015**, *293*, 2477–2483. [CrossRef]
190. Hassan, T.; Rangari, V.; Rana, R.; Jeelani, S. Sonochemical effect on size reduction of CaCO3 nanoparticles derived from waste eggshells. *Ultrason. Sonochem.* **2013**, *20*, 1308–1315. [CrossRef]
191. Hassan, T.; Rangari, V.; Jeelani, S. Sonochemical synthesis and characterisation of bio-based hydroxyapatite nanoparticles. *Int. J. Nano Biomater.* **2014**, *5*, 103–112. [CrossRef]
192. Hariharan, M.; Varghese, N.; Cherian, A.B.; Paul, J. Synthesis and characterisation of CaCO3 (Calcite) nano particles from cockle shells using chitosan as precursor. *Int. J. Sci. Res. Publ.* **2014**, *4*, 5.
193. Channappa, B.; Kambalagere, Y.; Mahadevan, K.M.; Narayanappa, M. Synthesis of Calcium Oxide Nanoparticles and Its Mortality Study on Fresh Water Fish Cyprinus Carpio. *IOSR J. Environ. Sci. Toxicol. Food Technol.* **2016**, *10*, 55–60.
194. Mirghiasi, Z.; Bakhtiari, F.; Darezereshki, E.; Esmaeilzadeh, E. Preparation and characterization of CaO nanoparticles from Ca(OH)$_2$ by direct thermal decomposition method. *J. Ind. Eng. Chem.* **2014**, *20*, 113–117. [CrossRef]
195. Roy, A.; Bhattacharya, J. Microwave-assisted synthesis and characterization of CaO nanoparticles. *Int. J. Nanosci.* **2011**, *10*, 413–418. [CrossRef]
196. Darcanova, O.; Beganskienė, A.; Kareiva, A. Sol-gel synthesis of calcium nanomaterial for paper conservation. *Chemija* **2015**, *26*, 25–31.
197. Ali, S.; Butt, A.; Ejaz, S.; Baron, C.J.; Ikram, D.M. CaO nanoparticles as a potential drug delivery agent for biomedical applications. *Dig. J. Nanomater. Biostructures* **2015**, *10*, 799.
198. Sadeghi, M.; Husseini, M.H. A Novel Method for the Synthesis of CaO Nanoparticle for the Decomposition of Sulfurous Pollutant. *J. Appl. Chem. Res.* **2013**, *7*, 10.
199. Liu, T.; Zhu, Y.; Zhang, X.; Zhang, T.; Zhang, J.; Li, X. Synthesis and characterization of calcium hydroxide nanoparticles by hydrogen plasma-metal reaction method. *Mater. Lett.* **2010**, *64*, 2575–2577. [CrossRef]
200. Yorug, A.H.; Ipek, Y. Sonochemical Synthesis of Hydroxyapatite Nanoparticles with Different Precursor Reagents. *Acta Phys. Pol.-Ser. A Gen. Phys.* **2012**, *121*, 230.
201. Mohd Abd Ghafar, S.L.; Hussein, M.Z.; Abu Bakar Zakaria, Z. Synthesis and Characterization of Cockle Shell-Based Calcium Carbonate Aragonite Polymorph Nanoparticles with Surface Functionalization. *J. Nanoparticles* **2017**, *2017*, 12. [CrossRef]
202. Mohadi, R.; Anggraini, K.; Riyanti, F.; Lesbani, A. Preparation Calcium Oxide from Chicken Eggshells. *Sriwijaya J. Environ.* **2016**, *1*, 32–35. [CrossRef]

203. Mulopo, L.; Radebe, V. Recovery of calcium carbonate from waste gypsum and utilization for remediation of acid mine drainage from coal mines. *Water Sci. Technol.* **2012**, *66*, 1296–1300. [CrossRef] [PubMed]
204. Render, D.; Samuel, T.; King, H.; Vig, M.; Jeelani, S.; Babu, R.J.; Rangari, V. Biomaterial-Derived Calcium Carbonate Nanoparticles for Enteric Drug Delivery. *J. Nanomater.* **2016**, *2016*, 8. [CrossRef] [PubMed]
205. Tangboriboon, N.; Kunanuruksapong, R.; Sirivat, A. Preparation and properties of calcium oxide from eggshells via calcination. *Mater. Sci.-Pol.* **2012**, *30*, 313–322. [CrossRef]
206. Jirimali, H.D.; Chaudhari, B.C.; Khanderay, J.C.; Joshi, S.A.; Singh, V.; Patil, A.M.; Gite, V.V. Waste Eggshell-Derived Calcium Oxide and Nanohydroxyapatite Biomaterials for the Preparation of LLDPE Polymer Nanocomposite and Their Thermomechanical Study. *Polym. Plast. Technol. Eng.* **2018**, *57*, 804–811. [CrossRef]

Review

Pyrolytic Conversion of Plastic Waste to Value-Added Products and Fuels: A Review

Sadegh Papari, Hanieh Bamdad and Franco Berruti *

Department of Chemical and Biochemical Engineering, Institute for Chemicals and Fuels from Alternative Resources (ICFAR), Western University, London, ON N6A 3K7, Canada; spapari@uwo.ca (S.P.); hbamdad@uwo.ca (H.B.)
* Correspondence: fberruti@uwo.ca

Abstract: Plastic production has been rapidly growing across the world and, at the end of their use, many of the plastic products become waste disposed of in landfills or dispersed, causing serious environmental and health issues. From a sustainability point of view, the conversion of plastic waste to fuels or, better yet, to individual monomers, leads to a much greener waste management compared to landfill disposal. In this paper, we systematically review the potential of pyrolysis as an effective thermochemical conversion method for the valorization of plastic waste. Different pyrolysis types, along with the influence of operating conditions, e.g., catalyst types, temperature, vapor residence time, and plastic waste types, on yields, quality, and applications of the cracking plastic products are discussed. The quality of pyrolysis plastic oil, before and after upgrading, is compared to conventional diesel fuel. Plastic oil yields as high as 95 wt.% can be achieved through slow pyrolysis. Plastic oil has a heating value approximately equivalent to that of diesel fuel, i.e., 45 MJ/kg, no sulfur, a very low water and ash content, and an almost neutral pH, making it a promising alternative to conventional petroleum-based fuels. This oil, as-is or after minor modifications, can be readily used in conventional diesel engines. Fast pyrolysis mainly produces wax rather than oil. However, in the presence of a suitable catalyst, waxy products further crack into oil. Wax is an intermediate feedstock and can be used in fluid catalytic cracking (FCC) units to produce fuel or other valuable petrochemical products. Flash pyrolysis of plastic waste, performed at high temperatures, i.e., near 1000 °C, and with very short vapor residence times, i.e., less than 250 ms, can recover up to 50 wt.% ethylene monomers from polyethylene waste. Alternatively, pyrolytic conversion of plastic waste to olefins can be performed in two stages, with the conversion of plastic waste to plastic oil, followed by thermal cracking of oil to monomers in a second stage. The conversion of plastic waste to carbon nanotubes, representing a higher-value product than fuel, is also discussed in detail. The results indicate that up to 25 wt.% of waste plastic can be converted into carbon nanotubes.

Keywords: pyrolysis; plastic waste; carbon nanotubes; plastic oil; fuels; monomer recovery; olefins

1. Introduction

Plastic products play a critical role in our lives and are being used in large quantities due to their durability, versatility, light weight, and low cost [1,2]. Plastic waste materials, generated in different sectors of the economy, such as agriculture, residential and commercial, automobiles, construction and demolition, packing materials, toys, and electrical equipment are growing rapidly and are either recycled, combusted (waste incineration), or disposed of [3]. Plastic waste consists mainly of low-density polyethylene (LDPE), high-density polyethylene (HDPE), polypropylene (PP), polyvinylchloride (PVC), polystyrene (PS), and polyethylene terephthalate (PET) [4]. Polyethylene and polypropylene constitute the greatest portion of plastic waste [4].

The increase in the world population and subsequent living standards have caused a rapid increase in municipal solid waste generation of to up to 1.3 billion tons per annum [3]. Reportedly, plastic waste is the third largest contributor of municipal solid waste [4]. The global production of plastic has increased, from 1.5 million tons in 1950 to approximately 359 million tons in 2018 and is attributed to a rapid rise in the packaging/wrapping sector [5]. Today, over 250 million tons per year are either landfilled or dispersed in the environment and an estimated 10 million tons per year end up in the oceans. Considering an increase of 9–13% of plastic waste per year [4], it is predicted that billions of tons of plastic could be produced by 2050, of which the greatest portion could go to landfills or be dispersed, both in the land environment and in the oceans.

An increase in daily demand of plastic materials, which are petroleum-based substances, can result in the depletion of non-renewable fossil resources. Approximately 4% of crude oil production is directly utilized in plastic production [6,7]. In addition to contributing to a global energy crisis, plastic waste can affect the environment and, therefore, either disposing or reusing/recycling is crucial. It is well known that plastics can persist in the environment for a prolonged period. The continuous disposal of plastic wastes is destructive to both terrestrial and marine ecosystem, as they are not readily biodegraded and can take several years to vanish [8]. Photo-degradation, auto-oxidation, thermo-oxidation, thermal degradation and biodegradation are plastic nature-based degradation mechanisms, however with a very slow rate [4]. On the other hand, the pollutants, such as toluenes, xylenes, benzenes, and phenols, released into the air, water, and soil as a result of plastic degradation cause undeniable issues, such as impacts on human and animal health and deterioration of soil fertility [4]. Photo-degradation converts plastic waste into fine pieces (micro-plastics), which float on the surfaces of rivers, ponds, lakes, end up in seas and oceans and can penetrate into the food chain and subsequently pass to humans [9].

In addition to landfilling, there are four distinctive approaches for plastic waste management. Primary mechanical recycling is a technique in which single-type uncontaminated and clean plastic wastes are reprocessed, resulting in a product without changing the basic structure and equivalent quality. Although this method is cost-effective, washing the waste materials generates a new waste stream and, more importantly, plastic wastes usually consist of mixtures of different plastics, often arranged as composites with other materials and are either difficult or impossible to recycle. Secondary recycling is another mechanical recycling approach that follows a decontamination process, remelting, remolding and re-extruding. Size reduction, contaminants removal, separation from other waste materials, make this approach less favorable in terms of operating cost. Tertiary recycling is a chemical or/and thermochemical recycling, which includes chemolysis/solvolysis (i.e., glycolysis, hydrolysis, methanolysis, and alcoholysis), gasification, partial oxidation, and pyrolysis. In this approach, large polymer molecules of plastic wastes are converted into shorter molecules through the use of heat and/or chemical reactions. This technique produces fuels or value-added chemicals that are useful for the synthesis of new plastics and other products. Quaternary recycling is waste combustion (incineration) of the waste material for energy recovery. It seems to be the simplest method; however, it generates pollution and may not meet the circular economy milestones [10,11].

According to the waste management hierarchy (from the most to the least preferred), we identify prevention, minimization, reuse, recycle, energy recovery, and disposal. Energy recovery through incineration is in a lower rank compared to reuse and recycling [12]. As mentioned, the primary and secondary recycling methods suffer some drawbacks, such as labor-intensive operation for the separation process prior to recycling, a high material loss, possible water contamination, and, overall, a high cost. More importantly, the recycled products are often more expensive than the virgin plastics and may not maintain the original properties. As such, the thermochemical conversion (i.e., tertiary recycling) can be an economically and environmentally friendly solution, leading to a high value fuel/chemical production from plastic waste.

The interest in thermochemical conversion of plastic waste, particularly pyrolysis, has increased considerably over the last few years, primarily since China stopped accepting post-consumer plastic waste in 2018, after having taken up to 45% of the world's plastic waste for recycling, landfilling and incineration [13–16]. Zhang et al. [17] conducted a comprehensive review on various advanced non-biodegradable plastic waste treatment technologies. They concluded that physical recycling methods are the most sustainable technologies with consideration of a decrease in the performance of plastic after several recycling cycles. It has been stated that pyrolysis is the most widely used thermal remediation, and gasoline/diesel yield is an indicator parameter which can reflect the actual valuable yield of the process and its industrial applicability. Degradation is another promising technology, but most studies focus on using selected kinds of microorganism to degrade specific polymers. As such, research on microorganisms able to degrade mixtures of various plastics is needed. Fojt et al. [18] critically reviewed the overlooked challenges associated with the accumulation of micro-plastics in the soil. Products made from biodegradable plastics are beginning to replace conventional plastics. Composting is highly suggested for bioplastic disposal; however, the compost formed could contain micro-bioplastic particles resulting from incomplete biodegradation, causing soil contamination. These authors addressed this problem by summarizing sample pre-treatments and analytical techniques. The analytical techniques include both thermo-analytical (i.e., pyrolysis) and non-thermoanalytical (i.e. pre-sorting and respective detection limits) approaches. They concluded that, due to the poor knowledge of the production rate of micro-plastics, fate, sorption properties and toxicity, a rapid and suitable approach is required for their determination. Yet, thermo-analytical approach is the most promising strategy. Murthy et al. [19] carried out an in-depth review study of the plastic pyrolysis process and discussed the influence of various operating parameters as well as the characterization of the liquid oil obtained from the process. The results revealed that most of the plastics produce oil with reasonable calorific values (i.e., approximately similar to conventional fuels). The plastic pyrolysis product distribution depends on the type of reactor used. Significant studies have been conducted on batch-style reactors due to the easy design, fabrication, operation, and control. On the other hand, continuous fluidized bed reactors can provide a uniform mixing of feedstock and heat carriers or catalysts during operation and, therefore, generate more stable products. Nanda and Berruti [20] systematically reviewed solid waste technologies, such as pyrolysis, liquefaction and gasification for converting waste plastic into fuels/chemicals. They stated that pyrolysis and hydrothermal liquefaction technologies are able to reduce the volume of plastics to landfills/oceans, reduce the overall carbon footprints, and, more importantly, have high conversion efficiencies and relatively lower costs when compared to higher temperature processes, such as gasification. Selectively, plastics can be converted either to bio-oil, bio-crude oil, synthesis gas, hydrogen and aromatic char. As such, the influence of process parameters, such as temperature, heating rate, feedstock concentration, reaction time, reactor type, and catalysts, have been discussed thoroughly. Damodharan et al. [21] conducted a review on the utilization of waste plastic oil in diesel engines. They used waste plastic oil obtained from the pyrolysis of mixed waste plastics in the presence of catalyst (e.g. silica, alumina, ZSM-5 and kaolin), with up to 80 wt.% yield. The pyrolysis oil had a lower cetane number than fossil diesel and, therefore, longer ignition delays and higher heat releases. NOx emissions are higher with plastic pyrolysis oil. Smoke emissions were chiefly low with plastic oil and could be further decreased to Euro levels by the use of oxygenated additives. They finally concluded that, plastic pyrolysis oil is a good candidate for fossil diesel replacement and found it to run smoothly in diesel engines. Williams [22] carried out a review on converting waste plastic to hydrogen and carbon nanotubes via pyrolysis coupled with catalytic steam reforming. This author investigated the influence of reactor designs, catalyst type, and operating conditions on the yield and quality of the carbon nanotubes. He concluded that the process temperature along with the type of catalyst are the prominent factors in plastic to hydrogen and carbon nanotubes pyrolysis. There is a balance between introduction of steam, which enhances hydrogen yield, and carbon nanotubes quality, since higher steam flowrates tend to oxidize the carbon nanotubes.

This review aims to thoroughly discuss the different types of plastic waste pyrolysis processes (i.e., slow, fast, and flash) with respect to the quality and quantity of the products. Furthermore, the application of pyrolysis plastic oil, as a fuel and/or material, is reviewed.

Upgrading plastic oil through different methods (e.g., thermal cracking for monomer recovery, hydrogenation, and blending) along with the conversion of plastic waste to carbon nanotubes is reviewed and discussed in detail. This paper contributes to the science of waste management and waste valorization, providing the most updated information and insight through a comprehensive study of the most advanced literature on the pyrolytic conversion of waste plastics.

2. Plastic Waste Properties

To achieve a very good heat/mass transfer during the pyrolysis process, plastic wastes are typically crushed, shredded and sieved to obtain small size flakes, i.e., less than 2 mm. Proximate and ultimate analysis of different plastic wastes are presented in Tables 1 and 2. A high volatile matter content (above 90 wt.%) along with a high carbon and hydrogen content make plastic waste an excellent candidate for the pyrolysis process, leading to a high conversion to the liquid/gas products.

Table 1. Ultimate analysis of different plastic wastes [23].

Plastic Types	Carbon	Hydrogen	Oxygen	Nitrogen	Sulfur
HDPE	78	13	4	0.06	0.08
PP	84	14	1	0.02	0.08
PS	90	9	1	0.07	0.08
PET	77	13	5	0.20	NA

Table 2. Proximate analysis of various plastic types [2].

Plastic Types	Moisture Content	Fixed Carbon	Volatile Matters	Ash Content	HHV (MJ/Kg)
HDPE	0.0	0.3	99.8	1.4	49.4
LDPE	0.3	0.0	99.7	0.4	46.4
PP	0.2	1.2	97.8	1.9	46.4
PS	0.3	0.2	99.6	0.0	41.9
PET	0.5	7.8	91.8	0.1	30.2

3. Pyrolysis Process

Pyrolysis is a versatile thermal cracking process that occurs in the absence of oxygen at temperatures above 400 °C. Typically, pyrolysis processes are classified as slow, fast, and flash [24,25]. This thermochemical process breaks down the long chain polymer molecules into smaller and less complex molecules through heat and chemical reactions. Slow pyrolysis is typically performed at temperatures between 350 and 550 °C, with 1 to 10 °C/min heating rates, and a prolonged vapor residence time. The major product of slow pyrolysis is a solid residue, called char, as a slow heating rate favors solid formation among various parallel-competitive reactions [25]. Fast pyrolysis often takes place at temperatures between 500 and 700 °C. The heating rate experienced by the feedstock is above 1000 °C/min, and vapor residence times are normally in the range of a few seconds [26]. Fast pyrolysis favors liquid production and, depending on the feedstock type, the liquid yield can surprisingly reach up to 90 wt.% for the pyrolysis of polyolefin materials [27]. In flash pyrolysis, the temperature is usually above 700 °C, the heating rate experienced by the feed is extremely fast, and vapor residence times are in the range of milliseconds [25]. Flash pyrolysis can produce higher yields of oil than fast pyrolysis for biomass feedstocks, while it differs for plastic waste, as the latter produces more gas compared to other products [28]. The products obtained from the pyrolysis of plastic wastes (all types, alone, or as mixtures) are categorized into liquid/wax, solid residues, and gas [29].

Pyrolysis is a robust technique and can be used for either fuel or monomer recovery, particularly while addressing plastic waste management challenges. For example, liquid pyrolysis oil obtained from fast pyrolysis is an excellent source of gasoline and diesel [29]. Unlike water-rich pyrolysis bio-based oils derived from biomass feedstocks, the plastic oils have a high heating value, almost three times more than bio-oils, and similar to diesel fuels, due to absence of highly oxygenated compounds and water [30]. The acid content of plastic oils also is dramatically lower than that of bio-oils and, therefore, no further upgrading may be required for fuel applications. The pyrolysis coproducts include solid (char), which can be used as an adsorbent [31–33], and gas, a valuable resource that can be used as an energy supplier for the pyrolysis process.

The produced plastic oil can be a liquid or a wax. The wax is yellowish with a high viscosity at room temperature and is predominantly composed of alkanes and alkenes hydrocarbons with a high boiling point (C20+) [34]. Wax is typically an intermediate product, and a further process, such as fluid catalytic cracking (FCC), is required to convert it into liquid fuels. Liquid plastic oil is comprised of mainly aliphatic compounds as well as mono- and polyaromatics [29]. In addition to being a promising precursor for fuel applications, the plastic oil can be used as an intermediate and converted into ethylene and propylene through further cracking at higher temperatures and extremely low contact times. The major gaseous species forming the "gas product" are methane, ethylene, ethane, propylene, butadiene, and butane [35]. The gas product can be used as an energy source to provide the required pyrolysis energy, making the process self-sustaining and independent from external energy sources. In addition, the valuable olefin components present in the gas stream can be separated and recovered for chemical recycling. The solid residue is the remaining pyrolysis product, mostly made of coke and ash [36].

3.1. Slow Pyrolysis

Table 3 summarizes slow pyrolysis trials conducted for the conversion of plastic wastes with/without catalyst utilization using various reactors and under a wide range of operating conditions. The results reveal that the liquid produced during slow pyrolysis is typically oily rather than waxy. The oil yield can reach up to 93 wt.% when LDPE is pyrolyzed at 550 °C, which implies a remarkable yield with a broad range of applications [21]. The plastic oil is versatile and can be used either directly in steam boilers for electricity generation or as a platform chemical for other applications, such as transportation fuels, monomer recovery, and carbon nanotubes (CNTS) production. The solid residue yield is significantly less than bio-based char, as a consequence of the lower fixed carbon and a higher volatile matter associated with plastic wastes compared to biomass (Table 1). The gasoline fraction, C6–C12, can constitute up to 90 wt.% of the liquid product, making it valuable for conventional gasoline replacement.

3.1.1. Influence of Plastic Types

Table 3 indicates that the pyrolysis of polyolefins, including LDPE, HDPE, and PP, typically produces a liquid oil with a significant fraction of aliphatic (alkanes and alkenes), specifically in the absence of catalyst. The impact of catalyst on the composition of plastic oil is discussed later, in Section 3.1.2. The desired pyrolysis temperature to achieve a high conversion of polyolefins is above 450 °C, since, below this temperature, the solid residue drastically increases. Polystyrene (PS), which is composed of styrene monomers, can generate a liquid with a remarkable amount of aromatic compounds, such as benzene, toluene, and ethyl benzene [2]. Although the pyrolysis of polyolefins and polystyrene leads to the formation of a liquid oil which can be an excellent precursor for fuels/chemicals, the pyrolysis of PET and PVC generates a significant amount of benzoic acid and hydrochloric acid, respectively, which are toxic and corrosive to the reactors [2,37]. As such, these two polymers are typically excluded from pyrolysis.

3.1.2. Influence of Catalyst

Among the typical plastic pyrolysis process catalysts (e.g., FCC, HZSM-5, MCM-41, HY, Hβ, HUSY, mordenite and amorphous silica-alumina), acidic zeolites have been widely investigated [38–43]. Zeolite catalysts have shown an excellent catalytic efficiency on cracking, isomerization and oligomerization/aromatization, attributed to their specific physicochemical properties, including a strong acidity and a micropore crystalline structure [44]. As illustrated in Table 3, the plastic pyrolysis in the presence of catalysts, particularly HZSM5, tends to produce remarkably more aromatic and polycyclic aromatic hydrocarbons, compared to the uncatalyzed pyrolysis process, therefore contributing to the gasoline fraction. Further, a significantly higher production of gases is typically observed in the presence of zeolite catalysts, due to the enhanced cracking reactions [45]. Amorphous silica-alumina catalysts significantly contribute to the production of light olefins, with no noticeable changes in the aromatics formation [46]. ZSM-5 and zeolite-Y promote the formation of both aromatics and branched hydrocarbons, along with a significant increase in the proportion of gaseous hydrocarbons. These results are consistent with other published reports in the literature [47,48]. Catalytic reforming over Al-MCM-41 proactively contributes to the gasoline production with a lower impact on gas generation, likely due to the weaker acid properties and larger pore dimensions of the catalyst [1]. In the presence of both Y-zeolite and ZSM-5 catalysts, the oil yield dramatically decreases in favor of gas production [49]. The superiority of Y-zeolite compared to the ZSM-5, in terms of aromatic compounds production, is rationalized by the differences in physical and chemical catalyst properties, i.e., pore size, surface area, and surface acidity [50].

Table 3. Slow pyrolysis of different plastic wastes.

Plastic Types, Temp., Cat., Ref.	Feed: Catalyst Ratio	Liquid/Wax Yield (wt.%)	Solid Residue Yield (wt.%)	Gas Yield (wt.%)	Gasoline (C6–C12) (wt.%)	Diesel (C13–C20) (wt.%)	Wax (C20+) (wt.%)	Monomer Recovery (wt.%)
HDPE-450 °C-None-[46]	-	84	3	13	47	32	5	-
HDPE-450 °C-ZSM-5-[46]	20	35	2	63	35	0	0	-
HDPE-450 °C-Silica-alumina-[46]	20	78	1	21	71	7	0	-
LDPE-425 °C-None-[1]	10	44	45	11	20	24	<1	3
LDPE-450 °C-None-[1]	10	74	10	16	34	39	1	7
LDPE-475 °C-None-[1]	10	69	4	27	28	36	<1	16
LDPE-425 °C-HZSM-5-[1]	10	7	48	45	7	<1	<1	26
LDPE-450 °C-HZSM-5-[1]	10	16	11	73	15	0	1	47
LDPE-475 °C-HZSM-5-[1]	10	22	4	74	21	<1	<1	53
LDPE-475 °C-Al-MCM-41-[1]	10	40	50	10	30	7	1	7
LDPE-475 °C-Al-MCM-41-[1]	10	34	18	58	31	3	2	31
LDPE-475 °C-Al-MCM-41-[1]	10	42	4	54	38	5	1.5	37
PE-500 °C-Y-zeolite-[49]	NR	80	10	10	NR	NR	NR	4
PE-500 °C-ZSM-5-[49]	NR	70	10	20	NR	NR	NR	5
LDPE-550 °C-None-[21]	10	93	-	14	NR	NR	NR	9
HDPE-550 °C-None-[45]	10	84	-	16	NR	NR	NR	11
LDPE-550 °C-LDPE-HZSM5-[45]	10	18	1	71	NR	NR	NR	59
HDPE-550 °C-LDPE-HZSM5-[45]	10	17	1	72	NR	NR	NR	53
LDPE-550 °C-HUSY-[45]	10	62	2	34	NR	NR	NR	22
LDPE-550 °C-HUSY-[45]	10	41	2	39	NR	NR	NR	31
HDPE-450 °C-None-[50]	34	74	19	6	15	60	25	21
HDPE-450 °C-MCM-[50]	34	78	15	6	15	60	25	22
HDPE-450 °C-FCC-[50]	34	82	11	6	25	65	10	21
HDPE-450 °C-HZSM-5-[50]	34	81	4	15	25	62	23	21
PS-550 °C-[51]	-	90	2	9	42	37	11	3
PET-550 °C-None-[51]	-	84	4	12	-	-	-	-
Mixed-550 °C-None-[51]	-	83	6	11	56	20	6	4

3.2. Fast Pyrolysis

Unlike slow pyrolysis, which is typically performed in the batch reactors, fast pyrolysis takes place in continuous systems. The faster char removal from the reactor space associated with continuous processes prevents undesirable catalytic effects leading to the excessive cracking of vapors, which, coupled with short vapor residence times, minimizing secondary cracking reactions, result in a higher liquid production. Table 4 indicates that fast pyrolysis can convert up to 95 wt.% of plastic waste into the liquid/wax product (e.g., pyrolysis of HDPE at 600 °C). In addition, PE yields a higher liquid/wax product compared to the PP. The waxy fraction (C20+) of fast pyrolysis is greater than that of slow pyrolysis, attributed to a shorter vapor residence time and, consequently, reduced cracking reactions. As previously mentioned, the waxy product, which is an excellent source of paraffins and olefins, can be used as a feedstock in FCC units for the production of transportation fuels and other valuable petrochemical compounds.

Table 4. Fast pyrolysis of different plastic wastes.

Plastic Type, Temp., Ref.	Liquid/Wax Yield (wt.%)	Solid Residues Yield (wt.%)	Gas Yield (wt.%)	Gasoline (C6–C12) (wt.%)	Diesel (C13–C20) (wt.%)	Wax (C20+) (wt.%)	Monomer Recovery (wt.%)
PP-668 °C-[52]	43	2	54	40	-	-	26
PP-703 °C-[52]	35	6	57	34	-	-	27
PP-746 °C-[52]	29	4	65	29	-	-	17
PE-728 °C-[52]	38	2	59	36	-	-	34
HDPE-600 °C-[53]	95	-	5	18	25	53	4
HDPE-650 °C-[53]	85	-	15	27	21	37	12
HDPE-700 °C-[53]	60	-	40	32	17	11	37
HDPE-428 °C-[53]	93	-	7	52	33	17	-
PP-409 °C-[53]	96	-	4	70	21	9	-
HDPE-650 °C-[54]	80	-	20	10	18	52	-
PVC-740 °C-[55]	28	49	15	-	-	-	-

Influence of Temperature

Temperature plays a key role in all pyrolysis processes, regardless of the feedstock type. In the pyrolysis of plastic wastes, as in any other pyrolysis process, the increase in temperature results in a rapid increase in gas yields from the enhanced cracking reactions and, correspondingly, in a decrease of the oil/wax yield (Table 4). In addition to the alteration of yields, temperature expectedly affects the products quality, due to its impacts on the pyrolysis kinetic mechanisms. Generally speaking, high temperature favors the production of less waxy and more oily compounds production, attributed to the conversion of long-chain paraffins/olefins to shorter molecules. Conversely, the solid residue yield decreases at elevated temperatures. A qualitative assessment of plastic oil shows high temperature favors an increase in gasoline production corresponding to a higher concentration of aromatics [52]. The yield of ethylene and propylene are found increase as the temperature rises.

3.3. Flash Pyrolysis

In order to avoid over-cracking reactions during pyrolysis, especially at high temperatures (above 700 °C), which converts a significant amount of liquid to gaseous products, flash pyrolysis taking place within milliseconds is a suitable option. Unlike the fast pyrolysis of biomass, which generates the highest yields of bio-oil, flash pyrolysis of plastic waste produces more gas rather than liquids (Table 5). As illustrated in Table 5, up to 75 wt.% of monomers, i.e., ethylene and propylene, can be recovered through flash pyrolysis. The byproduct, which is oil, can be used to provide the required energy for the process. Kannan et al. [56] performed a flash pyrolysis of LDPE in a microreactor with a minimal heat/mass transfer resistance at temperatures of 600–1000 °C, and vapor residence time of 250 ms to investigate the effect of temperature on monomer recovery (yield of olefins). They found that the 950–1000 °C temperature range is optimal to recover up to 68 wt.% of monomers.

Table 5. Flash pyrolysis of LDPE with different experimental parameters.

Plastic Type, Temp., Ref.	Vapor Residence Time (s)	Liquid/Wax Yield (wt.%)	Solid Residues Yield (wt.%)	Gas Yield (wt.%)	Monomer Recovery Yield (wt.%)
LDPE-900 °C-[57]	0.75	-	-	95.0	50
LDPE-850 °C-[54]	0.6	11.4	-	88.6	-
LDPE-825 °C-[58]	0.4	5	2	93	75
LDPE-790 °C-[59]	0.5	32.1	0.2	62.2	51.6
LDPE-1000 °C-[56]	0.25	-	-	99	68

Influence of Vapor Residence Time

Together with temperature, the vapor residence time in the hot zone (i.e. reactor) is the key parameter that significantly affects the pyrolysis products yields and compositions. Previous work [28,57,58] on the flash pyrolysis of different waste plastics in micro and large scale reactors indicated that short vapor residence time resulted in a high olefin content gas. Although the Internally Circulating Fluidized Bed (ICFB) reactor illustrated in Figure 1 has been shown to be suitable for thermochemical processes requiring short residence times, this capability is not achieved without cost [58]. The control of the residence time, particularly for large scale reactors, presents a significant challenge to the reactor designer. Nielsen et al. [60] invented a fenestrated centrifugal riser terminator for use in an ICFB and/or in conventional fluid bed reactor risers. This terminator can separate solids from gas in less than 20 ms with 99.5% separation efficiency. Such a fast separation is critically important when the objective is to minimize the vapor residence time at high temperature and rapidly quench the reaction (Table 5).

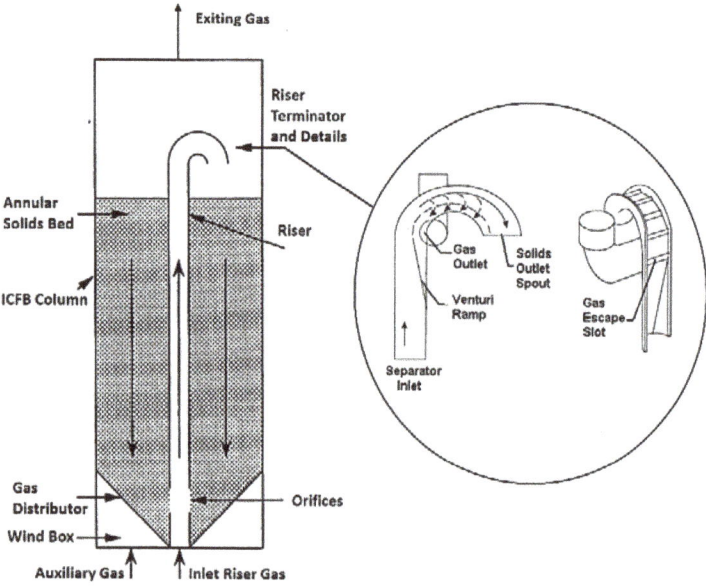

Figure 1. Internally circulating fluidized bed (ICFB) and riser terminator (adapted from [58,60]).

4. Plastic Oil Cracking

Tsuji et al. [35] utilized a two-stage unit, including a liquid–liquid extraction followed by a pyrolysis reactor, to investigate the thermal cracking of pyrolysis plastic oils containing considerable amounts of aromatic compounds, such as styrene. In the first stage, sulfolane solvent was used to remove the aromatic compounds prior to the pyrolytic cracking, in order to mitigate the coking effects, since stable aromatics (e.g., styrene) tend to be coked rather than cracked during pyrolysis. The results were promising, and the gas yield reached 85 wt.% at 750 °C, corresponding to a 20 wt.% increase compared to non-extracted oil. A schematic diagram of the pyrolysis oil cracking set-up is shown in Figure 2.

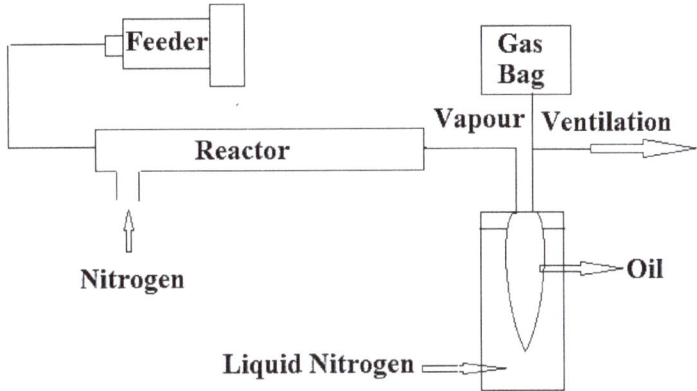

Figure 2. Schematic diagram of the experimental apparatus for cracking of raw plastic pyrolysis oil and extracted oil after separation of aromatics (adapted from Reference [35]).

The results of cracking of different plastic oils are summarized in Table 6. SL and SM are the light and medium fractions of pyrolysis plastic oil obtained from Sapporo Plastic Recycle Co. DH and DL stand for heavy and light fractions of plastic oil obtained from Dohoh Recycle Center Co. in Japan. Model plastics waste oil was obtained in the lab [35] from the pyrolysis of mixed plastics at 450 °C. The analysis of data presented in Table 6 suggests that a higher cracking temperature (above 850 °C) and a lower vapor residence time (less than 730 ms) are required to potentially achieve the best monomer recoveries.

Table 6. Pyrolysis of different plastic wastes [35].

Plastic Type and Temp.	Vapor Residence Time (s)	Liquid/Wax Yield (wt.%)	Solid Residues Yield (wt.%)	Gas Yield (wt.%)	Ethylene Yield (wt.%)	Propylene Yield (wt.%)	Total Olefin Yield (wt.%)
SL-700 °C	0.96	43.1	0.1	28.4	7	7	16
SL-850 °C	0.81	34.6	3.9	31.1	15	18	20
SM-700 °C	0.91	30.1	2.1	49.4	15	25	30
SM-850 °C	1.06	26.2	4	-	25	30	35
DL-700 °C	0.95	31.4	0.2	46.2	10	20	25
DL-850 °C	0.77	28.6	2.2	41.8	20	25	30
DH-700 °C	0.95	32.1	2	54.3	20	15	40
DH-850 °C	0.75	18.6	2.3	65	40	5	50
MO-700 °C	0.92	26.7	0.8	45	15	10	32
MO-850 °C	0.73	28.8	2.7	55.3	30	5	38

5. Upgrading of Pyrolysis Plastic Oils

The comparison between diesel fuel and a sample plastic oil obtained from the pyrolysis of mixed plastics, including 58 wt.% of HDPE and LDPE, 27 wt.% of PP, 9 wt.% of PS, 5 wt.% of PET [61] is shown in Table 7. The GC-MS results reveal that the carbon number distributions of the produced plastic oil is as follows: 35.41 wt.% of C6–C9, 48.40 wt.% of C10–C14, 13.21 wt.% of C15–C20, and 1.83 wt.% of C20+ [61]. The comparison between fuel properties of pyrolysis plastic oil and diesel indicates similar heating values and kinematic viscosity, while the plastic oil has a higher ash content and a lower cetane number compared to diesel fuel.

Table 7. A comparison between plastic oil and diesel physicochemical properties [61].

Properties	Plastic Oil *	Diesel
Density (kg/m^3)	734	820–850
Ash content (wt.%)	1	0.04
Calorific value (MJ/kg)	41	42
Kinematic viscosity (cSt)	2.9	3.05
Cetane number	49	55
Flash point (°C)	46	50
Fire point (°C)	51	56
Carbon residue (wt.%)	0.01	0.002
Sulphur content (wt.%)	<0.001	<0.035
Pour point (°C)	−3	−15
Cloud point (°C)	−27	-
Aromatic content (wt.%)	32	11–15

* Composition: 35.41 wt.% of C6-C9, 48.40 wt.% of C10-C14, 13.21 wt.% of C15-C20, and 1.83 wt.% of C20+.

5.1. Blending

As discussed earlier, raw pyrolysis plastic oil has a high heating value (40–55 MJ/kg), a low water content (<1 wt.%), and approximately neutral pH. Therefore, boilers can readily burn it as-is for the electricity generation. Damodharan et al. [21] stated that diesel engines can smoothly run plastic oil and no modification is required to the existing engine infrastructure. In contrast, there are several researchers [8,62–64] who believe improvements in plastic oil quality are needed to meet EN590 standards. In terms of drawbacks associated with the utilization of pyrolysis plastic oil in internal combustion engines, a high heat release and delayed ignition have been reported [61]. As such, a blend of conventional fuel and plastic oil can be a potential solution. The fuel trials using blends in conventional engines reveal a stable performance with less emission of NOx and SOx compared to diesel and gasoline fuels [65]. A reduced specific fuel consumption has also been reported [61]. Awasthi and Gaikwad [66] stated that the overall performance of the blend of diesel and plastic oil in a single cylinder four stroke VCR diesel engine was very satisfactory, particularly with 20 wt.% of pyrolysis plastic oil. Singh et al. [61] experimentally showed that the blend of plastic oil with diesel up to 50% can be easily utilized in conventional diesel engines.

5.2. Hydrogenation

Hydrogenation process takes place in the presence of three components: hydrogen, a catalyst, and an unsaturated compound. The transfer of hydrogen pairs to the unsaturated compound is facilitated via a heterogeneous catalyst which enables the reaction to occur at a lower temperature and pressure. For instance, hydrogenation converts alkenes to alkanes in plastic oil [67]. Due to the significant presence of unsaturated compounds in the plastic oils, some storage instability challenges may be experienced over time. Hydrogenation of pyrolysis oil occurring at temperatures above 700 °C, pressures around 70 bar, and in the presence of catalyst (such as ZSM-5) can alter unsaturated compounds into saturated and makes the oil more stable. A combination of hydrogenating and blending have been suggested to upgrade the plastic oil quality in order to meet the EN590 standard [67]. The fuel properties of plastic oil, diesel, and hydrogenated plastic oil along with the EN590 standard are compared in Table 8.

Table 8. Physicochemical properties of diesel, plastic oil and hydrogenated plastic oil [67].

Properties	Lower Limit Standards EN590	Upper Limit Standards EN590	Diesel	Plastic Pyrolysis Oil	Hydrogenated Plastic Pyrolysis Oil
Density (kg/m^3)	820	840	837	771	851
Pour Point (°C)	-	-	−15	−30	−20
Flash Point (°C)	55	-	72	20	65
Fire Point (°C)	-	-	82	30	72
Calculated Cetane Index	46	-	52	60	62
Kinematic Viscosity (mm^2/s)	2	4.5	2.31	1.78	3.5
Gross Calorific Value (MJ/kg)	-	-	46	45	45
Ash Content (wt.%)	-	0.1	0.01	0.01	0.01
Conradson Carbon Residue (wt.%)	-	-	0.18	0.1	0.1

5.3. Liquid-Liquid Extraction

The high aromatic content of plastic oils, particularly those obtained from pyrolysis of mixed plastics containing polystyrene, leads to a decrease in the engine performance and an increase in emissions due to a long ignition delay [68]. Generally, the low cetane number fuels, caused by high aromatic content, are not suitable for conventional diesel engines as they can cause unstable combustion. As such, the aromatic compounds can be separated and removed via solvent extraction prior to the utilization of plastic oil in diesel engines. Sulfolane as solvent was proven effective by Tsuji et al. [35] for the separation of aromatic compounds.

6. Carbon Nanotubes

Carbon nanotubes (CNTs) have gained recognition as very attractive materials due to their unique properties, including great electrical conductivity (100 times greater than copper), excellent mechanical strength (100 times greater than steel), high thermal conductivity, stable chemical properties, extremely high thermal stability, and an ideal one-dimensional (1D) structure with anisotropy [69–72]. Conventionally, methane, natural gas, acetylene, and benzene from nonrenewable resources have been utilized as a feedstock for CNTs production. Recently, the potential fabrication of CNTs from the pyrolysis of plastic waste has drawn researchers' attentions, adding a significant value to the plastic wastes. The process of converting plastic waste into CNTs is composed of two successive stages (Figure 3). In the first stage, the plastic waste is converted to the volatile vapor in the absence of oxygen and at a moderate temperature (approximately 550 °C). Then, the produced vapor is introduced into the second stage at a high pressure of 1 MPa and a temperature of 750 °C in the presence of Ni-based catalyst where it is converted into CNTs on the surface of the catalyst through the chemical vapor deposition mechanism. In this advanced process, CNT yields can reach up to 25 wt.% [73]. In the second stage, during pyrolysis at 750 °C, plastic waste vapors further decompose to the mixtures of their monomers (e.g., ethylene, propylene, and styrene). These light gases serve as carbon donors for CNTs formation. Moreover, the produced vapors from the first stage contain a significant amount of hydrogen, which plays an undeniable role in the formation of CNTs. Hydrogen moderates the rate of carbon deposition and prevents catalyst deactivation and poisoning by continuous surface cleansing of the catalyst surfaces [69–72]. A SEM image of CNT growth on Ni-based catalyst is shown in Figure 4.

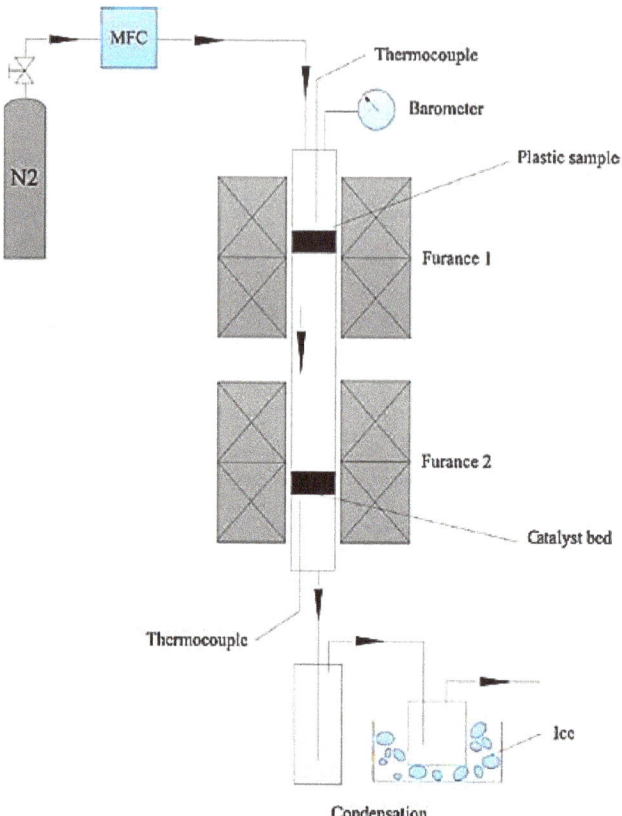

Figure 3. Schematic diagram of two-stage pyrolysis reactor system (adapted from [73]).

Figure 4. CNT growth on Ni-based catalyst (adapted from [73]).

7. Conclusions

The pyrolytic conversion of plastic waste into value added products and/or fuels is extensively reviewed in this paper. Plastic waste, which can be a source of detrimental problems to terrestrial and marine ecosystems, can be thermochemically converted into valuable products, such as gasoline, diesel, and wax. Fast pyrolysis leads to the production of waxy hydrocarbon mixtures, whereas slow pyrolysis typically produces more oil than wax. This is attributed to a difference in the vapor residence times, since longer residence times in slow pyrolysis allow for more cracking reactions breaking down the larger molecules into smaller and lighter fragments. The utilization of catalyst in plastic pyrolysis favors aromatic compounds in the liquid phase and gas production. Higher pyrolysis temperatures result in enhanced secondary cracking reactions, and, therefore, in a greater conversion of waxy compounds to oily and gaseous products. PE and PP produce pyrolysis oils with more aliphatic compounds, while PS generates higher aromatic hydrocarbons. In flash pyrolysis, conducted at 1000 °C and 250 ms of vapor residence time, up to approximately 70 wt.% of olefin monomers including 50 wt.% of ethylene can be recovered, making it a promising process for monomer recovery. The plastic oil can be blended with diesel and utilized as a fuel in conventional diesel engines. A two-stage process, including a pyrolysis unit followed by a fixed bed reactor with a nickel-based catalyst can be utilized to convert up to 25 wt.% of plastic waste into very valuable carbon nanotubes.

Author Contributions: S.P. contributed to the writing, data analysis, and organization of the literature review on slow, and fast pyrolysis. H.B. contributed to the literature review and data analysis/organization of CNTs and application of the section on pyrolysis plastic oil. F.B. supervised the overall project and contributed to developing and expanding the whole sections of this manuscript, particularly to the flash pyrolysis of plastic waste and plastic oil cracking. All authors have read and agreed to the published version of the manuscript.

Funding: This research was funded by the Natural Sciences and Engineering Research Council of Canada (NSERC) and the industry partners involved in the NSERC Industrial Research Chair program entitled "Thermochemical Conversion of Biomass and Waste to Bioindustrial Resources".

Institutional Review Board Statement: Not applicable.

Informed Consent Statement: Not applicable.

Data Availability Statement: Data sharing not applicable.

Acknowledgments: The authors are grateful to the Natural Sciences and Engineering Council of Canada and to the industry sponsors of the NSERC Industrial Research Chair in "Thermochemical Conversion of Biomass and Waste to Bioindustrial Resources" for the financial support for this project.

Conflicts of Interest: The authors declare no conflict of interest.

References

1. Aguado, J.; Serrano, D.; Miguel, G.S.; Castro, M.; Madrid, S. Feedstock Recycling of Polyethylene in a Two-Step Thermo-Catalytic Reaction System. *J. Anal. Appl. Pyrolysis* **2007**, *79*, 415–423. [CrossRef]
2. Al-Salem, S.M.; Antelava, A.; Constantinou, A.; Manos, G.; Dutta, A. A review on thermal and catalytic pyrolysis of plastic solid waste (PSW). *J. Environ. Manag.* **2017**, *197*, 177–198. [CrossRef]
3. Kunwar, B.; Cheng, H.; Chandrashekaran, S.R.; Sharma, B.K. Plastics to Fuel: A Review. *Renew. Sustain. Energy Rev.* **2016**, *54*, 421–428. [CrossRef]
4. Kumar, S.; Panda, A.K.; Singh, R. A Review on Tertiary Recycling of High-Density Polyethylene to Fuel. *Resour. Conserv. Recycl.* **2011**, *55*, 893–910. [CrossRef]
5. Tiseo, I. Global Plastic Production 1950–2019. Available online: https://www.statista.com/statistics/282732 (accessed on 16 May 2021).
6. The Compelling Facts about Plastics 2009—An Analysis of Eurropean Plastics Producction, Demand and Recovery for 2008. Available online: https://www.plasticseurope.org (accessed on 16 May 2021).
7. Ray, R.; Thorpe, R. A Comparison of Gasification With Pyrolysis for the Recycling of Plastic Containing Wastes. *Int. J. Chem. React. Eng.* **2007**, *5*, 1–14. [CrossRef]
8. Kumar, S.; Prakash, R.; Murugan, S.; Singh, R. Performance and Emission Analysis of Blends of Waste Plastic Oil Obtained by Catalytic Pyrolysis of Waste HDPE With Diesel in a CI Engine. *Energy Convers. Manag.* **2013**, *74*, 323–331. [CrossRef]

9. Butler, E.; Devlin, G.; McDonnell, K. Waste Polyolefins to Liquid Fuels via Pyrolysis: Review of Commercial State-of-the-Art and Recent Laboratory Research. *Waste Biomass-Valorization* **2011**, *2*, 227–255. [CrossRef]
10. Singh, A.; Sharma, T.C.; Kishore, P. Thermal Degradation Kinetics and Reaction Models of 1,3,5-Triamino-2,4,6-Trinitrobenzene-Based Plastic-Bonded Explosives Containing Fluoropolymer Matrices. *J. Therm. Anal. Calorim.* **2017**, *129*, 1403–1414. [CrossRef]
11. Achilias, D.; Roupakias, C.; Megalokonomos, P.; Lappas, A.; Antonakou, E. Chemical Recycling of Plastic Wastes Made from Polyethylene (LDPE and HDPE) and Polypropylene (PP). *J. Hazard. Mater.* **2007**, *149*, 536–542. [CrossRef]
12. Canopoli, L.; Fidalgo, B.; Coulon, F.; Wagland, S. Physico-Chemical Properties of Excavated Plastic from Landfill Mining and Current Recycling Routes. *Waste Manag.* **2018**, *76*, 55–67. [CrossRef]
13. Rajmohan, K.; Yadav, H.; Vaishnavi, S.; Gopinath, M.; Varjani, S. *Perspectives on Bio-Oil Recovery from Plastic Waste*; Elsevier BV: Amsterdam, The Netherlands, 2020; pp. 459–480. [CrossRef]
14. Chandrasekaran, S.R.; Sharma, B.K. *From Waste to Resources*; Elsevier BV: Amsterdam, The Netherlands, 2019; pp. 345–364. [CrossRef]
15. Al-Salem, S. *Feedstock and Optimal Operation for Plastics to Fuel Conversion in Pyrolysis*; Elsevier BV: Amsterdam, The Netherlands, 2019; pp. 117–146. [CrossRef]
16. Till, Z.; Varga, T.; Sója, J.; Miskolczi, N.; Chován, T. *Kinetic Modeling of Plastic Waste Pyrolysis in a Laboratory Scale Two-Stage Reactor*; Elsevier BV: Amsterdam, The Netherlands, 2018; Volume 43, pp. 349–354. [CrossRef]
17. Zhang, F.; Zhao, Y.; Wang, D.; Yan, M.; Zhang, J.; Zhang, P.; Ding, T.; Chen, L.; Chen, C. Current Technologies for Plastic Waste Treatment: A Review. *J. Clean. Prod.* **2021**, *282*, 124523. [CrossRef]
18. Fojt, J.; David, J.; Přikryl, R.; Řezáčová, V.; Kučerík, J. A Critical Review of the Overlooked Challenge of Determining Micro-Bioplastics in Soil. *Sci. Total. Environ.* **2020**, *745*, 140975. [CrossRef] [PubMed]
19. Murthy, K.; Shetty, R.J.; Shiva, K. Plastic Waste Conversion to Fuel: A Review on Pyrolysis Process and Influence of Operating Parameters. *Energy Sources Part A Recover. Util. Environ. Eff.* **2020**, 1–21. [CrossRef]
20. Nanda, S.; Berruti, F. Thermochemical Conversion of Plastic Waste to Fuels: A Review. *Environ. Chem. Lett.* **2021**, *19*, 123–148. [CrossRef]
21. Damodharan, D.; Kumar, B.R.; Gopal, K.; De Poures, M.V.; Sethuramasamyraja, B. Utilization of Waste Plastic Oil in Diesel Engines: A Review. *Rev. Environ. Sci. Bio/Technol.* **2019**, *18*, 681–697. [CrossRef]
22. Williams, P.T. Hydrogen and Carbon Nanotubes from Pyrolysis-Catalysis of Waste Plastics: A Review. *Waste Biomass-Valorization* **2021**, *12*, 1–28. [CrossRef]
23. Yao, D.; Yang, H.; Chen, H.; Williams, P.T. Co-Precipitation, Impregnation and so-Gel Preparation of Ni Catalysts for Pyrolysis-Catalytic Steam Reforming of Waste Plastics. *Appl. Catal. B Environ.* **2018**, *239*, 565–577. [CrossRef]
24. Papari, S.; Hawboldt, K. A Review on the Pyrolysis of Woody Biomass to Bio-Oil: Focus on Kinetic Models. *Renew. Sustain. Energy Rev.* **2015**, *52*, 1580–1595. [CrossRef]
25. Papari, S.; Hawboldt, K.; Helleur, R. Pyrolysis: A Theoretical and Experimental Study on the Conversion of Softwood Sawmill Residues to Biooil. *Ind. Eng. Chem. Res.* **2015**, *54*, 605–611. [CrossRef]
26. Papari, S.; Hawboldt, K. Development and Validation of a Process Model To Describe Pyrolysis of Forestry Residues in an Auger Reactor. *Energy Fuels.* **2017**, *31*, 10833–10841. [CrossRef]
27. Heydariaraghi, M.; Ghorbanian, S.; Hallajisani, A.; Salehpour, A. Fuel Properties of the Oils Produced from the Pyrolysis of Commonly-Used Polymers: Effect of Fractionating Column. *J. Anal. Appl. Pyrolysis* **2016**, *121*, 307–317. [CrossRef]
28. Lovett, S.; Berruti, F.; Behie, L.A. Ultrapyrolytic Upgrading of Plastic Wastes and Plastics/Heavy Oil Mixtures to Valuable Light Gas Products. *Ind. Eng. Chem. Res.* **1997**, *36*, 4436–4444. [CrossRef]
29. Williams, P.T.; Williams, E.A. Fluidised Bed Pyrolysis of Low Density Polyethylene to Produce Petrochemical Feedstock. *J. Anal. Appl. Pyrolysis* **1999**, *51*, 107–126. [CrossRef]
30. Papari, S.; Hawboldt, K.; Fransham, P. Study of Selective Condensation for Woody Biomass Pyrolysis Oil Vapours. *Fuel* **2019**, *245*, 233–239. [CrossRef]
31. Bamdad, H.; Hawboldt, K. Comparative Study Between Physicochemical Characterization of Biochar and Metal Organic Frameworks (MOFs) As Gas Adsorbents. *Can. J. Chem. Eng.* **2016**, *94*, 2114–2120. [CrossRef]
32. Bamdad, H.; Hawboldt, K.; MacQuarrie, S.; Papari, S. Application of Biochar for Acid Gas Removal: Experimental and Statistical Analysis Using CO_2. *Environ. Sci. Pollut. Res.* **2019**, *26*, 10902–10915. [CrossRef] [PubMed]
33. Bamdad, H.; Hawboldt, K.; MacQuarrie, S. A Review on Common Adsorbents for Acid Gases Removal: Focus on Biochar. *Renew. Sustain. Energy Rev.* **2018**, *81*, 1705–1720. [CrossRef]
34. Arabiourrutia, M.; Elordi, G.; Lopez, G.; Borsella, E.; Bilbao, J.; Olazar, M. Characterization of the Waxes Obtained by the Pyrolysis of Polyolefin Plastics in a Conical Spouted Bed Reactor. *J. Anal. Appl. Pyrolysis* **2012**, *94*, 230–237. [CrossRef]
35. Tsuji, T.; Hasegawa, K.; Masuda, T. Thermal Cracking of Oils from Waste Plastics. *J. Mater. Cycles Waste Manag.* **2003**, *5*, 102–106. [CrossRef]
36. Onwudili, J.A.; Insura, N.; Williams, P.T. Composition of Products from the Pyrolysis of Polyethylene and Polystyrene in a Closed Batch Reactor: Effects of Temperature and Residence Time. *J. Anal. Appl. Pyrolysis* **2009**, *86*, 293–303. [CrossRef]
37. Miandad, R.; Barakat, M.; Aburiazaiza, A.S.; Rehan, M.; Nizami, A. Catalytic Pyrolysis of Plastic Waste: A Review. *Process. Saf. Environ. Prot.* **2016**, *102*, 822–838. [CrossRef]

38. Lopez, G.; Artetxe, M.; Amutio, M.; Bilbao, J.; Olazar, M. Thermochemical Routes for the Valorization of Waste Polyolefinic Plastics to Produce Fuels and Chemicals. A Review. *Renew. Sustain. Energy Rev.* **2017**, *73*, 346–368. [CrossRef]
39. Zhang, X.; Lei, H.; Yadavalli, G.; Zhu, L.; Wei, Y.; Liu, Y. Gasoline-Range Hydrocarbons Produced from Microwave-Induced Pyrolysis of Low-Density Polyethylene over ZSM-5. *Fuel* **2015**, *144*, 33–42. [CrossRef]
40. Schirmer, J.; Kim, J.; Klemm, E. Catalytic Degradation of Polyethylene Using Thermal Gravimetric Analysis and a Cycled-Spheres-Reactor. *J. Anal. Appl. Pyrolysis* **2001**, *60*, 205–217. [CrossRef]
41. Boronat, M.; Corma, A. Are Carbenium and Carbonium Ions Reaction Intermediates in Zeolite-Catalyzed Reactions? *Appl. Catal. A Gen.* **2008**, *336*, 2–10. [CrossRef]
42. del Remedio Hernández, M.; Gómez, A.; García, Á.N.; Agulló, J.; Marcilla, A. Effect of the Temperature in the Nature and Extension of the Primary and Secondary Reactions in the Thermal and HZSM-5 Catalytic Pyrolysis of HDPE. *Appl. Catal. A Gen.* **2007**, *317*, 183–194. [CrossRef]
43. del Remedio Hernández, M.; García, Á.N.; Marcilla, A. Catalytic Flash Pyrolysis of HDPE in a Fluidized Bed Reactor for Recovery of Fuel-Like Hydrocarbons. *J. Anal. Appl. Pyrolysis* **2007**, *78*, 272–281. [CrossRef]
44. Mastral, J.; Berrueco, C.; Gea, M.; Ceamanos, J. Catalytic Degradation of High Density Polyethylene over Nanocrystalline HZSM-5 Zeolite. *Polym. Degrad. Stab.* **2006**, *91*, 3330–3338. [CrossRef]
45. Marcilla, A.; Beltrán, M.; Navarro, R. Thermal and Catalytic Pyrolysis of Polyethylene over HZSM5 and HUSY Zeolites in a Batch Reactor under Dynamic Conditions. *Appl. Catal. B Environ.* **2009**, *86*, 78–86. [CrossRef]
46. Seo, Y.-H.; Lee, K.-H.; Shin, D.-H. Investigation of Catalytic Degradation of High-Density Polyethylene by Hydrocarbon Group Type Analysis. *J. Anal. Appl. Pyrolysis* **2003**, *70*, 383–398. [CrossRef]
47. Sakata, Y.; Uddin, A.; Muto, A. Degradation of Polyethylene and Polypropylene into Fuel Oil by Using Solid Acid and Non-Acid Catalysts. *J. Anal. Appl. Pyrolysis* **1999**, *51*, 135–155. [CrossRef]
48. Mordi, R.C.; Dwyer, J.; Fields, R. H-ZSM-5 Catalysed Degradation of Low Density Polyethylene, Polypropylene, Polyisobutylene and Squalane: Influence of Polymer Structure on Aromatic Product Distribution. *Polym. Degrad. Stab.* **1994**, *46*, 57–62. [CrossRef]
49. Bagri, R.; Williams, P.T. Catalytic Pyrolysis of Polyethylene. *J. Anal. Appl. Pyrolysis* **2002**, *63*, 29–41. [CrossRef]
50. Miskolczi, N.; Bartha, L.; Deák, G.; Jóver, B.; Kalló, D. Thermal and Thermo-Catalytic Degradation of High-Density Polyethylene Waste. *J. Anal. Appl. Pyrolysis* **2004**, *72*, 235–242. [CrossRef]
51. Singh, R.; Ruj, B.; Sadhukhan, A.; Gupta, P. Thermal Degradation of Waste Plastics under Non-Sweeping Atmosphere: Part 1: Effect of Temperature, Product Optimization, and Degradation Mechanism. *J. Environ. Manag.* **2019**, *239*, 395–406. [CrossRef] [PubMed]
52. Jung, S.-H.; Cho, M.-H.; Kang, B.-S.; Kim, J.-S. Pyrolysis of a Fraction of Waste Polypropylene and Polyethylene for the Recovery of BTX Aromatics Using a Fluidized Bed Reactor. *Fuel Process. Technol.* **2010**, *91*, 277–284. [CrossRef]
53. Elordi, G.; Olazar, M.; Lopez, G.; Artetxe, M.; Bilbao, J. Product Yields and Compositions in the Continuous Pyrolysis of High-Density Polyethylene in a Conical Spouted Bed Reactor. *Ind. Eng. Chem. Res.* **2011**, *50*, 6650–6659. [CrossRef]
54. Berrueco, C.; Mastral, F.J.; Esperanza, E.; Ceamanos, J. Production of Waxes and Tars from the Continuous Pyrolysis of High Density Polyethylene. Influence of Operation Variables. *Energy Fuels* **2002**, *16*, 1148–1153. [CrossRef]
55. Kaminsky, W.; Predel, M.; Sadiki, A. Feedstock Recycling of Polymers by Pyrolysis in a Fluidised Bed. *Polym. Degrad. Stab.* **2004**, *85*, 1045–1050. [CrossRef]
56. Kannan, P.; Al Shoaibi, A.; Srinivasakannan, C. Temperature Effects on the Yield of Gaseous Olefins from Waste Polyethylene via Flash Pyrolysis. *Energy Fuels* **2014**, *28*, 3363–3366. [CrossRef]
57. Sodero, S.F.; Berruti, F.; Behie, L.A. Ultrapyrolytic Cracking of Polyethylene—A High Yield Recycling Method. *Chem. Eng. Sci.* **1996**, *51*, 2805–2810. [CrossRef]
58. Milne, B.J.; Behie, L.A.; Berruti, F. Recycling of Waste Plastics by Ultrapyrolysis Using an Internally Circulating Fluidized Bed Reactor. *J. Anal. Appl. Pyrolysis* **1999**, *51*, 157–166. [CrossRef]
59. Scott, D.S.; Czernik, S.R.; Piskorz, J.; Radlein, D.S.A.G. Fast Pyrolysis of Plastic Wastes. *Energy Fuels* **1990**, *4*, 407–411. [CrossRef]
60. Nielsen, B.B.; Berruti, F.; Behie, L.A. Riser Terminator for Internally Circulating Fluid Bed Reactor. U.S. Patent 5,665,130, 9 September 1997.
61. Singh, R.; Ruj, B.; Sadhukhan, A.; Gupta, P.; Tigga, V. Waste Plastic to Pyrolytic Oil and Its Utilization in CI Engine: Performance Analysis and Combustion Characteristics. *Fuel* **2020**, *262*, 116539. [CrossRef]
62. Ramesha, D.K.; Kumara, G.P.; Mohammed, A.V.; Mohammad, H.A.; Kasma, M.A. An Experimental Study on Usage of Plastic Oil and B20 Algae Biodiesel Blend As Substitute Fuel to Diesel Engine. *Environ. Sci. Pollut. Res.* **2015**, *23*, 9432–9439. [CrossRef] [PubMed]
63. Vu, P.H.; Nishida, O.; Fujita, H.; Harano, W.; Toyoshima, N.; Iteya, M. Reduction of NOx and PM from Diesel Engines by WPD Emulsified Fuel. *SAE Tech. Pap. Ser.* **2001**, *2001*, 5–7. [CrossRef]
64. Kaimal, V.K.; Vijayabalan, P. A Study on Synthesis of Energy Fuel from Waste Plastic and Assessment of Its Potential As an Alternative Fuel for Diesel Engines. *Waste Manag.* **2016**, *51*, 91–96. [CrossRef]
65. Sharuddin, S.D.A.; Abnisa, F.; Daud, W.M.A.W.; Aroua, M.K. A Review on Pyrolysis of Plastic Wastes. *Energy Convers. Manag.* **2016**, *115*, 308–326. [CrossRef]
66. Awasthi, P.K.; Gaikwad, A. Comparison of Power Output's of Different Blends of Pyrolysis Plastic Oil, Diesel With Pure Diesel on Single Cylinder 4-S (VCR) Diesel Engine. *Int. J. Mech. Prod. Eng. Res. Dev.* **2017**, *7*, 255–262. [CrossRef]

67. Mangesh, V.; Padmanabhan, S.; Tamizhdurai, P.; Narayanan, S.; Ramesh, A. Combustion and Emission Analysis of Hydrogenated Waste Polypropylene Pyrolysis Oil Blended With Diesel. *J. Hazard. Mater.* **2020**, *386*, 121453. [CrossRef]
68. Chandran, M.; Tamilkolundu, S.; Murugesan, C. Characterization Studies: Waste Plastic Oil and Its Blends. *Energy Sources Part A Recover. Util. Environ. Eff.* **2019**, *42*, 281–291. [CrossRef]
69. Gou, X.; Zhao, D.; Wu, C. Catalytic Conversion of Hard Plastics to Valuable Carbon Nanotubes. *J. Anal. Appl. Pyrolysis* **2020**, *145*, 104748. [CrossRef]
70. Shamsi, R.; Sadeghi, G.M.M.; Vahabi, H.; Seyfi, J.; Sheibani, R.; Zarrintaj, P.; Laoutid, F.; Saeb, M.R. Hopes Beyond PET Recycling: Environmentally Clean and Engineeringly Applicable. *J. Polym. Environ.* **2019**, *27*, 2490–2508. [CrossRef]
71. Moo, J.G.S.; Veksha, A.; Oh, W.-D.; Giannis, A.; Udayanga, W.C.; Lin, S.-X.; Ge, L.; Lisak, G. Plastic Derived Carbon Nanotubes for Electrocatalytic Oxygen Reduction Reaction: Effects of Plastic Feedstock and Synthesis Temperature. *Electrochem. Commun.* **2019**, *101*, 11–18. [CrossRef]
72. Panahi, A.; Wei, Z.; Song, G.; Levendis, Y.A. Influence of Stainless-Steel Catalyst Substrate Type and Pretreatment on Growing Carbon Nanotubes from Waste Postconsumer Plastics. *Ind. Eng. Chem. Res.* **2019**, *58*, 3009–3023. [CrossRef]
73. Wang, J.; Shen, B.; Lan, M.; Kang, D.; Wu, C. Carbon Nanotubes (CNTs) Production from Catalytic Pyrolysis of Waste Plastics: The Influence of Catalyst and Reaction Pressure. *Catal. Today* **2020**, *351*, 50–57. [CrossRef]

Review

Influence of Design Parameters on Fresh Properties of Self-Compacting Concrete with Recycled Aggregate—A Review

Rebeca Martínez-García [1,*], P. Jagadesh [2], Fernando J. Fraile-Fernández [1], Julia M. Morán-del Pozo [3] and Andrés Juan-Valdés [3]

1. Department of Mining Technology, Topography, and Structures, University of León, Campus de Vegazana s/n, 24071 León, Spain; fjfraf@unileon.es
2. Department of Civil Engineering, Coimbatore Institute of Technology, Coimbatore 641014, Tamil Nadu, India; jaga.86@gmail.com
3. Department of Agricultural Engineering and Sciences, University of León, Avenida de Portugal 41, 24071 Léon, Spain; julia.moran@unileon.es (J.M.M.-d.P.); andres.juan@unileon.es (A.J.-V.)
* Correspondence: rmartg@unileon.es; Tel.: +34-987-291-000 (ext. 5421); Fax: +34-987-291-787

Received: 18 November 2020; Accepted: 14 December 2020; Published: 16 December 2020

Abstract: This article presents an overview of the bibliographic picture of the design parameter's influence on the mix proportion of self-compacting concrete with recycled aggregate. Design parameters like water-cement ratio, water to paste ratio, and percentage of superplasticizers are considered in this review. Standardization and recent research on the usage of recycled aggregates in self-compacting concrete (SCC) exploit its significance in the construction sector. The usage of recycled aggregate not only resolves the negative impacts on the environment but also prevents the usage of natural resources. Furthermore, it is necessary to understand the recycled aggregate property's role in a mixed design and SCC properties. Design parameters are not only influenced by a mix design but also play a key role in SCC's fresh properties. Hence, in this overview, properties of SCC ingredients, calculation of design parameters in mix design, the effect of design parameters on fresh concrete properties, and the evolution of fresh concrete properties are studied.

Keywords: recycled aggregates; self-compacting concrete; design parameters; fresh concrete properties; mix design

1. Introduction

In recent decades, the construction industry has experienced exponential growth worldwide and throughout the territory of the European Union (EU). This growth has caused a very important increase in the generation (among others) of so-called construction and demolition waste (C&DW). The developed world, or first world, is the great consumer of raw materials on the planet and also the largest generator of waste. According to data from the European Statistical Office, Eurostat, each EU citizen produces an average of 2000 kg of waste per year, without counting waste from mining (including the latter, the figure would exceed 5000 kg/person/year) [1]. Of this set of waste, more than a third corresponds to the construction sector.

After water, aggregates (basically sand and gravel) are the raw material most consumed by human beings and their main use is a key part of construction materials such as concrete, mortar, bituminous mixtures, etc. It is necessary to promote the recovery of waste generated by these activities, thus reducing the extraction of natural aggregates and the efficient means of waste management [2,3]. Concrete is the world's second most consumed material by human beings next to water. Due to the modernization of the world, the demolition of concrete is increasing day by day [4]. Hence, usage of such demolished waste is more focused by researchers across the globe and even several countries like Italy and Denmark are making standards and principles regarding it. Few countries (like Denmark and Germany) accomplish reuse percentages of over 80% [1] but several other countries across the globe have reuses percentage below 10% [5], so reuses problems remain the same. Usage of recycled aggregate (RA) not only provides the solution for landfill deposition but also preserves natural resources for natural aggregate quarrying.

Apart from this, it also reduces the overall footprint of concrete by a reduction in carbon dioxide (CO_2) emission [6,7], and energy associated with it. Usage of waste materials leads to a circular economy [8] which will respond to environmental challenges as shown in Figure 1. Usage of RA in concrete leads to approximately a 20% drop in CO_2 emissions and 60% preservation of natural resources due to aggregate quarrying [9]. Most industry across the globe undergoes a sustainable process, likewise the construction industry [10]. Over the past few decades, there have been a lot of changes occurring in the concrete sector to traditional ways of its production, notably the addition of minerals, chemical admixture, new supplementary cementitious materials (SCM), etc. To accomplish the sustainability in the concrete production process, the following measures are to be taken:

- Reduction in carbon dioxide emission produced during raw material quarrying and processing [11,12];
- Reduction in natural resource utilization [12,13];
- Reduction in cost [14,15];
- Enhancement of fresh and hardened properties of concrete [9,16,17];
- Less energy consumption during the production process [18];
- Reduction in noise level due to vibration caused by compaction of concrete [14].

Figure 1. Concept of circular economy.

Today, local and state administrations, the scientific community, and the general population are increasingly aware of the depletion of natural resources and advocate for sustainable development.

As a result of this development, the need arises to recycle waste, where resources are part of a circular economy and can be sustained from generation to generation as depicted in Figure 1. Society as a whole has become conscious of the need to combine economic development with sustainability and protection of the environment. It is in this context where the need arises to study new applications and developments for waste from C&DW. Obtaining cement-based materials made with RA (obtained from the treatment of C&DWs) has been one of the most studied applications. Although, indeed, there are currently some reliable studies that support the use of C&DW in conventional concrete [16–22], in recent years some studies on the use of RA have begun to be published with self-compacting concrete (SCC), although they are still rare [23–39]. Most of these studies compare the various fresh and hardened properties of SCC with RA and without RA (natural aggregates), apart from SCC mix design. Most of the studies only substitute the coarse fraction of the aggregate, but there are some (the fewest) that also substitute the fine fraction [27,29,30]. Numerous tests have concluded that the incorporation of RA into concrete as alternatives for natural aggregates in small proportions (≤20%) does not cause a reduction in the performance of these recycled concretes. These studies have made it possible for these types of recycled materials to be included in regulations and/or recommendations all over the world. In general, the guidelines for its use are usually the following: limitation of the recycled coarse aggregate (RCA) content to 20 wt.% on the total coarse aggregate content, opening the possibility of experimental studies for larger substitutions; the use of recycled fine aggregates (RFA) is not usually allowed, regardless of their nature, although some studies [27,29,30] show that substitutions of up to 10% do not produce significant variations in concrete characteristics; the use of mixed RA is excluded (that is, with the presence of ceramic materials in percentages by weight of around 30% or higher) [38]. Regulations and codes usually limit the composition and physical and mechanical characteristics of recycled aggregates for concrete for structural use. For example, in Europe, the UNE-EN 12,320 [40] and UNE-EN 933 [41] standards limit the content of ceramic materials to 5%, the gypsum content to 1%, the absorption capacity of aggregates below 5%, or the Flakiness index below 35% among others. These limits may vary from country to country.

However, the fact that international standards do not include the possibility of using mixed RA, the limitation on the total content substituted, and the impossibility of using fine fraction aggregates from CD&W, make it a challenge for the scientific community to develop studies and techniques capable of achieving these objectives, increasing the reuse capacity of these aggregates and promoting their use and recovery.

One of the advantages of SCC over conventional concrete is the handling process and the problem associated with it. One of the possible solutions to achieve the measure stated above is the fresh concrete properties of SCC with RA. Fresh properties of SCC play a key role in the behavior of hardened properties [42]. SCC composition is characterized by a high content of cement and fines, a reduction in the proportion of coarse aggregates, and the use of next-generation additives [43]. Due to the self-flowing ability of SCC, it requires a greater powder content, which leads to an increase in cost [44]. The presence of Ordinary Portland cement in a higher quantity of mix not only increases the cost but also liberates the high heat of hydration that is associated with concrete cracking [45]. To reduce the cost and heat of hydration of SCC, several investigators have introduced the usage of mineral admixtures [14,23] and the usage of RA [14]. In addition, the maximum volume of occupation in concrete is an aggregate that will occupy 60% to 70% [46] depending upon usage. Hence, any small change in the aggregate has a significant effect not only on the concrete properties but also on the cost. Hence, several researchers have examined the use of RA as a replacement for fine and coarse aggregate [14] in recent years. For the past two decades, extensive studies have been carried out on the effective usage of RA [32].

Ingredients used for SCC are similar to that of conventional concrete but include the addition of additives such as superplasticizers (SPs) (or viscosity modifiers) that are essential to avoid segregation and exudation of the mixture. This type of concrete requires special attention regarding the choice of materials, guaranteeing their uniformity and consistency, which makes them more complex when it comes to being dosed with substitute RA, due to their initially possible heterogeneity [43]. Several studies confirm the greater absorption capacity of RA, around 10% extra water, a premise to take into account. This higher water demand can be offset by the use of an SP [33,47–51]. Other authors also use other valid techniques such as pre-saturating the aggregates [52].

Adhered mortar on RA had a greater influence on the properties of RA [53] apart from the crushing type used [54]. The crushing type of aggregate determines the shape and size of the recycled product produced [54]. There is a possibility of four interfacial transition zone (ITZ) found in RA concrete: 1. ITZ (formed due to hydration reaction occurring between RA and new mortar) [55]; 2. ITZ (formed due to the reaction occurring between new aggregate and new mortar) [55]; 3. ITZ (formed due to reaction between new aggregate and old mortar) [56]; and 4. ITZ (formed due to reaction between old mortar and RA). The development of ITZ from old mortar and RA is very slow. And already-formed old ITZ by old mortars has microcracks and voids in it, which is responsible for a reduction in the hardened properties of RA concrete [23,56]. Properties of RA depend upon the source from which it is extracted [56] and has a direct influence on the RA concrete properties. Several researchers observed that the RA resulting from high strength concrete can be used as a 100% replacement for natural aggregate for production of RA concrete [4,57,58]. One of the main disadvantages in using RA as a commercialization of the product is due to the presence of this excess mortar adhering to it. But few investigations on SCC incorporated with RCA show mechanical and durable properties similar to that of the SCC with NCA by increasing cement content and SCMs, adjusting water to the cement (W/C) ratio and the size of RCA [59–63].

Quantity of RFA generated from C&DW occupies 20 to 50% of total C&DW [63]. Even several authors [64,65] recommended using a grinding technique to convert this fine part of fine aggregate to reactive powder (RP). RP consists of reactive (silicon dioxide, calcium hydroxide (CH) and calcium silicate hydrate) materials making it a higher pozzolanic activity. The reactive form of minerals is confirmed with help of microscope studies [66]. Even the surface area of RP particles is found to be higher than the cement [66], with an average particle size of 0.0154 mm. Usage of RP in SCC increases the strength performance due to the possible accessibility of extra pozzolanic material in RP and the possibility of a reaction with unreacted CH [66]. Due to the shape and texture of RP, it possesses better characteristics than other pozzolanic materials, which further enhance the bonding and interlocking of the cement matrix [66]. RP also occupies the pores in the cement matrix and provides better packing density due to its fine particle size.

The use of RA in the production of SCC has been gradually increasing due to its economic and environmental advantages derived from the fusion of the two techniques. On the one hand, the advantages of SCC: optimal compaction without vibrating, fluidity, complex and specially reinforced works, increased quality of finish in exposed concrete, greater adherence, etc., and on the other the advantages of RA: less demand for extraction of new quarry aggregates, reduction of pollution produced by C&DW, savings in transport costs, low energy consumption, reduction of investments since a good quality material is produced at a low cost, resources are optimized, etc. However, information on the quality of SCCs with RCA is still scarce. The quality of SCC depends on design parameters that influence the properties of concrete. Some of the design parameters considered in these studies are W/C ratio, water to powder (W/P) ratio, the ratio of coarse aggregate to total aggregate, percentage of powder content in total volume, and weight of SP in terms of percentage of cement. The preliminary objective of this article is to increase the usage of waste materials in the construction industry, especially SCC in order to understand the effect of design parameters of SCC mix on the fresh concrete properties. This study provides very useful information on design parameters, the influence of design parameters on mix design and fresh concrete properties, and comparison with available standards, which are helpful for practical use in concrete production.

Figure 2 shows the search criteria, the selection criteria, and the subsequent processing of the articles studied in this review.

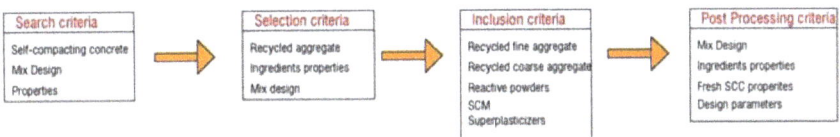

Figure 2. Flow diagram for selection criteria.

2. Material Properties of SCC Ingredients

The origin of SCC can be diverse, but there is a definite role of ingredient properties on the SCC fresh concrete properties. Hence, it is necessary to understand the properties of ingredients used in SCC. Tuyan et al. [14] used a laboratory-made aggregate from concrete waste, crushed with a jaw crusher. Pereira et al. [32] used two types of RA from a C&DW recycling plant. Señas et al. [49] used an RA from construction sites. The material was crushed with a jaw crusher. It was sieved to achieve a size similar to that of NCA and washed before adding it to the SCC. Duan et al. [67] used an RA from a C&DW recycling plant. The fine fraction was obtained by further grinding the RCA. Tang et al. [68] used an RA of unwashed, crushed concrete from a C&DW waste recycling plant. The aggregates were immersed in water 24 h before and air-dried for 1 h before adding them to the concrete.

In Djelloul et al. [69], the RA were obtained by crushing 1 m × 1 m concrete slabs with a thickness of 10 cm, manufactured in the laboratory, and kept for 28 days in water. They were first subjected to manual crushing and later with a mechanical crusher. Sasanipour et al. [56] used laboratory crushed concrete aggregates. They were first crushed with a small machine to produce RCA and then manually crushed using a hammer to produce the RFA. Aslani et al. [70] used aggregates that came from the demolition of buildings and concrete infrastructures. The rubber aggregates were obtained from the mechanical grinding of tires at the end of their useful life. Ouldkhaoua et al. [45] used metakaolin as obtained from the calcination of kaolin. Nieto [71] used an aggregate from crushed structural concrete. Dapena [72] used RA from structural concrete, crushing plants with a strength between 15 and 35 MPa.

The average value of ingredients used for SCC other than RA by various researchers is tabulated in Table 1. It is observed that the average specific gravity is obtained as 3.108. Specific gravity for fine and coarse aggregate investigated by several authors are 2.651 and 2.646. From Tables 1 and 2, it is observed that the physical properties of natural aggregate are higher than that of the RA due to old mortar adhering to it. The specific gravity of RA is lower than that of natural aggregate and is observed by most researchers, this leads to a density of RCA mixture as it it lower than that of the natural aggregate mixture. Higher water absorption of RA is observed from Table 2 when compared to Table 1 and it is reported by most authors.

Table 1. Average value for the properties of ingredients other than recycled aggregates used in the literature.

Ingredients	Specific Gravity	Fineness Modulus	Blaine's Surface Area (kg/cm²)	Water Absorption (%)	Bulk Density (kg/m³)	References
Cement	3.108		3495.5			Tuyan et al., 2014 [14], Pereira-de-Oliveria et al., 2014 [32], Señas et al., 2016 [49], Duan et al., 2020 [67], Tang et al., 2016 [68], Djelloul et al., 2018 [69], Sasanipour et al., 2019 [56], Aslani et al., 2018 [70], Ouldkhaoua et al., 2020 [45], Nieto et al., 2019 [71], Gupta et al., 2020 [73], Singh et al., 2019 [66], Abed et al., 2020 [74], Xavier et al., 2020 [75]
Fine aggregate	2.651	2.76		1.35	1632.5	Dapena et al., 2011 [72],
Coarse aggregate	2.646	6.05		1.24	1440.25	Rajhans et al., 2011 [76]

Table 2. Recycled aggregate properties used in the literature.

Source	Types	AS [1] (mm)	FM [2] (mm)	CF [3] (%)	SG [4]	W [5] (%)	BD [6] (kg/m³)	References
Waste laboratory made concrete	RCA	-	-	-	2.48	4.80	1410	Tuyan et al., 2014 [14]
Construction and demolition waste	RCA RCA	9.5 19	5.78 6.92	- -	- -	4.10 4.05	1509 1485	Pererira-de-Olivera et al., 2014 [32]
Construction work	RCA	12.5	3.77	-	-	-	-	Señas et al., 2016 [49]
Waste infrastructure component	RCA	12	2.59	-	-	6.53	1220	Duan et al., 2020 [67]
Construction and demolition waste facility	RCA	10	-	-	-	7.75	1450	Tang et al., 2016 [68]
Waste from laboratory concrete	RFA RCA RCA	- 9.5 12.5	3.8 - -	- - -	2.27 2.39 2.4	8.87 7.39 3.21	1258 1172 1154	Djelloul., 2018 [69]
Residential and sanitary buildings	RCA RFA	- -	- 3.67	- -	2.36 2.23	5.4 14.8	- -	Sasanipour et al., 2019 [56]
Demolition waste	RCA	-	-	-	-	5.60	1612	Abed et al., 2020 [74]
Concrete, bricks, asphalt, glass, others, aggregates	RCA	16	-	-	-	6.61	1584	Nieto et al., 2019 [71]
Demolished building	RCA	-	-	-	2.54	2.33	1220	Gupta et al., 2020 [73]
Cathode ray tube funnel glass	RFA	-	2.16	-	2.75	-	-	Ouldkhaoua et al., 2020 [45]

[1] Aggregate size (mm); [2] Fineness modulus; [3] Content of fineness (%); [4] Specific gravity; [5] Water absorption (%); [6] Bulk Density (kg/m³).

RA size investigated by most researchers less than 16 mm is observed in Table 2. It is also observed that the specific gravity and bulk density are lesser than natural aggregate. And also, higher water absorption than natural aggregate with an inconsistent form of fineness modulus of RA is noted. The quality and quantity of cement paste adhering to the surface RA affects the properties of it [15,32]. Accordingly, the RA has a lower density, higher water absorption, and lower mechanical strength properties [15]. From Table 2 and in past studies it can be observed that the density of RA is lower than that of natural aggregate due to the coating of cement paste on the surface of RA. This excess amount of cement paste results in an increase in the water absorption capacity of RA [15,32]. Coarse aggregate particle size is largely a result of the adhering of excess cement paste on it when compared to fine aggregate. Hence, most researchers have used an RCA size less than 16 mm. Coarser RCA shows a more negative effect than finer RFA [77–79]. To achieve the same workability compared to control concrete, concrete with RA requires a greater quantity of water due to the presence of excess cement paste [78]. This result in a higher water to binder ratio leads to the formation of more porosity in RA concrete and reduction in durability properties. Hence, usage of mineral admixtures may reduce this effect, since they provide the same workability to the mixture without increasing the water to binder ratio [15].

Salesa et al. [6] observed that the introduction of RCA in SCC leads to drops in fresh concrete properties [6]. This drop in fresh concrete property is due to two factors influenced by RCA. Firstly, RCA has a greater surface roughness to react and secondly, RCA has more fine particles adhering to the surface than natural aggregates and because of this they required more water to achieve similar fresh concrete properties.

Zhang et al., 2016 [80] observed that old ITZ in RCA did not develop with an increase in curing time, irrespective of surface treatment on RCA using nanomaterials (nano silica + nano calcium slurries or cement + nano silica slurries). Nanomaterials are not able to infiltrate into old ITZ to react with a few unhydrated particles present in it. New ITZs formed in RCA (surface treated by nano materials) were enhanced significantly. ITZ improvement is directly related to the elastic modulus of that particular ITZ. Old mortar adhering to RCA was surface-strengthened by nano surface treatment, resulting in enhanced RCA properties (water absorption, crushing value apparent density).

The higher water absorption nature of RCA results from the higher absorption of cement mortar attached to aggregate particles [81]. Cui et al., 2015 [81] further found that RCA with a higher water absorption nature can impact the fresh and hardened properties of concrete. Higher water absorption of RCA is due to the porous nature of old paste and the cracks present in it. RCA with surface modifications using an alkaline organosilicon modifier was effective in reducing the initial water absorption and it does not alter the mechanical properties of concrete when used in it.

Kou and Poon, 2010 [82] found that there are enhanced properties of RCA that can be observed when RCA is impregnated with Polyvinyl alcohol (PVA). There were improved mechanical and durability properties of concrete with RCA impregnated with PVA.

The enhancement of RA properties is observed after treatment when compared to untreated RA [15]. Enhanced properties are decreased in water absorption and an increase in the specific gravity of treated RA. Treated aggregate shows an improvement in fresh and hardened properties when compared to untreated aggregates.

As stated earlier, several works have been carried out to enhance the properties of RCA. Two methods are adopted in the literature either to remove the adhered mortar or to improve the adhered mortar quality [23,83]. In the first method, RCA is pre-soaked in different types of acids like hydrochloric acid and sulphuric acid at different concentrations to remove adhered mortar [84]. For the second method, a surface treatment like coating is carried out to improve the quality of RCA without removing the old mortar adhering to it. Surface treatment materials like microbial carbonate [85], silane-based water repellent [86], PVA [82], alkaline organosilicon modifier [81], silicate-based solution [87], and pozzolanic materials [88] have been used with different concentrations to improve the concrete properties.

3. Mix Design

The development of the SCC reference mix based on the strength achieved at 28 days as the average value of authors are shown in Table 3. Dosages of ingredients without RA as proposed by several authors are compiled for the production of control SCC in order to identify the range of ingredients required for it. It is also observed that the unit weight of the mix is found to be less than the standard weight of concrete. The proportion of coarse aggregate to total aggregate from the reference mix is 0.477, which is suitable for SCC as prescribed by subsequent mixes by several authors. Design constants for control mix from Table 3 are water to binder ratio as 0.34, water to powder ratio as 0.26, % of SP as 0.94%, % of SCM as 0.5, and filler to binder ratio as 0.29.

Table 3. Average of reference mix proportions (control mix in weight basis) used by several authors from 2010–2020.

Ingredients (Kg/m^3)	Average Values	Standard Deviation	References
Cement	386.31	83.64	Tuyan et al., 2014 [14], Pereira-de-Oliveria et al., 2014 [32],
SCM	191.4	80.84	Señas et al., 2016 [49], Duan et al., 2020 [67],
Filler	166.5	182.37	Tang et al., 2016 [68], Djelloul et al., 2018 [69],
Fine aggregate	823.95	214.93	Sasanipour et al., 2019 [56], Aslani et al., 2018 [70],
Coarse Aggregate	749.26	113.77	Ouldkhaoua et al., 2020 [45], Nieto et al., 2019 [71],
Water	194.07	23.52	Gupta et al., 2020 [73], Singh et al., 2019 [66],
SP	3.62	1.67	Abed et al., 2020 [74], Xavier et al., 2020 [75],
Unit weight	2335.61	46.12	Dapena et al., 2011 [72], Rajhans et al., 2011 [76]

We have selected the most representative studies of the use of SCC concrete with RA published in recent years and summarized in three tables (Tables 4–6) based on the weight basis of ingredients.

Table 4. Mix proportion (weight basis) adopted from Tuyan et al., 2014 [14], Pereira-de-Oliveria et al., 2014 [32], Señas et al., 2016 [49] and Duan et al., 2020 [67].

RCA Content (%)	FRA Content (%)	Cement (kg/m³)	W/C [1]	% SP [2]	Other Waste	Unit Weight (kg/m³)	W/B [3]	W/P [4]	References
0	-	315	0.62	1.55	Fly ash	2294	0.42	0.32	
0	-	315	0.69	1.04		2260	0.48	0.36	
0	-	315	0.76	0.69		2224	0.53	0.40	
20	-	315	0.62	1.74		2295	0.42	0.32	
20	-	315	0.69	1.07		2260	0.48	0.36	Tuyan et al., 2014 [14]
20	-	315	0.76	0.76		2224	0.53	0.40	
40	-	315	0.62	1.18		2295	0.42	0.32	
40	-	315	0.69	1.11		2260	0.48	0.36	
40	-	315	0.76	0.85		2225	0.53	0.40	
60	-	315	0.62	1.96		2295	0.42	0.32	
60	-	315	0.69	1.17		2260	0.48	0.36	
60	-	315	0.76	0.95		2225	0.53	0.40	
0	-	284.9	0.57	1.19	-	2358.5	0.25	0.129	
20	-	284.9	0.57	1.68		2349.8	0.25	0.130	Pereira-de-Oliveria et al.,
40	-	284.9	0.56	1.61		2335.3	0.25	0.128	2014 [32]
100	-	284.9	0.56	2.1		2282.2	0.25	0.130	
0	0	332	0.4	1.25		2400	0.4	0.5	
50	0	332	0.4	1.50		2359	0.4	0.5	
50	20	332	0.4	1.75		2341	0.4	0.5	
0	0	332	0.4	0.87		2398	0.4	0.5	Señas et al., 2016 [49]
50	0	332	0.4	1.62		2359	0.4	0.5	
50	20	332	0.4	1.87		2342	0.4	0.5	
0	0	430.5	0.57	0.355		2301.9	0.4	0.4	
0	10	430.5	0.57	0.355		2301.9	0.44	0.4	
25	10	430.5	0.57	0.355		2301.9	0.44	0.4	
50	10	430.5	0.57	0.355	Fly ash	2301.9	0.44	0.4	Duan et al., 2020 [67]
100	10	430.5	0.57	0.355	Silica	2301.9	0.44	0.4	
0	20	430.5	0.57	0.355	fume	2301.9	0.50	0.4	
25	20	430.5	0.57	0.355		2301.9	0.50	0.4	
50	20	430.5	0.57	0.355		2301.9	0.50	0.4	
100	20	430.5	0.57	0.355		2301.9	0.50	0.4	

[1] water/cement ratio; [2] % of superplasticizer to cement; [3] water/binder ratio; [4] water/powder ratio.

Table 5. Mix proportion adopted from Tang et al., 2020 [68], Djelloul et al., 2018 [69], Sasanipour et al., 2019 [56].

RCA Content (%)	FRA Content (%)	Cement (kg/m³)	W/C [1]	% SP [2]	Other Waste	Unit Weight (kg/m³)	W/B [3]	W/P [4]	References
0	-	445	0.49	1.011		2329.5	0.35	0.35	
25	-	445	0.49	1.011	Fly ash	2316.5	0.35	0.35	
50	-	445	0.49	1.011	Silica	2304.5	0.35	0.35	Tang et al., 2020 [68]
75	-	445	0.49	1.011	fume	2292.5	0.35	0.35	
100	-	445	0.49	1.011		2279.5	0.35	0.35	
0	0	507	0.37	1.5	GGBFS	2352	0.37	0.38	
25	25	507	0.37	1.5		2307	0.37	0.38	
50	50	507	0.37	1.5		2261	0.37	0.38	
75	75	507	0.37	1.5		2265	0.37	0.38	Djelloul et al., 2018 [69]
100	100	507	0.37	1.5		2170	0.37	0.38	
0	0	434	0.37	1.5		2353	0.37	0.38	
25	25	434	0.37	1.5		2307	0.37	0.38	
50	50	434	0.37	1.5		2261	0.37	0.38	
75	75	434	0.37	1.5		2265	0.37	0.38	
100	100	434	0.37	1.5		2170	0.37	0.38	
0	0	359	0.37	1.5		2353	0.37	0.38	
25	25	359	0.37	1.5		2307	0.37	0.38	
50	50	359	0.37	1.5		2261	0.37	0.38	
75	75	359	0.37	1.5		2265	0.37	0.38	
100	100	359	0.37	1.5		2170	0.37	0.38	
0	0	420	0.4	0.9		2322	0.4	0.28	
0	0	386.4	0.4	1	Silica	2322	0.4	0.28	
25	0	420	0.4	1.06	fume	2317	0.4	0.28	
50	0	420	0.4	1.09		2294	0.4	0.28	Sasanipour et al., 2019 [56]
75	0	420	0.4	1.05		2287	0.4	0.28	
100	0	420	0.4	1		2274	0.4	0.28	
25	0	386.4	0.4	1		2317	0.4	0.28	
50	0	386.4	0.4	1.09		2299	0.4	0.28	
75	0	386.4	0.4	1		2287	0.4	0.28	
100	0	386.4	0.4	1		2274	0.4	0.28	
0	25	420	0.4	0.92		2287	0.4	0.28	
0	25	420	0.4	0.9		2320	0.4	0.28	

[1] water/cement ratio; [2] % of SP to cement; [3] water/binder ratio; [4] water/powder ratio.

Table 6. Mix proportion adopted from Aslani et al., 2018 [71], Ouldkhaoua et al., 2020 [46], Nieto et al., 2019 [72], Dapena et al., 2011 [73].

RCA Content (%)	FRA Content (%)	Cement (kg/m^3)	W/C [1]	% SP [2]	Other Waste	Unit Weight (kg/m^3)	W/B [3]	W/P [4]	References
0	0	180	1.125	1.667		2332	0.45	0.25	
10	10	180	1.125	1.222		2316	0.45	0.26	Aslani et al., 2018 [70]
20	20	180	1.125	1.333		2301	0.45	0.27	
30	30	180	1.125	1.444		2286	0.45	0.29	
40	40	180	1.125	1.556	Fly ash	2270	0.45	0.30	
0	0	180	1.125	1.333	GGBFS	2245	0.45	0.25	
0	10	180	1.125	1.472	Silica	2238	0.45	0.26	
0	20	180	1.125	1.556	fume	2231	0.45	0.27	
0	30	180	1.125	1.667	Crumbled	2223	0.45	0.29	
0	40	180	1.125	1.778	rubber	2215	0.45	0.30	
0	0	180	1.125	1.111	Scoria	2168	0.45	0.25	
0	10	180	1.125	1.222		2160	0.45	0.26	
0	20	180	1.125	1.333		2152	0.45	0.27	
0	30	180	1.125	1.444		2144	0.45	0.29	
0	40	180	1.125	1.556		2136	0.45	0.30	
0	0	446	0.4	0.8	MK	2380	0.4	0.4	
0	0	446	0.42	0.85	CRTG	2380	0.4	0.4	Ouldkhaoua et al., 2020 [45]
0	10	446	0.42	0.85		2380	0.4	0.4	
0	20	446	0.42	0.85		2380	0.4	0.4	
0	30	446	0.42	0.83		2380	0.4	0.4	
0	40	446	0.42	0.83		2380	0.4	0.4	
0	50	446	0.42	0.8		2380	0.4	0.4	
0	0	422	0.44	1.1		2380	0.4	0.41	
0	10	422	0.44	1.1		2380	0.4	0.41	
0	20	422	0.44	1.1		2380	0.4	0.41	
0	30	422	0.44	1.05		2380	0.4	0.41	
0	40	422	0.44	1		2380	0.4	0.41	
0	50	422	0.44	0.95		2380	0.4	0.41	
0	0	399	0.47	1.2		2380	0.4	0.41	
0	10	399	0.47	1.2		2380	0.4	0.41	
0	20	399	0.47	1.2		2380	0.4	0.41	
0	30	399	0.47	1.15		2380	0.4	0.41	
0	40	399	0.47	1.15		2380	0.4	0.41	
0	50	399	0.47	1.1		2380	0.4	0.41	
0	-	367	0.55	1.5	-	2354	0.55	0.42	Nieto et al., 2019 [71]
20	-	367	0.55	1.5		2349	0.55	0.42	
40	-	367	0.55	1.5		2344	0.55	0.42	
60	-	367	0.55	1.5		2339	0.55	0.42	
80	-	367	0.55	1.5		2334	0.55	0.42	
100	-	367	0.55	1.5		2329	0.55	0.42	
0	-	386	0.5	1.5		2371	0.5	0.38	
20	-	386	0.5	1.5		2366	0.5	0.38	
40	-	386	0.5	1.5		2361	0.5	0.38	
60	-	386	0.5	1.5		2356	0.5	0.38	
0	-	408	0.45	1.5		2390	0.45	0.35	
20	-	408	0.45	1.5		2385	0.45	0.35	
40	-	408	0.45	1.5		2380	0.45	0.35	
60	-	408	0.45	1.5		2375	0.45	0.35	
0	-	408	0.45	1.5		2390	0.45	0.35	
20	-	408	0.45	1.5		2391	0.45	0.36	
40	-	408	0.45	1.5		2391	0.45	0.37	
60	-	408	0.45	1.5		2392	0.45	0.38	
0	-	408	0.45	1.5		2390	0.45	0.35	
20	-	408	0.45	1.5		2391	0.45	0.36	
40	-	408	0.45	1.5		2391	0.45	0.37	
60	-	408	0.45	1.5		2392	0.45	0.38	
0	0	380	0.5	0.7	-	2290	0.5	0.5	
20	0	380	0.5	0.7		2330	0.5	0.5	Dapena et al., 2011 [72]
20	5	380	0.5	0.7		2263	0.5	0.5	
20	10	380	0.5	0.7		2263	0.5	0.5	
50	0	380	0.5	0.7		2224	0.5	0.5	
50	5	380	0.5	0.7		2224	0.5	0.5	
50	10	380	0.5	0.7		2224	0.5	0.5	
100	0	380	0.5	0.7		2161	0.5	0.5	
100	5	380	0.5	0.7		2161	0.5	0.5	
100	10	380	0.5	0.7		2161	0.5	0.5	

[1] water/cement ratio; [2] % of SP to cement; [3] water/binder ratio; [4] water/powder ratio.

Tuyan et al., 2014 [14] design four types of mixes with substitutions of 0%, 20%, 40%, and 60% of NCA with RCA. The cement was class C with a content of 315 kg/m^3. Three W/C ratios are used, 0.62,

0.69, and 0.76. The SP content was adjusted about the percentage of replacement from 0.95% to 1.97% of the cement weight. The water absorption of the RA was 4.80%.

Pereira de Oliveira et al. [32] designed four types of mixes with substitutions of 0%, 20%, 40%, and 100% of NCA with RCA employing two types of RA, using CEM I 42.5-R at 284.9 kg/m^3. The W/C ratio was 0.56 and 0.57. The SP was adjusted about the replacement ratio from 1.193% to 2.106% of the cement weight. The water absorption of the two used RA was 4.10% and 4.05%.

Señas et al., 2016 [49] design six types of mixes with substitutions of 50% of NCA with RCA and 20% of natural fine aggregate (NFA) with RFA. The cement was Portland type I with a content of 415 kg/m^3. W/C ratio was used as 0.50. They used two different high range water reducing admixtures as "S" and "H" types. The SP content was adjusted from 0.7% to 1.50% of the cement weight.

Duan et al., 2020 [67] designed nine types of mixes with a combination of RFA and RCA substitution. The C1 series with a 10% substitution of NFA for RFA and substitutions of 25%, 50%, and 100% of NCA for RCA. The C2 series with a 20% substitution of NFA for RFA and substitutions of 25%, 50%, and 100% of NCA for RCA. The cement used is PO425 with a content of 430.5 kg/m^3. The W/C ratio is constant at 0.57. A constant SP of value 0.25% of the cement weight. The water absorption of the RA was 6.53% and was previously submerged in water for 24 h.

Tang et al., 2020 [68] design five types of mixes with substitutions of 0%, 25%, 50%, 75%, and 100% of NCA with RCA. The cement was ordinary Portland cement, type I, with a content of 445 kg/m^3. Three W/C ratios used 0.35. The SP content was 0.1 of the cement weight in all mixtures. The water absorption of the RA was 7.75% and was previously submerged in water for 24 h.

Djelloul et al., 2018 [69] design fifteen types of mixes with substitutions of 0%, 25%, 50%, 75% and 100% combined coarse and fine recycle aggregates. The cement was CEM II/A 42.5, with three different contents, 359, 434 and 507 kg/m^3. The W/C ratio and SP dosage were kept constant at 0.38 and 1.5% by weight of cement, respectively. The water absorption of the RA was 3.21 and 7.39 for RCA and 6.3 for RFA.

Sasanipour et al., 2019 [56] designed twelve types of mixes with substitutions of 0%, 25%, 50%, 75%, and 100% combined RCA and RFA. The cement was Portland cement type II, with 420 kg/m^3 content kg/m^3. The W/C ratio was kept constant at 0.4 and SP content was adjusted from 0.9% to 1.15% of the cement weight. The RA was previously submerged in water for 24 h.

Aslani et al., 2018 [70] designed three series with 5 mixes each. Series I with substitutions of 0%, 10%, 20%, 30%, and 40% combined coarse and fine recycle aggregates. Series II crumb rubber replacement is kept constant at 20% of the volume of coarse aggregates, and NFA is then replaced by RFA at 0%, 10%, 20%, 30%, and 40% of fine aggregate volume respectively. Series II kept constant coarse aggregate replacement consisting of 50% scoria lightweight aggregates, NFA is then replaced by RFA at 0%, 10%, 20%, 30%, and 40% of fine aggregate volume, respectively. The cement was type GP, with 450 kg/m^3 content kg/m^3. The W/C ratio was kept constant at 0.45 and SP content was adjusted from 1.222% to 1.778% of the cement weight.

Ouldkhaoua et al., 2020 [45] designed nineteen mixes, natural sand has been replaced with recycled cathode ray tube glass (CRTG) and metakaolin at levels of 0%, 10%, 20%, 30%, 40%, and 50% by weight and the cement has been partially replaced by MK at substitution ratios of 5%, 10%, and 15% by weight. The cement was Portland CEMI 42.5, the content ranges from 399.15 kg/m^3 to 469.59 kg/m^3. The W/C ratio from 0.4 to 0.47 and SP content was adjusted from 0.8% to 1.1% of the cement weight.

Nieto et al., 2019 [71] prepared five different mixes, each divided into two groups. The first group contained concrete with different water-cement ratios W/C 0.55, 0.50, and 0.45, with different quantities of cement 367, 386, and 408 kg/m^3, respectively and different substitution rates of NCA with RCA 0%, 20%, 40%, 60%, 80%, and 100%. The cement was CEMI 52.5 N/SR. The SP content was kept constant at 1.5% of the weight of cement. The water absorption of the RA was 5.35%, 5.83%, and 6.61% and was previously submerged in water.

Dapena et al., 2011 [72] designed ten mixes with substitutions of 20%, 50%, and 100% of coarse RA combined with substitutions of 5%, 10%, and 15% of fine RA. Prepared mixes with a constant W/C ratio of 0.50. The cement was type CEMI 42.5 R/SR, with 380 kg/m^3 content kg/m^3. The SP ratio was kept constant at 0.70% of the weight of cement. The water absorption of the RA was 4.12%.

To study one of the objectives, the authors grouped the mixed design in the literature into the following four families based on the design parameters.

- Family IA, W/C ratio as a varying parameter and its influence on the ingredients. Family IB, W/C ratio as a constant parameter;
- Family II, W/P ratio has varying parameters and has an influence on the ingredients;
- Family III, the influence of superplasticizer dosage on the ingredients.

The following features were observed from the above Table 7:

- For family IA mix, an increase in W/C ratio shows a decrease in the percentage of cement content in terms of unit weight;
- For the family IA mix, an increase in the W/C ratio causes an increase in the SCM content of the mixes;
- For the family IB mix, an increase in the W/C ratio is inconsistent with cement and SCM content;
- For a family I mix, the filler content varies concerning the W/C ratio but the sequence shows that a greater amount of filler material results in a higher water content of mixes;
- For the family, I mix, increase in the percentage of aggregate in mixes results in an increase in water required to achieve the same workability of the mix

Table 7. Grouping of mixed design concerning water-cement ratio (design parameters) based on the weight of ingredients (all ingredients are in terms of percentage).

F^a	W/C	C (%)	SCM (%)	Filler (%)	FA (%)	CA (%)	RFA (%)	RCA (%)	SP (%)	W (%)	References
IA	0.42	18.67	0.79	-	34.27–19.04	34.32	3.81–19.03	-	0.17	7.87	Ouldkhaoua et al., 2020 [45]
	0.44	17.71	1.57	-	34.32–19.10	34.50	3.81–19.07	-	0.18–0.21	7.88	Ouldkhaoua et al., 2020 [45]
	0.45	17.18–17.11	-	5.18–5.16	40.0–39.83	24.42–12.21	-	17.68–0	0.26	7.75–7.71	Nieto et al., 2019 [71]
	0.47	16.76	2.35	-	34.38–19.07		3.82–19.10	-	0.21–0.23	7.89	Ouldkhaoua et al., 2020 [45]
	0.50	16.38–16.31	-	4.92–4.90	40.32–40.15	24.51–12.31	-	17.83–0	0.25	8.19–8.16	Nieto et al., 2019 [71]
	0.55	15.76–15.62	-	4.72–4.68	40.79–40.44	24.69–0	-	30.06–0	0.24	8.67–8.60	Nieto et al., 2019 [71]
	0.62	13.73	5.85	7.06	32.17	25.9712.98	-	19.47–6.49	0.27–0.24	8.45	Tuyan et al., 2014 [14]
	0.69	13.93	5.97	6.95	31.55	24.56 12.74	-	19.12–6.37	0.16–0.15	9.56	Tuyan et al., 2014 [14]
	0.76	14.16	6.07	6.79	30.92	24.94–12.49	-	18.70–6.25	0.13–0.11	10.71	Tuyan et al., 2014 [14]
IB	0.4	13.89–14.18	-	3.54	45.15–36.17	15.89–16.01	8.33	14.11–14.22	0.21–0.26	7.09–7.04	Señas et al., 2016 [49]
	0.49	19.21–19.52	7.99–8.12	-	35.18–35.75	0–21.37		0–26.33	0.20–0.19	9.50–9.65	Tang et al., 2016 [68]
	0.56	12.20–12.48	15.85–16.22	25.59–26.19	5.70–5.84	20.76		12.82–31.90	0.20–0.26	6.88–7.12	Pereira-de-Oliveria et al., 2014 [32]
	0.57	12.12	15.75	25.43	5.67	27.50		6.37	0.19–0.26	6.94	Pereira-de-Oliveria et al., 2014 [32]
	0.57	18.70	2.67–5.34	2.6–5.34	28.67	16.94–25.41	-	8.47–33.89	0.07	28.67	Duan et al., 2020 [67]

F^a = Family; W/C = Water/cement ratio; C = Cement; SCM = Supplementary cementitious materials: FA = Fine aggregate; CA = Coarse aggregate; RFA = Recycled fine aggregate; RCA = Recycled coarse aggregate; W = Water.

The following features are observed from the above Table 8

- With an increase in W/P ratio, powder content of mix is decreased;
- With an increase in the W/P ratio, the proportion of fine aggregate content is decreased and coarse aggregate content is increased;
- With an increase in the W/P ratio, the proportion of replacement of fine aggregate by RFA is decreased;
- Water and SP content in the mix has an indirect relationship with the total powder content of the mix.

Table 8. Grouping of mix design concerning water to powder ratio (design parameters) based on the weight of ingredients (all ingredients are in terms of percentage).

Fª	W/C	C (%)	SCM (%)	Filler (%)	FA (%)	CA (%)	RFA (%)	RCA (%)	SP (%)	W (%)	References
	0.28	18.12–16.99	9.04–7.42	-	45.39–44.54	23.00–5.75	9.75–9.62	21.02–5.16	0.20–0.16	7.38–7.25	Sasanipour et al., 2019 [56]
	0.32	13.73	5.85	7.06	32.17	25.97–12.98	-	19.47–6.49	0.27–0.24	8.45	Tuyan et al., 2014 [14]
	0.35	17.18–17.11	-	5.18–5.16	40.00–39.83	24.42–12.21	-	17.68–0	0.26	7.75–7.71	Nieto et al., 2019 [71]
	0.35	19.21–19.52	7.99–8.12	-	35.18–35.75	0–21.37	-	0–26.33	0.20–0.19	9.50–9.65	Tang et al., 2016 [68]
II	0.36	13.93	5.97	6.95	31.55	24.56–12.74	-	19.12–6.37	0.16–0.15	9.56	Tuyan et al., 2014 [14]
	0.38	16.38–16.31	-	4.92–4.90	40.32–40.15	24.51–12.31	-	17.83–0	0.25	8.19–8.16	Nieto et al., 2019 [71]
	0.38	18.81–16.54	6.82–3.16	-	29.58–0	23.90–0	37.56–8.83	29.82–11.17	0.33–0.35	8.89–8.36	Djelloul et al., 2018 [69]
	0.40	18.70	2.67–5.34	2.67–5.34	28.67	16.94–25.41	-	8.47–33.89	0.07	28.67	Duan et al., 2020 [67]
	0.40	14.16	6.07	6.79	30.92	24.94–12.49	-	18.70–6.25	0.13–0.11	10.71	Tuyan et al., 2014 [14]
	0.41	18.67–16.76	2.35–0.79	-	34.38–19.04	34.50–34.32	19.10–3.81	-	0.23–0.17	7.89–7.87	Ouldkhaoua et al., 2020 [45]
	0.42	15.76–15.62	-	4.72–4.68	40.79–40.44	24.69–0	-	30.06–0	0.24	8.67–8.60	Nieto et al., 2019 [71]
	0.50	13.89–14.18	-	3.54	45.15–36.17	15.89–16.01	8.33	14.11–14.22	0.21–0.26	7.09–7.04	Señas et al., 2016 [49]
	0.50	16.79–16.30	-	-	32.89–31.94	33.44–32.48	0.83–0.42	8.12–7.52	0.12–0.11	8.39–8.15	Dapena et al., 2011 [72]

Fª = Family; W/C = Water/cement ratio; C = Cement; SCM = Supplementary cementitious materials; FA = Fine aggregate; CA = Coarse aggregate; RFA = Recycled fine aggregate; RCA = Recycled coarse aggregate; W = Water.

The following features are observed from the above Table 9:

- An increase in the replacement of aggregate by RA results in an increase in SP to achieve the same consistency and is observed in mixed designs in the literature;
- Increases in SP result in an increase in SCM content of the mixture;
- Increase in SP results in an increase in the filler quantity of the mix;
- An increase in SP content increases the total aggregate content of the mixture.

Table 9. Grouping of mix design to a percentage of superplasticizer (design parameters) based on the weight of ingredients (all ingredients are in terms of percentage).

Fª	SP (%)	C (%)	SCM (%)	Filler (%)	FA (%)	CA (%)	RFA (%)	RCA (%)	W (%)	References
	0.35	18.70	2.67–5.34	2.67–5.34	28.67	16.94–25.41	-	8.47–33.89	28.67	Duan et al., 2020 [67]
	0.8–1.2	18.67–16.76	2.35–0.79	-	34.38–19.04	34.50–34.32	19.10–3.81	-	7.89–7.87	Ouldkhaoua et al., 2020 [45]
III	0.9–1.09	18.12–16.99	9.04–7.42	-	45.39–44.54	23.00–5.75	9.75–9.62	21.02–5.16	7.38–7.25	Sasanipour et al., 2019 [56]
	0.95–0.76	14.16	6.07	6.79	30.92	24.94–12.49	-	18.70–6.25	10.71	Tuyan et al., 2014 [14]
	1.011	19.21–19.52	7.99–8.12	-	35.18–35.75	0–21.37	-	0–26.33	9.50–9.65	Tang et al., 2016 [68]
	1.18–1.08	13.93	5.97	6.95	31.55	24.56–12.74	-	19.12–6.37	9.56	Tuyan et al., 2014 [14]
	1.5	15.76–15.62	-	4.72–4.68	40.79–40.44	24.69–0	-	30.06–0	8.67–8.60	Nieto et al., 2019 [71]
	1.5	16.38–16.31	-	4.92–4.90	40.32–40.15	24.51–12.31	-	17.83–0	8.19–8.16	Nieto et al., 2019 [71]
	1.5	17.18–17.11	-	5.18–5.16	40.00–39.83	24.42–12.21	-	17.68–0	7.75–7.71	Nieto et al., 2019 [71]
	1.5	18.81–16.54	6.82–3.16	-	29.58–0	23.90–0	37.56–8.83	29.82–11.17	8.89–8.36	Djelloul et al., 2018 [69]
	1.5–1.88	13.89–14.18	-	3.54	45.15–36.17	15.89–16.01	8.33	14.11–14.22	7.09–7.04	Señas et al., 2016 [49]
	1.97–1.75	13.73	5.85	7.06	32.17	25.97–12.98	-	19.47–6.49	8.45	Tuyan et al., 2014 [14]
	2.11–1.19	12.48–12.12	16.22–15.75	26.18–25.43	5.83–5.67	27.50–0	-	31.89–6.37	7.12–6.95	Pereira-de-Oliveria et al., 2014 [32]
	2.66	16.79–16.30	-	-	32.89–31.94	33.44–32.48	0.83–0.42	8.12–7.52	8.39–8.15	Dapena et al., 2011 [72]

Fª = Family; W/C = Water/cement ratio; C = Cement; SCM = Supplementary cementitious materials; FA = Fine aggregate; CA = Coarse aggregate; RFA = Recycled fine aggregate; RCA = Recycled coarse aggregate; W = Water.

Apart from the above common observations, other works of literature show the most valid features of SCC properties. Most previous research has confirmed that the reduction percentage of recycled concrete elastic modulus is higher than the strength reduction [89,90]. An increase in the flowability of recycled SCC is achieved by adding a greater quantity of admixtures [70]. Due to an increase in SP content and porous nature of RA, it leads to instability and segregation of mix, which was controlled by adding a viscous modifying agent (VMA) [70,91]. The addition of GGBFS in SCC enhances the durability properties of SCC [69]. With an increase in the replacement percentage of NCA by RCA, the fluidity of SCC is reduced [71] due to the higher water absorption characteristics of RCA. This leads to an increase in the utilization of SP in SCC to maintain the same fluidity [71]. From the material and economic optimization perspective, there is a more focus on the reduction of SP quantity in the mix [71]. The addition of marble powder results in a decrease in void content due to its filler effect [75] in the mixture.

4. Influence of Design Parameters on Fresh Concrete Properties

For each fresh SCC property and every author, the results/findings obtained are presented in the form of a table/graph and the new tendencies are identified and studied based on the design parameters obtained from the mix design. A comparison of similar kinds of experiments is made from various authors to understand the RCA trends on fresh properties.

4.1. Effect of the Water to Cement Ratio on Fresh Concrete Properties

The researchers concluded that several parameters are used to evaluate the slump flow, which is defined as free fluidity and flow in the absence of barriers. The parameters considered are the time required by SCC takes to form a 500 mm circle, called flow time (T_{500}), and the slump flow (SF) diameter. SF ratio is defined as the ratio between the slump flow by RA SCC to the slump flow by natural aggregate SCC. A common observation made by most researchers is that increase in the replacement of coarse aggregate by RA results in T_{500} increases with SF decreases.

EFNARC [92] classified SF into three categories viz. SF1 (550–650 mm), SF2 (660–750 mm), and SF3 (760–850 mm). This means that the slump flow can differ from 550 mm to 850 mm. Table 10 shows that most of the RA SCC shows that most of the literature shows SF1 and SF2, which is also confirmed by Figure 3. EFNARC [92] categorized T500 mm spread flow into two classes, namely VS1 (≤2 s) and VS2 (>2 s). Similarly, V funnel flow time is also classified as two classes, namely VF1 (≤8 s) and VF2 (9 to 25 s). The minimum criteria for passing ability classes by L box ratio is identified as PA1 (≥0.80) and segregation resistance is of two classes, SR1 (15 to 20) and SR2 (≤15).

Table 10. Classification of the family I as per EFNARC standards [92] and the effect of the W/C ratio on fresh concrete properties.

F^a	W/C	T500SF (s)	SF (mm)	T500 J-Ring (s)	d_{max} J-Ring (mm)	L-Box H2/H1	V-f (s)	Sieve Seggreg. (%)	References
IA	0.42	-	-	-	-	0.87–0.97 (PA1)	6.1–8.6 (VF1–VF2)	0–15 (SR2)	[45]
	0.44	-	-	-	-	0.84–0.95 (PA1)	6.2–9 (VF1–VF2)	6–14 (SR2)	[45]
	0.45	2–6 (VS1–VS2)	568–715 (SF1)	10–248	578–675	-	5–103 (VF1–VF2)	-	[71]
	0.47	-	-	-	-	0.81–0.91 (PA1)	9.2–10 (VF2)	5–13 (SR2)	[45]
	0.50	2–5 (VS1–VS2)	628–715 (SF2)	2–41	445–683	-	5–25 (VF1–VF2)	-	[71]
	0.55	1–5 (VS1–VS2)	570–723 (SF1–SF2)	2–40	590–720	-	5–28 (VF1–VF2)	-	[71]
	0.62	2–4.6 (VS1–VS2)	665–690 (SF2)	-	-	0.89–0.95 (PA1)	13.9–28.2 (VF2)	-	[14]
	0.69	1.4–1.8 (VS1)	655–700 (SF2)	-	-	0.81–0.90 (PA1)	6.7–8.2 (VF1–VF2)	-	[14]
	0.76	1.2–1.7 (VS1)	650–700 (SF2)	-	-	0.60–0.7	2.7–6.2 (VF1)	-	[14]
IB	0.4	2.7–10.5 (VS2)	610–710 (SF1–SF2)	6.1–13.8	570–670	-	7.5–14 (VF1–VF2)	-	[49]
	0.49	2.9–4.3 (VS2)	700–720 (SF2)	-	-	0.80–0.89 (PA1)	-	5.20–9.90 (SR2)	[68]
	0.56	-	670 (SF2)	-	-	-	17.2 (VF2)	-	[32]
	0.57	1.8–3.4 (VS1–VS2)	660–750 (SF2)	-	-	0.78–0.92 (PA1)	4.1–6.1 (VF1)	-	[67]
	0.57	-	650–675 (SF2)	-	-	-	13–14.4 (VF2)	-	[32]

F^a = Family; W/P = Water/power ratio; SF = Slump Flow; V-f = V-funnel.

Corinaldesi and Moriconi [93], when studying SCC with RA and Municipal Solid Waste Ash (MSWA) (as NA substitutes), found that the slump flow is similar and the SF lies in SF3 as per EFNARC standards [92]. Thus, the authors concluded that the NFA and MSWA contribute identically to the

SCC's fluidity. As for the flow time, Corinaldesi and Moriconi [93] obtained identical results with NFA and MSWA and the results lie within the limit.

Grdic et al. [94] obtained that the SF decreased as the RCA integration ratio increased since higher ratios mean more water is absorbed by the RCA. In terms of flow time, Grdic et al. [94] found that it increases with the RCA integration ratio. This is due to the RCA being more angular and having a rougher surface than the CAN.

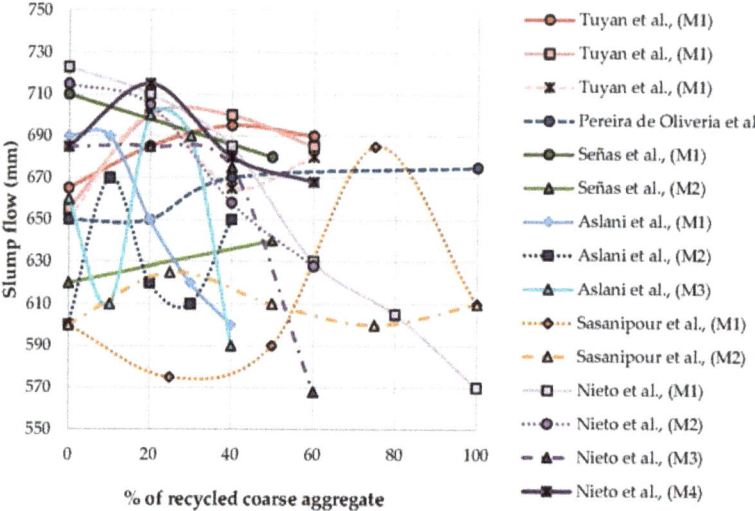

Figure 3. Effect of recycled coarse aggregate on slump Flow and its classification based on EFNARC [95].

The addition of silica fume and RP tends to decrease the flowability by absorbing excess free water in the mix [59]. Fineness and shape of the powder particles are two important factors in keeping the initial slump under control [59]. This is also one of the reasons for the addition of SCM in SCC. RP possesses a greater surface area and an irregular microstructure and hence it absorbs more water [33]. RP reduces flow over the time of SCC [33].

Table 10 indicates that the family IA slump flow mostly lies in the class SF2 as per the EFNARC standard [92] and a few in SF1. Similar to family IA, the family IB slump flow lies in the class SF1 and SF2. This indicates that either an increase or a constant level of W/C ratio influences the classification of the slump flow. For family IA, an increase in W/C ratio results in the conversion of slump flow and class SF1 to SF2 is observed but for family IB, an increase in W/C ratio result in a change of slump flow class SF1 and SF2 to SF2. T500 of slump flow from the literature indicates that most of it lies above 2 s, indicating that the flow time lies in class VS2 for both families IA and IB as per EFNARC standard [92]. For family IA, an increase in W/C ratio, resulted in a change of flow time classes from VS1 and VS2 to VS1 but for family IB, an increase in W/C ratio resulted in a change of flow time classes from VS2 to VS1 and VS2.

From Figure 3 it is observed that most SCCs investigated by previous researchers are lying between SF1 and SF2 as per the EFNARC standard [92]. Most slump flows are lying in the range of 600 mm to 720 mm, indicating much flow for most replacement ratio of coarse aggregate by RA is flowability. From Figure 4 it is observed that the slump flow ratio goes wider with an increase in replacement of NCA by RCA. The slump flow ratio varies by ± 0.05, up to 40% of the replacement of NCA by RA by most researchers is observed.

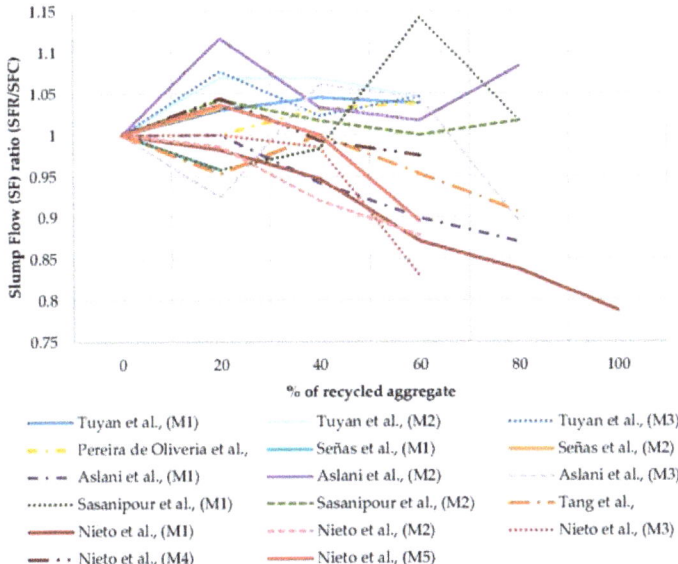

Figure 4. Effect of W/C ratio on slump flow.

Señas et al. [49] and Topcu et al. [62], observed that for the same W/C ratio (Family IA), the workability of RCA SCC is decreased and this is also observed in Table 10. Due to the water absorption nature of RCA, the water requirement of the mix is increased to achieve the same workability. Depending upon the quantity and quality of old mortar in RA, the water requirement varies. This is the reason for the lower slump flow for family IA and in most cases, it lies in SF1 as per the EFNARC standard [92]. V funnel time increase with an increase in the W/C ratio is observed for family IA.

Señas et al. [49] found, when fine aggregates are replaced by RFA, there is a need for a greater quantity of admixture to obtain a similar fluidity compared to the reference mixture. This is due to the rough texture and porous nature of RFA.

Usage of fine filler material in a recycled SCC mix fills the pores, which contributes to the greater density, making a homogeneous material and reducing internal bleeding [33]. It is also confirmed by microanalytical studies [95].

Safiuddin et al. [33] observed that an increase in RCA content of the mix results in an increase in V funnel flow time due to RCA possessing greater rugosity and angularity. It is also reported that the increase in slump flow time is directly related to V funnel flow time because RCA possesses a low flowing ability. Passing ability by L box ratio for most of the results in literature lies in class PA1, as per the EFNARC standard [92].

Pereira-de-Oliveira et al. [32] observed that the V-funnel flow time increased for the RCA as replacement ratios like 20% and 40% for NCA, and 40% of the RCA mix had the highest flow time and has the lowest W/C (0.56). All the others have an identical W/C ratio (0.57). Therefore, the authors mention that the V-funnel flow time is powerfully influenced by the W/C ratio.

It is observed that an increase in W/C ratio either in family IA or IB results in the V-funnel flow lying in either class VF1 or VF2. An increase in water content of mix results in increased in-flow time in V-funnel but with the addition of RCA results in a decrease in V-funnel time. The segregation percentage lies below 15% for both families.

Salesa et al. [6] reported that the quantity of cement paste from RA has a greater influence on the water requirement of the mix. An increase in old cement paste increases the quantity of unhydrated cement paste volume, which requires a greater water content to achieve the same workability. The fineness of RCA also affects the water requirement of the mix because the finer particles had a larger surface area to wet.

Pereira-de-Oliveira et al. [32] observed the influence of the W/C ratio and the percentage of SP content on the fresh concrete properties of RCA SCC. A gradual increase in SP dosage is associated with an increase in RCA, it is observed. An increase in RCA in SCC is associated with an increase in the W/C ratio, which can be controlled by the addition of SP. Higher water absorption behavior of RCA results in higher water requirement for lubricating the mixture to flow.

4.2. Effect of Water to Powder Ratio on Fresh Concrete Properties

Safiuddin et al. [33] reported that the increase in incorporation ratio results in increased segregation resistance up to lower replacement levels of NCA. For higher replacement levels of NCA, the segregation resistance decreases because the increase in RCA content leads to an increase in the fineness content of the mixture. Grdic et al. [94] found that the increase in RCA content leads to an increase in the W/C ratio, resulting in a decrease in segregation resistance.

From Table 11, the following key observations are made from works of literature:

- An increase in the W/P ratio results in T500 flow time to decrease because the flowability of the mix is better when the suspension of solid materials in liquid media increases;
- Slump flow increases with an increase in W/P ratio;
- SCC flow regarding obstruction (J-ring) increases with an increase in W/P ratio;
- L-Box ratio has no relationship concerning W/P ratio;
- V-funnel initially decreases and then increases to W/P ratio.

Table 11. Classification of family II as per EFNARC standard [92] and the effect of the W/P ratio on fresh concrete properties.

F^a	W/P	T500 SF(s)	SF (mm)	T500 J-Ring (s)	dmax J-Ring (mm)	L-Box H2/H1	V-f (s)	Sieve Segregation (%)	References
II	0.28	3.1–7.1 (VS2)	570–670 (SF1–SF2)	4–11	545–656	0.87–0.97 (PA1)	-	-	[56]
	0.32	2–4.5 (VS2)	665–695 (SF2)	-	-	0.84–0.95 (PA1)	2.3–28.2 (VF1–VF2)	-	[68]
	0.35	2–15(VS2)	560–685 (SF1–SF2)	5–8	643–675	-	7–25 (VF1–VF2)	-	[71]
	0.35	2.9–4.3 (VS2)	700–720 (SF2)	-	-	0.81–0.91 (PA1)	-	5.20–9.90 (SR2)	[68]
	0.36	1.4–1.8 (VS1)	655–700 (SF1–SF2)	-	-	-	2.7–8 (VF1)	-	[68]
	0.38	1–6 (VS1–VS2)	613–715 (SF1–SF2)	2–41	365–683	-	5–25 (VF1–VF2)	-	[71]
	0.38	2.9–4.9 (VS2)	713–792 (SF2–SF3)	-	-	0.89–0.95 (PA1)	6.16–23.20 (VF1–VF2)	5.10–16.50 (SR1–SR2)	[69]
	0.40	1.8–3.4 (VS1–VS2)	640–750 (SF1–SF2)	-	-	0.81–0.90 (PA1)	4.1–6.1 (VF1)	-	[67]
	0.40	1.2–1.7 (VS1)	650–700 (SF1–SF2)	-	-	0.60–0.7	2.7–7.2 (VF1)	-	[67]
	0.41	-	-	-	-	-	6.5–9 (VF1–VF2)	6–14 (SR2)	[45]
	0.42	1–6(VS1–VS2)	570–723 (SF1–SF2)	2–40	475–720	0.80–0.89 (PA1)	5–28 (VF1–VF2)	-	[71]
	0.50	2.7–10.5 (VS2)	610–710 (SF1–SF2)	6.1–13.8	570–670	-	7.5–14 (VF1–VF2)	-	[49]

F^a = Family; W/P = Water/power ratio; SF = Slump Flow; V-f = V-funnel.

An increase in replacement percentage of natural aggregate by the coarse aggregate resulting in an increase in flow time is observed in Figure 5.

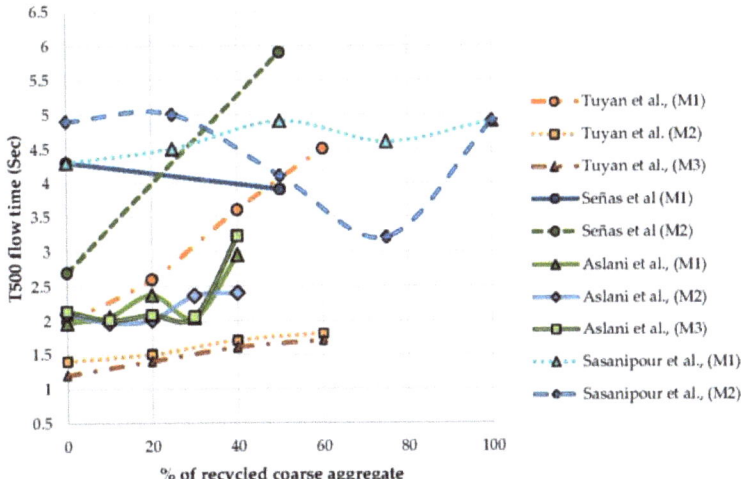

Figure 5. Effect of T500 flow time on the percentage of recycled coarse aggregate.

The time flow ratio is the ratio of T500 time flow of SCC with RCA to SCC with NCA. The time flow ratio increase with an increase in the percentage of recycled content in the SCC mix is observed. To keep the ratio constant, several parameters like the dosage of SP can be adjusted as observed by Aslani et al. [70] From Figure 6, it is observed that up to 40% of the replacement of NCA by RCA results in not much variation in the ratio. The ratio is varied by altering the design parameters, it is observed.

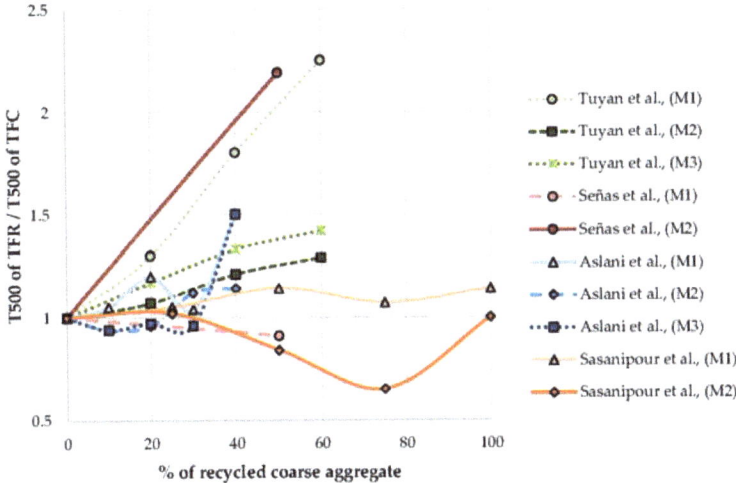

Figure 6. Effect of T500 time flow ratio on the percentage of recycled coarse aggregate.

4.3. Effect of the Percentage of Superplasticizer on Fresh Concrete Properties

Señas et al. [49] observed that by varying percentages and types of SP, the flow parameters lie within the EFNARC standard [92]. Flow class is also not modified when coarse aggregate or fine aggregate is replaced by RA. Admixture with a lower density influences the viscosity with an increase in RA content in mixture, and admixture with a higher density does not influence viscosity, it is observed. This was confirmed by the V-funnel test and it is observed that viscosity nature increases with an increase in RA content. Incorporation of SP to the mixture shows a better presence without any sign of segregation or bleeding. From Figure 7, it is observed that the larger proportions of RA in SCC mixes shows a variation in V-slump flow. Most of the V-slump flow in the range of 5 to 30 sec is observed. An increase with the replacement of NCA by RCA results in an increase in V-slump flow, it is noted.

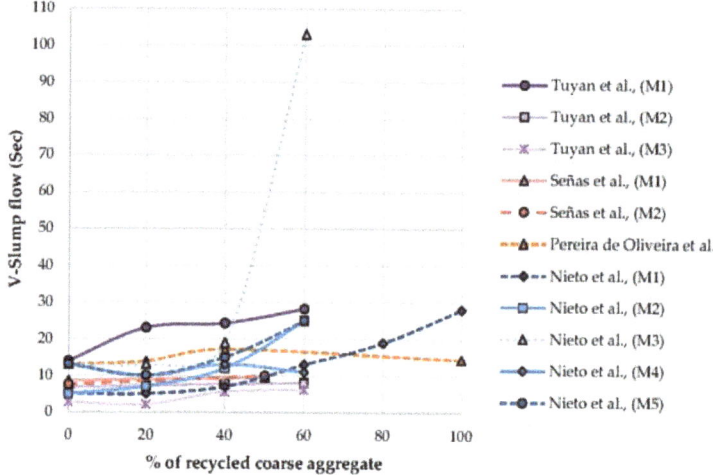

Figure 7. Effect of V-slump flow time on the percentage of recycled coarse aggregate.

Salesa et al., [6] 2017 reported that the SCC with RCA has greater flowability compared to conventional SCC and it is influenced by two factors. First, the particle size distribution of RCA and secondly due to the activation of SP from RA apart from SP in the current mix leads to achieving better flowability. The percentage of RCA does not influence air content, as observed by Salesa et al. [6]. Gridic et al. [94] observed that the passing ability lies in the range of 0.94 to 0.98, which increases with increases in the RCA content. Tuyan et al. [14] found that the increase in SP content increases the passing ability of mix with an increase in RCA. Modani and Mohitkar [93] observed that 100% incorporation of RCA content results in a lower passing ability with a higher amount of SP in it. Tuyan et al. [14] reported that the V-funnel flow time increased with an increase in RCA ratio with a lower SP content in it. The inclusion of metakaolin results in a decrease in workability, which is compensated by increasing the percentage of SP [45]. Although there is a decrease in SCC workability, it increases the mechanical strength of SCC [45]. An increase in metakaolin decreases the alkali-silica reaction (ASR) of aggregates in concrete [96]. An increase in Cathode Ray Tunnel Glass (CRTG) in the mixture increases the workability [97].

From Table 12, the following features are observed:

- An increase in the percentage of SP increases in T500 flow time;
- No constant relationship between slump flow and percentage of SP;
- An increase in the percentage of SP increases J-ring flow;
- An increase in the percentage of SP up to 1%, increases bypassing ability, i.e., L-box ratio;
- An increase in the percentage of SP increases segregation resistance.

Table 12. Classification of family III as per EFNARC standard [95] and effect of % of SP on fresh concrete properties.

Fª	SP (%)	T500 SF(s)	SF (mm)	T500 J-Ring (s)	d_{max} J-Ring (mm)	L-Box H2/H1	V-f (s)	Sieve Seggreg. (%)	References
III	0.35	1.8–3.4 (VS1–VS2)	640–750 (SF1–SF2)	-	-	0.78–0.925	4.1–6.1 (VF1)	-	[67]
	0.8–1.2	-	-	-	-	0.87–0.99 (PA1)	6.1–8.6 (VF1–VF2)	7–15 (SR2)	[45]
		-	-	-	-	0.84–0.95 (PA1)	6.2–9 (VF1–VF2)	6–14 (SR2)	[45]
		-	-	-	-	0.81–0.91 (PA1)	9.2–10 (VF2)	5–13 (SR2)	[45]
	0.9–1.09	3.1–7.1 (VS2)	570–685 (SF1–SF2)	4–11	545–656	-	-	-	[56]
	0.9–0.76	1.2–1.7 (VS1)	650–700 (SF1–SF2)	-	-	0.60–0.70 (PA1)	2.7–6.2 (VF1)	-	[68]
	1.01	2.9–4.3 (VS2)	700–720 (SF2)	-	-	0.80–0.89 (PA1)	-	5.20–9.90 (SR2)	[68]
	1.18–1.08	1.4–1.8 (VS1)	665–700 (SF2)	-	-	0.89–0.90 (PA1)	7.2–13.9 (VF1–VF2)	-	[68]
	1.5	1–6 (VS1–VS2)	570–723 (SF1–SF2)	2–40	475–720	-	5–28 (VF1–VF2)	-	[71]
	1.5	1–5 (VS1–VS2)	628–715 (SF1–SF2)	2–41	445–683	-	5–25 (VF1–VF2)	-	[71]
	1.5	2–15 (VS2)	568–685 (SF1–SF2)	5–8	353–675	-	13–103 (VF2)	-	[71]
	1.5	2.9–4.9 (VS2)	713–792 (SF2)	-	-	0.75–1 (PA1)	6.16–23.20 (VF1–VF2)	5.10–16.50 (SR1–SR2)	[69]
	1.55–1.88	2.7–10.5 (VS2)	610–710 (SF1–SF2)	6.1–13.8	570–660	-	8.5–14 (VF2)	-	[49]
	1.97–1.75	2.6–4.5 (VS2)	685–695 (SF2)	-	-	0.95–0.95 (PA1)	23–28.2 (VF2)	-	[68]
	2.11–1.19	-	650–675 (SF1–SF2)	-	-	-	13–14.4 (VF2)	-	[32]

Fª = Family; W/P = Water/power ratio; SF = Slump Flow; V-f = V-funnel.

5. Conclusions

The literature results analyzed in this research show that recycled aggregate (RA) can be considered as a feasible alternative material for natural aggregates and is feasible for use in self-compacting concrete (SCC). The fresh properties of SCC with recycled coarse aggregate (RCA) show fully satisfactory results for structural purposes and are classified as per EFNRARC standards. RA would make a significant contribution to sustainability in concrete production, and could directly have an impact on the construction sector and in the circular economy. The following conclusions have been drawn from the development and results of this research study:

- For the family, I mix, An increase in the water to cement (W/C) ratio shows a decrease in the percentage of cement content in terms of unit weight. An increase in the W/C ratio causes an increase in the supplementary cementitious material (SCM) content of the mixes. An increase in the percentage of aggregate in mixes results in the increase in water required to achieve the same workability of the mix.
- For family II mix, Increase in water to powder (W/P) ratio, powder content of mix is decreased; Increase in W/P ratio, the proportion of fine aggregate content is decreased and coarse aggregate content is increased; Increase in W/P ratio, the proportion of replacement of fine aggregate by RFA is decreased.
- For family III mix, An increase in the replacement of aggregate by RA increases SP to achieve the same consistency; Increases in SP increase the SCM content of the mixture; An increase in SP increases the filler quantity of the mix.
- An increase in the W/P ratio results in T500 flow time to decrease because the flowability of the mix is better when the suspension of solid materials in liquid media increases. Slump flow increases

with an increase in the W/P ratio. SCC flow about obstruction (J-ring) increases with an increase in the W/P ratio. L-Box ratio has no relationship concerning the W/P ratio. V-funnel initially decreases and then increases to the W/P ratio.

- An increase in the percentage of SP increases in T500 flow time. There is no constant relationship between slump flow and percentage of SP. An increase in the percentage of SP increases J-ring flow. An increase in the percentage of SP up to 1%, increases by passing ability, i.e., L-box ratio. An increase in the percentage of SP increases segregation resistance.

The good results of SCC with RA obtained by some authors should be an incentive to reconsider the use of this type of waste as RA in concrete, which would also help to make the construction process more sustainable. However, more studies would be necessary for its properties as well as the development of standards and guides on the use of RA in concrete.

Author Contributions: Conceptualization, P.J., R.M.-G. and A.J.-V.; methodology, P.J., R.M.-G. and A.J.-V.; investigation, R.M.-G., P.J. and J.M.M.-d.P.; writing—original draft preparation, P.J., R.M.-G., F.J.F.-F.; writing—review and editing, P.J., R.M.-G., F.J.F.-F. and A.J.-V.; supervision, J.M.M.-d.P., and A.J.-V.; All authors have read and agreed to the published version of the manuscript.

Funding: This research received no external funding.

Conflicts of Interest: The authors declare no conflict of interest.

References

1. Eurostat. Recycling Rate of Waste Excluding Major Mineral Wastes. Available online: https://ec.europa.eu/eurostat/tgm/refreshTableAction.do?tab=table&plugin=1&pcode=ten00106&language=en (accessed on 24 April 2019).
2. Eusrostat. Generation of Waste by Economic Activity. Available online: https://ec.europa.eu/eurostat/tgm/table.do?tab=table&tableSelection=1&labeling=labels&footnotes=yes&layout=time,geo,cat&language=en&pcode=sdg_12_50&plugin=1 (accessed on 3 February 2020).
3. Anefa. *Informe de Situación Económica Sectorial*; Madrid, Spain, 2018. Available online: https://www.aridos.org/wp-content/uploads/2018/10/Informe-sectorial-Asamblea-General-ANEFA-2018.pdf (accessed on 10 February 2020).
4. Padmini, A.K.; Ramamurthy, K.; Mathews, M.S. Influence of Parent Concrete on the Properties of Recycled Aggregate Concrete. *Constr. Build. Mater.* **2009**, *23*, 829–836. [CrossRef]
5. Fischer, C.; Werge, M. *EU as a Recycling Society Present Recycling Levels of Municipal Waste and Construction & Demolition Waste in the Europe*; European Environmental Agency (EEA), European Topic Centre on Susatainable Consumption and Production: Copenhagen, Denmark, 2009.
6. Salesa, Á.; Pérez-Benedicto, J.Á.; Esteban, L.M.; Vicente-Vas, R.; Orna-Carmona, M. Physico-Mechanical Properties of Multi-Recycled Self-Compacting Concrete Prepared with Precast Concrete Rejects. *Constr. Build. Mater.* **2017**, *153*, 364–373. [CrossRef]
7. Sandanayake, M.; Zhang, G.; Setunge, S. Estimation of Environmental Emissions and Impacts of Building Construction—A Decision Making Tool for Contractors. *J. Build. Eng.* **2019**, *21*, 173–185. [CrossRef]
8. Esquinas, A.R.; Ledesma, E.F.; Otero, R.; Jiménez, J.R.; Fernández, J.M. Mechanical Behaviour of Self-Compacting Concrete Made with Non-Conforming Fly Ash from Coal-Fired Power Plants. *Constr. Build. Mater.* **2018**, *182*, 385–398. [CrossRef]
9. Guo, H.; Shi, C.; Guan, X.; Zhu, J.; Ding, Y.; Ling, T.C.; Zhang, H.; Wang, Y. Durability of Recycled Aggregate Concrete—A Review. *Cem. Concr. Compos.* **2018**, *89*, 251–259. [CrossRef]
10. Bostanci, S.C.; Limbachiya, M.; Kew, H. Use of Recycled Aggregates for Low Carbon and Cost Effective Concrete Construction. *J. Clean. Prod.* **2018**, *189*, 176–196. [CrossRef]
11. Gao, M.; Beig, G.; Song, S.; Zhang, H.; Hu, J.; Ying, Q.; Liang, F.; Liu, Y.; Wang, H.; Lu, X.; et al. The Impact of Power Generation Emissions on Ambient PM 2.5 Pollution and Human Health in China and India. *Environ. Int.* **2018**, *121*, 250–259. [CrossRef]
12. Braga, A.M.; Silvestre, J.D.; de Brito, J. Compared Environmental and Economic Impact from Cradle to Gate of Concrete with Natural and Recycled Coarse Aggregates. *J. Clean. Prod.* **2017**, *162*, 529–543. [CrossRef]

13. Jagadesh, P.; Ramachandramurthy, A.; Murugesan, R. Evaluation of Mechanical Properties of Sugar Cane Bagasse Ash Concrete. *Constr. Build. Mater.* **2018**, *176*, 608–617. [CrossRef]
14. Tuyan, M.; Mardani-aghabaglou, A.; Ramyar, K. Freeze—Thaw Resistance, Mechanical and Transport Properties of Self-Consolidating Concrete Incorporating Coarse Recycled Concrete Aggregate. *Mater. Des.* **2014**, *53*, 983–991. [CrossRef]
15. Güneyisi, E.; Gesoğlu, M. Properties of Self-Compacting Portland Pozzolana and Limestone Blended Cement Concretes Containing Different Replacement Levels of Slag. *Mater. Struct. Constr.* **2011**, *44*, 1399–1410. [CrossRef]
16. Tabsh, S.W.; Abdelfatah, A.S. Influence of Recycled Concrete Aggregates on Strength Properties of Concrete. *Constr. Build. Mater.* **2009**, *23*, 1163–1167. [CrossRef]
17. Evangelista, L.; de Brito, J. Mechanical Behaviour of Concrete Made with Fine Recycled Concrete Aggregates. *Cem. Concr. Compos.* **2007**, *29*, 397–401. [CrossRef]
18. Ghanbari, M.; Monir Abbasi, A.; Ravanshadnia, M. Economic and Environmental Evaluation and Optimal Ratio of Natural and Recycled Aggregate Production. *Adv. Mater. Sci. Eng.* **2017**, *2017*. [CrossRef]
19. Juan-Valdés, A.; Rodríguez-Robles, D.; García-González, J.; Guerra-Romero, M.I.; Morán-del Pozo, J.M. Mechanical and Microstructural Characterization of Non-Structural Precast Concrete Made with Recycled Mixed Ceramic Aggregates from Construction and Demolition Wastes. *J. Clean. Prod.* **2018**, *180*, 482–493. [CrossRef]
20. Pacheco, J.; de Brito, J.; Chastre, C.; Evangelista, L. Experimental Investigation on the Variability of the Main Mechanical Properties of Concrete Produced with Coarse Recycled Concrete Aggregates. *Constr. Build. Mater.* **2019**, *201*, 110–120. [CrossRef]
21. Mas, B.; Cladera, A.; del Olmo, T.; Pitarch, F. Influence of the Amount of Mixed Recycled Aggregates on the Properties of Concrete for Non-Structural Use. *Constr. Build. Mater.* **2012**, *27*, 612–622. [CrossRef]
22. Etxeberria, M.; Vázquez, E.; Marí, A.; Barra, M. Influence of Amount of Recycled Coarse Aggregates and Production Process on Properties of Recycled Aggregate Concrete. *Cem. Concr. Res.* **2007**, *37*, 735–742. [CrossRef]
23. Güneyisi, E.; Gesoglu, M.; Algın, Z.; Yazıcı, H. Effect of Surface Treatment Methods on the Properties of Self-Compacting Concrete with Recycled Aggregates. *Constr. Build. Mater.* **2014**, *64*, 172–183. [CrossRef]
24. Herbudiman, B.; Saptaji, A.M. Self-Compacting Concrete with Recycled Traditional Roof Tile Powder. *Procedia Eng.* **2013**, *54*, 805–816. [CrossRef]
25. González-Taboada, I.; González-Fonteboa, B.; Eiras-López, J.; Rojo-López, G. Tools for the Study of Self-Compacting Recycled Concrete Fresh Behaviour: Workability and Rheology. *J. Clean. Prod.* **2017**, *156*, 1–18. [CrossRef]
26. Esquinas, A.R.; Ramos, C.; Jiménez, J.R.; Fernández, J.M.; de Brito, J. Mechanical Behaviour of Self-Compacting Concrete Made with Recovery Filler from Hot-Mix Asphalt Plants. *Constr. Build. Mater.* **2017**, *131*, 114–128. [CrossRef]
27. Carro-López, D.; González-Fonteboa, B.; Martínez-Abella, F.; González-Taboada, I.; De Brito, J.; Varela-Puga, F. Proportioning, Microstructure and Fresh Properties of Self-Compacting Concrete with Recycled Sand. In *Procedia Engineering*; Elsevier Ltd.: Amsterdam, The Netherlands, 2017; Volume 171, pp. 645–657. [CrossRef]
28. Silva, P.; de Brito, J. Experimental Study of the Mechanical Properties and Shrinkage of Self-Compacting Concrete with Binary and Ternary Mixes of Fly Ash and Limestone Filler. *Eur. J. Environ. Civ. Eng.* **2017**, *21*, 430–453. [CrossRef]
29. Santos, S.A.; da Silva, P.R.; de Brito, J. Mechanical Performance Evaluation of Self-Compacting Concrete with Fine and Coarse Recycled Aggregates from the Precast Industry. *Materials* **2017**, *10*, 904. [CrossRef]
30. Carro-López, D.; González-Fonteboa, B.; Martínez-Abella, F.; González-Taboada, I.; de Brito, J.; Varela-Puga, F. Proportioning, Fresh-State Properties and Rheology of Self-Compacting Concrete with Fine Recycled Aggregates. *Hormigón y Acero* **2018**, *69*, 213–221. [CrossRef]
31. Ghalehnovi, M.; Roshan, N.; Hakak, E.; Shamsabadi, E.A.; de Brito, J. Effect of Red Mud (Bauxite Residue) as Cement Replacement on the Properties of Self-Compacting Concrete Incorporating Various Fillers. *J. Clean. Prod.* **2019**, *240*, 118213. [CrossRef]
32. Pereira-de-Oliveira, L.A.; Nepomuceno, M.C.S.; Castro-Gomes, J.P.; Vila, M.F.C. Permeability Properties of Self-Compacting Concrete with Coarse Recycled Aggregates. *Constr. Build. Mater.* **2014**, *51*, 113–120. [CrossRef]
33. Safiuddin, M.; Salam, M.A.; Jumaat, M.Z. Effects of Recycled Concrete Aggregate on the Fresh Properties of Self-Consolidating Concrete. *Arch. Civ. Mech. Eng.* **2011**, *11*, 1023–1041. [CrossRef]

34. Kebaïli, O.; Mouret, M.; Arabi, N.; Cassagnabere, F. Adverse Effect of the Mass Substitution of Natural Aggregates by Air-Dried Recycled Concrete Aggregates on the Self-Compacting Ability of Concrete: Evidence and Analysis through an Example. *J. Clean. Prod.* **2015**, *87*, 752–761. [CrossRef]
35. Khaleel, O.R.; Abdul Razak, H. Mix Design Method for Self Compacting Metakaolin Concrete with Different Properties of Coarse Aggregate. *Mater. Des.* **2014**, *53*, 691–700. [CrossRef]
36. Santos, S.; da Silva, P.R.; de Brito, J. Self-Compacting Concrete with Recycled Aggregates—A Literature Review. *J. Build. Eng.* **2019**, *22*, 349–371. [CrossRef]
37. Martínez-García, R.; Guerra-Romero, I.M.; Morán-del Pozo, J.M.; de Brito, J.; Juan-Valdés, A. Recycling Aggregates for Self-Compacting Concrete Production: A Feasible Option. *Materials* **2020**, *13*, 868. [CrossRef] [PubMed]
38. Rodríguez-Robles, D.; García-González, J.; Juan-Valdés, A.; Morán-del Pozo, J.; Guerra-Romero, M. Quality Assessment of Mixed and Ceramic Recycled Aggregates from Construction and Demolition Wastes in the Concrete Manufacture According to the Spanish Standard. *Materials* **2014**, *7*, 5843–5857. [CrossRef] [PubMed]
39. García-González, J.; Barroqueiro, T.; Evangelista, L.; de Brito, J.; De Belie, N.; Morán-del Pozo, J.; Juan-Valdés, A. Fracture Energy of Coarse Recycled Aggregate Concrete Using the Wedge Splitting Test Method: Influence of Water-Reducing Admixtures. *Mater. Struct. Constr.* **2017**, *50*, 1–15. [CrossRef]
40. UNE-EN 12620:2009. *Áridos Para Hormigón*; Aenor: Madrid, España, 2009.
41. UNE-EN 933-3:2012. *Ensayos Para Determinar Las Propiedades Geométricas de Los Áridos. Parte 3: Determinación de La Forma de Las Partículas. Índice de Lajas*; Aenor: Madrid, España, 2012.
42. Okamura, H.; Ouchit, M. *Self-Compacting High Performance Concrete Development of Self-Compacting Concrete*; SAGE Publications Sage: Los Angeles, CA, USA, 1998; Volume 998.
43. Ouchi, M.; Hibino, M.; Okamura, H. Effect of Superplasticizer on Self-Compactability of Fresh Concrete. *Transp. Res. Rec. J. Transp. Res. Board* **1997**, *1574*, 37–40. [CrossRef]
44. Mehta, P.K.; Monteiro, P.J.M. *Concrete. Microstructure, Properties and Materials*; McGraw-Hill: New York, NY, USA, 2006.
45. Ouldkhaoua, Y.; Benabed, B.; Abousnina, R.; Kadri, E.H.; Khatib, J. Effect of Using Metakaolin as Supplementary Cementitious Material and Recycled CRT Funnel Glass as Fine Aggregate on the Durability of Green Self-Compacting Concrete. *Constr. Build. Mater.* **2020**, *235*, 117802. [CrossRef]
46. Okamura, H.; Ouchi, M. Self-Compacting Concrete. *J. Adv. Concr. Tech.* **2003**, *1*, 5–15. [CrossRef]
47. Pereira-de Oliveira, L.A.; Nepomuceno, M.; Rangel, M. An Eco-Friendly Self-Compacting Concrete with Recycled Coarse Aggregates. *Inf. la Construcción* **2013**, *65*, 31–41. [CrossRef]
48. Tertre Torán, J.I.; Moreno Burriel, A. *Hormigón Con Árido Reciclado*; Cemex: Madrid, Spain, 2010.
49. Señas, L.; Priano, C.; Marfil, S. Influence of Recycled Aggregates on Properties of Self-Consolidating Concretes. *Constr. Build. Mater.* **2016**, *113*, 498–505. [CrossRef]
50. González Taboada, I.; González Fonteboa, B.; Martínez Abella, F.; Rojo López, G. Influencia de Las Variaciones En Los Materiales Sobre La Reología de Hormigones Autocompactantes Reciclados. In *V Congreso Iberoamericano de Hormigón Autocompactante y Hormigones Especiales*; Editorial Universitat Politecnica de València: Valencia, Spain, 2018. [CrossRef]
51. Guerrero Vilches, I.M.; Rodriguez Jeronimo, G.; Rodriguez Montero, J. Valorización Como Árido Reciclado Mixto de Un Residuo de Construcción y Demolición En La Confección de Hormigones Autocompactantes Durables En Terrenos Con Yesos. In *V Congreso Iberoamericano de Hormigón Autocompactante y Hormigones Especiales*; Editorial Universitat Politecnica de València: Valencia, Spain, 2018. [CrossRef]
52. García-González, J.; Rodríguez-Robles, D.; Juan-Valdés, A.; Morán-del Pozo, J.; Guerra-Romero, M. Pre-Saturation Technique of the Recycled Aggregates: Solution to the Water Absorption Drawback in the Recycled Concrete Manufacture. *Materials* **2014**, *7*, 6224–6245. [CrossRef]
53. Nagataki, S.; Gokce, A.; Saeki, T. Effects of Recycled Aggregate Characteristics on Performance Parameters of Recycled Aggregate Concrete. *Spec. Publ.* **2000**, *192*, 53–72. [CrossRef]
54. Rashwan, M.S.; Abourizk, S. The Properties of Recycled Concrete. *Concr. Int.* **1997**, *19*, 56–60.
55. Tam, V.W.Y.; Gao, X.F.; Tam, C.M. Microstructural Analysis of Recycled Aggregate Concrete Produced from Two-Stage Mixing Approach. *Cem. Concr. Res.* **2005**, *35*, 1195–1203. [CrossRef]
56. Sasanipour, H.; Aslani, F. Durability Properties Evaluation of Self-Compacting Concrete Prepared with Waste Fine and Coarse Recycled Concrete Aggregates. *Constr. Build. Mater.* **2020**, *236*, 117540. [CrossRef]

57. Kou, S.C.; Poon, C.S. Effect of the Quality of Parent Concrete on the Properties of High Performance Recycled Aggregate Concrete. *Constr. Build. Mater.* **2015**, *77*, 501–508. [CrossRef]
58. Akbarnezhad, A.; Ong, K.C.G.; Tam, C.T.; Zhang, M.H. Effects of the Parent Concrete Properties and Crushing Procedure on the Properties of Coarse Recycled Concrete Aggregates. *J. Mater. Civ. Eng.* **2013**, *25*, 1795–1802. [CrossRef]
59. Limbachiya, M.C.; Leelawat, T.; Dhir, R.K. Use of Recycled Concrete Aggregate in High-Strength Concrete. *Mater. Struct. Constr.* **2000**, *33*, 574–580. [CrossRef]
60. Beltrán, M.G.; Barbudo, A.; Agrela, F.; Galvín, A.P.; Jiménez, J.R. Effect of Cement Addition on the Properties of Recycled Concretes to Reach Control Concretes Strengths. *J. Clean. Prod.* **2014**, *79*, 124–133. [CrossRef]
61. Butler, L.; West, J.S.; Tighe, S.L. Effect of Recycled Concrete Coarse Aggregate from Multiple Sources on the Hardened Properties of Concrete with Equivalent Compressive Strength. *Constr. Build. Mater.* **2013**, *47*, 1292–1301. [CrossRef]
62. Topçu, I.B.; Şengel, S. Properties of Concretes Produced with Waste Concrete Aggregate. *Cem. Concr. Res.* **2004**, *34*, 1307–1312. [CrossRef]
63. Vázquez, E. (Ed.) *Progress of Recycling in the Built Environment*; Final Report of the RILEM Technical Committee 217-PRE; Springer: Dordrecht, The Netherlands, 2013. [CrossRef]
64. Yu, X.-X.; Li, R.-Y.; Xiang, D.; Gen, L.; Song, Z. Effect of Mechanical Force Grinding on the Properties of Recycled Powder. Available online: http://en.cnki.com.cn/Article_en/CJFDTOTAL-RGJT201704020.htm (accessed on 21 August 2020).
65. Lu, J.X.; Zhan, B.J.; Duan, Z.H.; Poon, C.S. Using Glass Powder to Improve the Durability of Architectural Mortar Prepared with Glass Aggregates. *Mater. Des.* **2017**, *135*, 102–111. [CrossRef]
66. Singh, A.; Arora, S.; Sharma, V.; Bhardwaj, B. Workability Retention and Strength Development of Self-Compacting Recycled Aggregate Concrete Using Ultrafine Recycled Powders and Silica Fume. *J. Hazard. Toxic Radioact. Waste* **2019**, *23*, 04019016. [CrossRef]
67. Duan, Z.; Singh, A.; Xiao, J.; Hou, S. Combined Use of Recycled Powder and Recycled Coarse Aggregate Derived from Construction and Demolition Waste in Self-Compacting Concrete. *Constr. Build. Mater.* **2020**, *254*, 119323. [CrossRef]
68. Tang, W.C.; Ryan, P.C.; Cui, H.Z.; Liao, W. Properties of Self-Compacting Concrete with Recycled Coarse Aggregate. *Adv. Mater. Sci. Eng.* **2016**, *2016*. [CrossRef]
69. Djelloul, O.K.; Menadi, B.; Wardeh, G.; Kenai, S. Performance of Self-Compacting Concrete Made with Coarse and Fine Recycled Concrete Aggregates and Ground Granulated Blast-Furnace Slag. *Adv. Concr. Constr.* **2018**, *6*, 103–121. [CrossRef]
70. Aslani, F.; Ma, G.; Wan, D.L.Y.; Muselin, G. Development of High-Performance Self-Compacting Concrete Using Waste Recycled Concrete Aggregates and Rubber Granules. *J. Clean. Prod.* **2018**, *182*, 553–566. [CrossRef]
71. Nieto, D.; Dapena, E.; Alaejos, P.; Olmedo, J.; Pérez, D. Properties of Self-Compacting Concrete Prepared with Coarse Recycled Concrete Aggregates and Different Water:Cement Ratios. *J. Mater. Civ. Eng.* **2019**, *31*, 04018376. [CrossRef]
72. Dapena, E.; Alaejos, P.; Lobet, A.; Pérez, D. Effect of Recycled Sand Content on Characteristics of Mortars and Concretes. *J. Mater. Civ. Eng.* **2011**, *23*, 414–422. [CrossRef]
73. Gupta, P.K.; Rajhans, P.; Panda, S.K.; Nayak, S.; Das, S.K. Mix Design Method for Self-Compacting Recycled Aggregate Concrete and Its Microstructural Investigation by Considering Adhered Mortar in Aggregate. *J. Mater. Civ. Eng.* **2020**, *32*, 04019371. [CrossRef]
74. Abed, M.; Nemes, R.; Lublóy, É. Performance of Self-Compacting High-Performance Concrete Produced with Waste Materials after Exposure to Elevated Temperature. *J. Mater. Civ. Eng.* **2020**, *32*, 05019004. [CrossRef]
75. Xavier, B.C.; Verzegnassi, E.; Bortolozo, A.D.; Alves, S.M.; Cecche Lintz, R.C.; Andreia Gachet, L.; Osório, W.R. Fresh and Hardened States of Distinctive Self-Compacting Concrete with Marble- and Phyllite-Powder Aggregate Contents. *J. Mater. Civ. Eng.* **2020**, *32*, 04020065. [CrossRef]
76. Rajhans, P.; Gupta, P.K.; Ranjan, R.K.; Panda, S.K.; Nayak, S. EMV Mix Design Method for Preparing Sustainable Self Compacting Recycled Aggregate Concrete Subjected to Chloride Environment. *Constr. Build. Mater.* **2019**, *199*, 705–716. [CrossRef]
77. Kou, S.C.; Poon, C.S. Enhancing the Durability Properties of Concrete Prepared with Coarse Recycled Aggregate. *Constr. Build. Mater.* **2012**, *35*, 69–76. [CrossRef]

78. Lovato, P.S.; Possan, E.; Molin, D.C.C.D.; Masuero, Â.B.; Ribeiro, J.L.D. Modeling of Mechanical Properties and Durability of Recycled Aggregate Concretes. *Constr. Build. Mater.* **2012**, *26*, 437–447. [CrossRef]
79. Poon, C.S.; Kou, S.C.; Lam, L. Influence of Recycled Aggregate on Slump and Bleeding of Fresh Concrete. *Mater. Struct.* **2007**, *40*, 981–988. [CrossRef]
80. Zhang, H.; Zhao, Y.; Meng, T.; Shah, S.P. Surface Treatment on Recycled Coarse Aggregates with Nanomaterials. *J. Mater. Civ. Eng.* **2016**, *28*, 04015094. [CrossRef]
81. Cui, H.Z.; Shi, X.; Memon, S.A.; Xing, F.; Tang, W. Experimental Study on the Influence of Water Absorption of Recycled Coarse Aggregates on Properties of the Resulting Concretes. *J. Mater. Civ. Eng.* **2015**, *27*, 04014138. [CrossRef]
82. Kou, S.C.; Poon, C.S. Properties of Concrete Prepared with PVA-Impregnated Recycled Concrete Aggregates. *Cem. Concr. Compos.* **2010**, *32*, 649–654. [CrossRef]
83. Ismail, S.; Ramli, M. Engineering Properties of Treated Recycled Concrete Aggregate (RCA) for Structural Applications. *Constr. Build. Mater.* **2013**, *44*, 464–476. [CrossRef]
84. Bui, N.K.; Satomi, T.; Takahashi, H. Mechanical Properties of Concrete Containing 100% Treated Coarse Recycled Concrete Aggregate. *Constr. Build. Mater.* **2018**, *163*, 496–507. [CrossRef]
85. Qiu, J.; Tng, D.Q.S.; Yang, E.H. Surface Treatment of Recycled Concrete Aggregates through Microbial Carbonate Precipitation. *Constr. Build. Mater.* **2014**, *57*, 144–150. [CrossRef]
86. Zhu, Y.G.; Kou, S.C.; Poon, C.S.; Dai, J.G.; Li, Q.Y. Influence of Silane-Based Water Repellent on the Durability Properties of Recycled Aggregate Concrete. *Cem. Concr. Compos.* **2013**, *35*, 32–38. [CrossRef]
87. Spaeth, V.; Djerbi Tegguer, A. Improvement of Recycled Concrete Aggregate Properties by Polymer Treatments. *Int. J. Sustain. Built Environ.* **2013**, *2*, 143–152. [CrossRef]
88. Kong, D.; Lei, T.; Zheng, J.; Ma, C.; Jiang, J.; Jiang, J. Effect and Mechanism of Surface-Coating Pozzalanics Materials around Aggregate on Properties and ITZ Microstructure of Recycled Aggregate Concrete. *Constr. Build. Mater.* **2010**, *24*, 701–708. [CrossRef]
89. Xiao, J.; Li, J.; Zhang, C. Mechanical Properties of Recycled Aggregate Concrete under Uniaxial Loading. *Cem. Concr. Res.* **2005**, *35*, 1187–1194. [CrossRef]
90. Katz, A. Properties of Concrete Made with Recycled Aggregate from Partially Hydrated Old Concrete. *Cem. Concr. Res.* **2003**, *33*, 703–711. [CrossRef]
91. Bandi, S.M.; Patel, Y.J.; Vyas, V.H. Study on Fresh and Hardened Properties of Self Compacted Concrete Using Recycled Concrete Aggregate. *Int. J. Innov. Res. Sci. Eng. Technol.* **2007**, 3297. [CrossRef]
92. EFNARC. *Specification and Guidelines for Self-Compacting Concrete*; EFNARC: Norfolk, UK, 2002; Volume 44, p. 32. ISBN 0-9539733-4-4.
93. Moriconi, G.; Corinaldesi, V. Self-Compacting Concrete: A Great Opportunity for Recycling Materials. In Proceedings of the International RILEM Conference on the Use of Recycled Materials in Building and Structures, Barcelona, Spain, 8–11 November 2004.
94. Grdic, Z.J.; Toplicic-Curcic, G.A.; Despotovic, I.M.; Ristic, N.S. Properties of Self-Compacting Concrete Prepared with Coarse Recycled Concrete Aggregate. *Constr. Build. Mater.* **2010**, *24*, 1129–1133. [CrossRef]
95. Collepardi, M.; Collepardi, S.; Troli, R. *Properties of SCC and Flowing Concrete*. 2006. Available online: http://www.encosrl.it/OLDSITE/pubblicazioni-scientifiche/pdf/scc/19.pdf (accessed on 10 February 2020).
96. Lee, G.; Ling, T.C.; Wong, Y.L.; Poon, C.S. Effects of Crushed Glass Cullet Sizes, Casting Methods and Pozzolanic Materials on ASR of Concrete Blocks. *Constr. Build. Mater.* **2011**, *25*, 2611–2618. [CrossRef]
97. Hui, Z.; Sun, W. Study of Properties of Mortar Containing Cathode Ray Tubes (CRT) Glass as Replacement for River Sand Fine Aggregate. *Constr. Build. Mater.* **2011**, *25*, 4059–4064. [CrossRef]

Publisher's Note: MDPI stays neutral with regard to jurisdictional claims in published maps and institutional affiliations.

© 2020 by the authors. Licensee MDPI, Basel, Switzerland. This article is an open access article distributed under the terms and conditions of the Creative Commons Attribution (CC BY) license (http://creativecommons.org/licenses/by/4.0/).

MDPI\
St. Alban-Anlage 66\
4052 Basel\
Switzerland\
Tel. +41 61 683 77 34\
Fax +41 61 302 89 18\
www.mdpi.com

Materials Editorial Office\
E-mail: materials@mdpi.com\
www.mdpi.com/journal/materials

www.ingramcontent.com/pod-product-compliance
Lightning Source LLC
LaVergne TN
LVHW070747100526
838202LV00013B/1320

9 7 8 3 0 3 6 5 3 0 6 2 8